Laser Additive Manufacturing: Design, Materials, Processes and Applications, 2nd Edition

Laser Additive Manufacturing: Design, Materials, Processes and Applications, 2nd Edition

Jie Yin
Yang Liu
Linda Ke
Kai Guan

MDPI

Basel • Beijing • Wuhan • Barcelona • Belgrade • Novi Sad • Cluj • Manchester

Jie Yin
Gemological Institute
China University of Geosciences
Wuhan
China

Yang Liu
Faculty of Mechanical
Engineering & Mechanics
Ningbo University
Ningbo
China

Linda Ke
Shanghai Engineering
Technology Research Center
of Near-Net-Shape Forming
for Metallic Materials
Shanghai
China

Kai Guan
TSC Laser Technology
Development (Beijing) Co., Ltd.
Beijing
China

Editorial Office
MDPI AG
Grosspeteranlage 5
4052 Basel, Switzerland

This is a reprint of articles from the Special Issue published online in the open access journal *Micromachines* (ISSN 2072-666X) (available at: www.mdpi.com/journal/micromachines/special_issues/2A090U7J2Q).

For citation purposes, cite each article independently as indicated on the article page online and using the guide below:

Lastname, A.A.; Lastname, B.B. Article Title. *Journal Name* **Year**, *Volume Number*, Page Range.

ISBN 978-3-7258-1720-7 (Hbk)
ISBN 978-3-7258-1719-1 (PDF)
https://doi.org/10.3390/books978-3-7258-1719-1

Cover image courtesy of Jie Yin, Yang Liu, Linda Ke and Kai Guan
The complex part is produced through 3D printing by laser additive manufacturing.

Contents

About the Editors

Jie Yin

Prof. Dr. Jie Yin was born in November 1984 and is currently a professor and doctoral supervisor at the Gemological Institute and Advanced Manufacturing Research Center of the China University of Geosciences (CUG). He received his Ph.D. degree from Huazhong University of Science and Technology (HUST) in 2014, then joined Wuhan National Laboratory for Optoelectronics (WNLO) as a postdoctoral and associate researcher for six years. His research focuses on laser-advanced manufacturing and laser-matter interaction. He is the recipient of the Young Outstanding Talents of CUG Scholars; of the Best Paper Award from the International Conference on Research Advances in Additive Manufacturing (RAAM) in 2019; of the Excellent Report Award by the 4D Printing Technology Forum in 2022; and of the Second Prize for Science and Technology Progress of the China National Light Industry Council in 2023. He has led more than 10 projects, including the National Natural Science Foundation of China (NSFC), the Shanghai Aerospace Science and Technology Innovation Key Fund (SAST), and the Fundamental Research Funds for the Central Universities. He has served as an editorial board member for international academic journals such as *Micromachines*, *Additive Manufacturing Frontiers*, and the *Chinese Journal of Lasers*, and a reviewer for 30 academic journals. Furthermore, he has published more than 50 academic papers, with more than 3000 citations in Scopus (H-index = 21), 4 ESI/F5000 highly cited papers, and 3 cover papers. He was listed among the "World's Top 2% Scientists" in 2023.

Yang Liu

Prof. Dr. Yang Liu is currently an associate professor in the Department of Mechanical Engineering at Ningbo University. He received his Ph.D. degree from South China University of Technology (SCUT) in 2015, and he worked as a visiting scholar at Cardiff University from 2018 to 2019. Now, he is a core member of the MoE Key Laboratory of Impact and Safety Engineering and the Stanford 2% Global Top Scientists for the year 2022. Dr. Liu is the head of the additive manufacturing research group at the NBU and one of the early scholars to study the properties of additive manufacturing alloys such as Ti, Al, and HEAs under extreme work environments (high pressure, high velocity, or high/low temperatures). His work mainly aims to reveal the dynamic deformation behaviors and failure mechanisms of additive manufacturing metals and structures under dynamic load, which will facilitate the widening of the application of additive manufacturing technologies and products. To this date, Dr. Liu is the PI or Co-I of over 20 scientific projects supported by the NSFC, Zhejiang Province, Ningbo City, enterprises, etc. He has published over 50 peer-reviewed articles in various top academic journals, including *Additive Manufacturing*, *Virtual and Physical Prototyping*, *Materials & Design*, *Materials Science & Engineering A.*, etc., with more than 2400 citations.

Linda Ke

Prof. Dr. Linda Ke was born in October 1986 and is currently a professor at the Research Center of the Shanghai Institute of Aerospace Precision Machinery. He received his Ph.D. degree from Huazhong University of Science and Technology (HUST) in 2014, and he worked as a visiting scholar at Bauman Moscow State Technical University in 2018. His research focuses on laser additive manufacturing and wire-arc additive manufacturing.

His team has successfully realized the application of more than 10 innovative design products based on additive manufacturing, represented by the "Tiandu-2" satellite integral tank in satellites and rockets, and received three Ministerial and Provincial-Level Science and Technology Awards. To date, Dr. Ke is the PI or Co-I of over 30 scientific projects supported by the National Natural Science Foundation of China (NSFC), the Science and Technology Commission of Shanghai Municipality, etc., and has published more than 40 academic papers in various top academic journals, including *Additive Manufacturing, Materials Design, Materials Science & Engineering, International Journal of Impact Engineering*, etc.

Kai Guan

Dr. Guan Kai is currently the general manager of Tianjin LiM Laser Technology Co., Ltd. He received his Ph.D. degree from Huazhong University of Science and Technology (HUST). He has been engaged in the research of additive manufacturing technology and laser processing technology for more than 10 years, conducting extensive research in basic research, application research, and engineering research. He presided over a key research and development project for the Tianjin Science and Technology Bureau, participated in numerous national and provincial projects, published over 20 peer-reviewed papers (including 3 ESI/Cover papers), applied for and was granted more than 60 patents, and his research results have won three Ministerial and Provincial-Level Science and Technology Awards. He led the development of more than 20 types of metal additive manufacturing equipment. The selective laser melting (SLM) equipment has a printing area ranging from ϕ140 mm to 1500 mm$\times$1500 mm and a molding height from 150 mm to 1500 mm. In the most advanced case, a single device can operate with 20 lasers simultaneously. The molding size of the laser-engineered net shaping (LENS) equipment can reach up to 4000 mm. This type of equipment provides a new solution for the molding of large titanium alloy parts, significantly reducing the production cycle and manufacturing costs. The research and development of this equipment in aerospace, biomedical, automotive mold, and other fields has achieved significant economic and social benefits.

micromachines

MDPI

Editorial

Editorial for the Special Issue on Laser Additive Manufacturing: Design, Materials, Processes, and Applications, 2nd Edition

Jie Yin [1,2,*], Yang Liu [3,*], Linda Ke [4,*] and Kai Guan [5,*]

1 Gemological Institute, China University of Geosciences, Wuhan 430074, China
2 Advanced Manufacturing Research Institute, China University of Geosciences, Wuhan 430074, China
3 Faculty of Mechanical Engineering & Mechanics, Ningbo University, Ningbo 315211, China
4 Shanghai Engineering Technology Research Center of Near-Net-Shape Forming for Metallic Materials, Shanghai Spaceflight Precision Machinery Institute, Shanghai 201600, China
5 TSC Laser Technology Development (Beijing) Co., Ltd., Beijing 100076, China
* Correspondence: yinjie@cug.edu.cn (J.Y.); liuyang1@nbu.edu.cn (Y.L.); kelinda_casc@163.com (L.K.); guankai@tsc-bj.com (K.G.)

Citation: Yin, J.; Liu, Y.; Ke, L.; Guan, K. Editorial for the Special Issue on Laser Additive Manufacturing: Design, Materials, Processes, and Applications, 2nd Edition. *Micromachines* **2024**, *15*, 787. https://doi.org/10.3390/mi15060787

Received: 29 May 2024
Accepted: 6 June 2024
Published: 15 June 2024

1. Introduction for Special Issue of Laser additive Manufacturing

Laser-based additive manufacturing (LAM) represents one of the most forward-thinking transformations in how we conceive, design, and bring to life engineered solutions. By fusing digital design with material layering processes, LAM transcends the limitations of traditional subtractive manufacturing and equal material manufacturing, enabling the realization of material–micro/macrostructure–performance integration previously considered unattainable [1].

The strategic significance of LAM extends beyond mere production. It embodies a confluence of innovation, sustainability, and efficiency, essential for securing technological advancement and maintaining a competitive edge in the global industrial landscape. With sustainability at its core, LAM minimizes waste through precision use of materials, contributing to a greener manufacturing paradigm [2].

However, to fully harness the power of LAM, we must delve into the intricacies of its many facets. This includes mastering the nuances of material behavior under the influence of laser energy [3,4], optimizing design for additive processes [5,6], understanding the thermodynamics of the laser–material interaction [7], and refining the reliability of the parts produced [8]. Qualified LAM parts must meet stringent requirements, often surpassing those fabricated through traditional means [9].

The pursuit of defect-free, structurally robust parts via LAM is not merely an academic goal but a practical necessity. As we push the boundaries of what can be created, from novel functional devices to high-performance components, we must ensure that the principles of design for additive manufacturing (DfAM) are well integrated with the material science and laser processing techniques that underpin this technology [10]. It is through this synergy that LAM will continue to redefine what is possible in manufacturing, offering a pathway to a new era of industrial capability and innovation.

Building on the work in the first edition [11] and sincere cooperation with the Guest Editors [12], the second edition will continue to concentrate on laser additive manufacturing, including macro- to micro-scale additive manufacturing with lasers, including structure design/material design [13], fabrication [14], modeling and simulation; in situ characterization of additive manufacturing processes [15]; and ex situ material characterization and performances [16], with an overview of various applications [17] in aerospace, biomedicine, optics, transportation, energy, etc.

As depicted in Figure 1, this Special Issue featured a diverse array of topics, publishing a total of 11 contributions, comprising 10 original research articles and 1 review paper. After the high-quality reviewing process, four articles [18–21] were selected as Editor's Choice.

Each of them is briefly introduced below according to four aspects (i.e., design, materials, processes, and applications) of the laser additive manufacturing in this Special Issue.

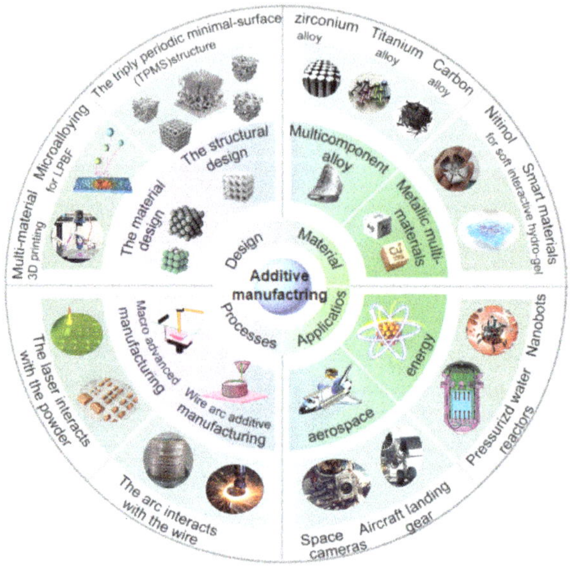

Figure 1. Topics covered in the Special Issue titled "Laser Additive Manufacturing: Design, Materials, Processes, and Applications, 2nd Edition".

2. Design of Laser Additive Manufacturing in Special Issue

The design of laser additive manufacturing covered in this Special Issue includes the material design (e.g., the high reflectivity and thermal conductivity multi-material 3D printing [22] and the microalloying for LPBF [20]) and the structural design (e.g., the triply periodic minimal-surface (TPMS) structure [18], the thin-walled main support structure [23]).

For the high reflectivity and thermal conductivity multi-material fabrication, Chen et al. [22] reported that the formation mechanism, interface characteristics, and molten pool behavior of the Ag7.5Cu/Cu10Sn (A/C) and Cu10Sn/Ag7.5Cu (C/A) interfaces were printed by laser powder bed fusion (LPBF) for the first time to reveal the influence of different building strategies. They found that the high thermal conductivity of the substrate enhanced the Marangoni convection, which strengthened interfacial bonding strength and reduced the defects.

For the particle-reinforced titanium matrix composites (PRTMCs), Li et al. [20] studied the impact of LaB6 content on the microstructure, quasi-static characteristics, and dynamic properties of titanium matrix composites, focusing primarily on the TC4 and TC4/LaB6 composites manufactured using LPBF. The sample with 0.5 wt.% LaB6 was found to have the best strength–toughness synergy among the three groups of composites due to having the smallest grain size. This study was able to provide a theoretical basis for an in-depth understanding of the compressive properties of additive manufacturing of PRTMCs under high-speed loading conditions.

For the TPMS structure high-precision fabrication, Qu et al. [18] reported the critical roles of energy density and minimal feature size on superelastic NiTi functional components by high-precision LPBF. In this work, extensive parameter studies were conducted to reveal their influences on the microstructure and mechanical properties of as-printed components with different minimal feature sizes (thin wall structure, bulk samples, TPMS structures, etc.). TPMS structures were proposed and verified to be a good candidate as the standard test for parameter study of printing NiTi intricate components and investigation of NiTi thin wall structure mechanical property. Furthermore, a robotic cannula tip with fine minimal feature size and high superelasticity was fabricated using HP-LPBF as a case study.

3. Materials of Laser Additive Manufacturing in Special Issue

This Special Issue covers a wide range of forming materials, including zirconium alloys [24], titanium alloys [20,23], steel [21,25,26], composites (e.g., electroless copper and carbon fiber [27]), and in particular, metallic multi-materials (e.g., copper/silver [22]), 4D printing materials (e.g., NiTi, smart materials for soft interactive hydrogel [19]).

Lei et al. [21] reported that AerMet100 steel was produced using an in situ rolling hybrid with wire arc additive manufacturing. In their work, the microstructure, tensile properties, and fracture toughness of as-deposited and heat-treated AerMet100 steel were evaluated in different directions. The results reveal that the manufacturing process leads to grain fragmentation and obvious microstructural refinement of the AerMet100 steel and weakens the anisotropy of the mechanical properties. After heat treatment, the microstructure of the AerMet100 steel is mainly composed of lath martensite and reversed austenite. Alloy carbides are precipitated within the martensite matrix, and a high density of dislocations is the primary strengthening mechanism. The existence of film-like austenite among the martensite matrix enhances the toughness of AerMet100 steel, which coordinates stress distribution and restrains crack propagation, resulting in an excellent balance between strength and toughness. The AerMet100 steel with in situ rolling is isotropy and achieves the following values: an average ultimate strength of 1747.7 ± 16.3 MPa, yield strength of 1615 ± 40.6 MPa, elongation of $8.3 \pm 0.2\%$ in the deposition direction, and corresponding values in the building direction are 1821.3 ± 22.1 MPa, 1624 ± 84.5 MPa, and $7.6 \pm 1.7\%$.

Wang et al. [27] investigated the effects of building directions on the crystallographic anisotropy, phase composition, superelastic properties, microhardness, geometrically necessary dislocation (GND) density, and impurity element content of NiTi shape memory alloys (SMAs) fabricated by LPBF. The main idea of this study was to change the introduction of impurity elements in NiTi alloys and alter their microstructures by adjusting the LPBF building direction, which in turn affects their phase transformation temperatures and superelasticity. In conclusion, for engineering applications, the 45° sample had a lower phase transformation temperature, which was more favorable for superelastic recovery under certain temperatures (austenite start temperature (As) < testing temperature < austenite finish temperature (Af)), while the preferred orientation of the 90° sample ensured a greater superelastic recovery strain when tested at Af + 10 °C.

Wang et al. [28] use a nanosecond ultraviolet laser to study the effects of laser process parameters on the adhesion strength between electroless copper and CFCs. To achieve good adhesion strength, four key process parameters, namely, the laser power, scanning line interval, scanning speed, and pulse frequency, were optimized experimentally using response surface methodology, and a central composite design was utilized to design the experiments. The numerical analysis indicated that the optimized laser power, scanning line interval, scanning speed, and pulse frequency were 5.5 W, 48.2 μm, 834.0 mm/s, and 69.5 kHz, respectively. A validation test confirmed that the predicted results were consistent with the actual values; thus, the developed mathematical model can adequately predict responses within the limits of the laser process parameters being used.

4. Processes of Laser Additive Manufacturing in Special Issue

This Special Issue focuses on macro-advanced manufacturing (e.g., laser powder bed fusion (L-PBF) [18], wire arc additive manufacturing (WAAM) [21]), but also on micro-advanced manufacturing (e.g., laser photochemical synthesis [19]). Furthermore, the process optimizations on the adhesion strength [28] and the mechanical properties [24] (in-process) and the influence of aging treatment regimens on the microstructure and mechanical properties [25] (post-process) were investigated.

Song et al. [24] reported the influences of process parameters on the printability, surface roughness, and mechanical properties of the LPBFed nuclear Zr-4 alloy. The results showed that the relative density of the Zr-4 alloy samples was greater than 99.3% with the laser power range of 120–160 W and the scanning speed range of 600–1000 mm/s. Under a moderate laser power in the range of 120–140 W, the printed Zr-4 alloy possessed excellent surface molding quality with a surface roughness less than 10 μm. The microstructure of the printed Zr-4 alloy was an acicular α phase with an average grain size of about 1 μm. The Zr-4 alloy printed with a laser power of 130 W and a scanning speed of 400 mm/s exhibited the highest compression strength of 1980 MPa and the highest compression strain of 28%. The findings demonstrate the potential in the fabrication of complex Zr-4 alloy parts by LPBF for industrial applications.

Liu et al. [26] investigated the surface morphologies and corrosion behaviors of LPBFed 316L stainless steel polished with different laser pulse widths. The experimental results show that, compared to the nanosecond (NS) and femtosecond (FS) lasers, the surface material's sufficient remelting realized by the continuous wave (CW) laser results in a significant improvement in roughness. The surface hardness is increased, and the corrosion resistance is the best. The microcracks on the NS laser-polished surface led to a decrease in the microhardness and corrosion resistance. The FS laser does not significantly improve surface roughness. The ultrafast laser-induced micro-nanostructures increase the contact area of the electrochemical reaction, resulting in a decrease in corrosion resistance.

Dong et al. [25] studied the effects of aging temperature and time on the microstructure and mechanical properties of SLM 17-4 PH steel. They found that the aging treatment induced the formation of the austenite phase with a face-centered cubic (FCC) structure. With prolonged aging treatment, the volume fraction of the austenite phase increased, which agreed with the EBSD phase mappings. The ultimate tensile strength (UTS) and yield strength gradually increased with increasing aging times at 482 °C. The UTS reached its peak value after aging for 3 h at 482 °C, which was similar to the trend of microhardness (i.e., UTS = 1353.4 MPa). However, the ductility of the SLM 17-4 PH steel decreased rapidly after aging treatment. This work reveals the influence of heat treatment on SLM 17-4 steel and proposes an optimal heat-treatment regime for the SLM high-performance steels.

5. Applications of Laser Additive Manufacturing in Special Issue

The findings of this Special Issue are expected to provide guidance and reference for scientists and engineers in the fields of aerospace (e.g., space camera and aircraft landing gear [23]), energy (e.g., fuel envelope positioning lattices of pressurized water reactors, component boxes, and heat exchangers [24]), and micromachines (e.g., four-dimensional microrobots/nanorobots [19]).

Tao et al. [19] reviewed the literature on the four-dimensional micro/nanorobots via laser photochemical synthesis towards the molecular scale. Miniaturized four-dimensional (4D) micro/nanorobots denote a forerunning technique associated with interdisciplinary applications, such as in embeddable labs-on-chip, metamaterials, tissue engineering, cell manipulation, and tiny robotics. With emerging smart interactive materials, static micro/nanoscale architectures have upgraded to the fourth dimension, evincing time-dependent shape/property mutation. Molecular-level 4D robotics promises complex sensing, self-adaptation, transformation, and responsiveness to stimuli for highly valued functionalities. To precisely control 4D behaviors, current–laser-induced photochemical additive manufacturing, such as digital light projection, stereolithography, and two-photon polymerization, is pursuing high-freeform shape-reconfigurable capacities and high-resolution spatiotemporal programming strategies, which challenge multi-field sciences while offering new opportunities. Herein, this review summarizes the recent development of micro/nano 4D laser photochemical manufacturing, incorporating active materials and shape-programming strategies to provide an envisioning of these miniaturized 4D micro/nanorobots. A comparison with other chemically/physically fabricated micro/nanorobots further explains the advantages and potential usage of laser-synthesized micro/nanorobots.

Peng et al. [23] reported the design and optimization of a thin-walled main support structure for a space camera based on additive manufacturing. They proposed a solution for designing and optimizing a large-scale complex thin-walled structure using additive manufacturing. Firstly, we devise an integrated thin-walled structure and test material for the main support. Secondly, shape optimization is achieved via the optimization of the lateral slope angle of the primary support based on Timoshenko cantilever beam theory. Additionally, an active fitting optimization algorithm is proposed for the purpose of refining the wall thickness of the thin-walled structure. Then, we determine the structural design of the main support. This primary support is manufactured via selective laser melting (SLM). Following processing, the structure size is 538 mm × 400 mm × 384 mm, and the mass is 7.78 kg. Finally, frequency scanning experiments indicate that, in the horizontal direction, there is a natural frequency of 105.97 Hz with an error rate of approximately 3% compared to finite element analysis results. This research confirms that our large-scale complex, thin-walled main support structure design meets all design requirements.

To conclude, we would like to acknowledge all the authors for their contributions to the success of this Special Issue in *Micromachines*, as well as the reviewers whose feedback helped to improve the quality of the published papers.

Acknowledgments: The Guest Editors are very grateful to Zihan Liu from China University of Geosciences for her visualization contribution, and Min Su from the *Micromachines* publishing office for her great guidance, assistance, help, and support that led to the success of this Special Issue.

Conflicts of Interest: Kai Guan employed by TSC Laser Technology Development (Beijing) Co., Ltd., The authors declare that the research was conducted in the absence of any commercial or financial relationships that could be construed as a potential conflict of interest.

References

1. Gu, D.; Shi, X.; Poprawe, R.; Bourell, D.L.; Setchi, R.; Zhu, J. Material-structure-performance integrated laser-metal additive manufacturing. *Science* **2021**, *372*, eabg1487. [CrossRef] [PubMed]
2. Wei, H.L.; Mukherjee, T.; Zhang, W.; Zuback, J.S.; Knapp, G.L.; De, A.; DebRoy, T. Mechanistic models for additive manufacturing of metallic components. *Prog. Mater. Sci.* **2020**, *116*, 100703. [CrossRef]
3. Wang, D.; Huang, J.; Tan, C.; Yang, Y. Review on Effects of Cyclic Thermal Input on Microstructure and Property of Materials in Laser Additive Manufacturing. *Acta Metall. Sin.* **2022**, *10*, 1221–1235.
4. Xiao, Y.; Yang, Y.; Wang, D.; Zhou, H.; Liu, Z.; Liu, L.; Wu, S.; Song, C. In-situ synthesis of spatial heterostructure Ti composites by laser powder bed fusion to overcome the strength and plasticity trade-off. *Int. J. Mach. Tools Manuf.* **2024**, *196*, 104117. [CrossRef]
5. Baghi, A.H.; Mansour, J. Development of a Genetic Algorithm—Artificial Neural Network model to optimize the Dimensional Accuracy of parts printed by FFF. *Rapid Prototyp. J.* **2024**, *30*, 840–857. [CrossRef]
6. Jin, J.; Yang, H.; Liu, Y.; Yang, J.; Li, K.; Yi, Y.; Chen, D.; Zhang, W.; Zhou, S. Microstructure and enhanced strength-ductility of TiNbCu alloys produced by laser powder bed fusion. *Mater. Sci.Eng. A* **2024**, *890*, 145889. [CrossRef]
7. Yin, J.; Wang, D.; Yang, L.; Wei, H.; Dong, P.; Ke, L.; Wang, G.; Zhu, H.; Zeng, X. Correlation between forming quality and spatter dynamics in laser powder bed fusion. *Addit. Manuf.* **2020**, *31*, 100958. [CrossRef]
8. Liu, Y.; Meng, J.; Zhu, L.; Chen, H.; Li, Z.; Li, S.; Wang, D.; Wang, Y.; Kosiba, K. Dynamic compressive properties and underlying failure mechanisms of selective laser melted Ti-6Al-4V alloy under high temperature and strain rate conditions. *Addit. Manuf.* **2022**, *54*, 102772. [CrossRef]
9. Chaugule, P.S.; Du, W.; Kamath, R.R.; Barua, B.; Messner, M.C.; Singh, D. Reliability comparisons between additively manufactured and conventional SiC–Si ceramic composites. *J. Am. Ceram. Soc.* **2024**, *107*, 3117–3133. [CrossRef]
10. Fischer, L.A.S.; Foggiatto, J.A.; Marcondes, P.V.P.; Lajarin, S.F. Design for Additive Manufacturing (DFAM) applied in the manufacture of Master Sample for the automotive industry. *Proc. Inst. Mech. Eng. Part B J. Eng. Manuf.* **2024**. [CrossRef]
11. Yin, J.; Liu, Y.; Zhao, P. Editorial for the Special Issue on Laser Additive Manufacturing: Design, Processes, Materials and Applications. *Micromachines* **2022**, *13*, 2057. [CrossRef] [PubMed]
12. Li, Z.; Li, H.; Yin, J.; Li, Y.; Nie, Z.; Li, X.; You, D.; Guan, K.; Duan, W.; Cao, L.; et al. A Review of Spatter in Laser Powder Bed Fusion Additive Manufacturing: In Situ Detection, Generation, Effects, and Countermeasures. *Micromachines* **2022**, *13*, 1366. [CrossRef] [PubMed]
13. Li, K.; Yang, J.; Yi, Y.; Liu, X.; Liu, Y.; Zhang, L.; Zhang, W.; Li, W.; Chen, D.; Zhou, S. Enhanced strength-ductility synergy and mechanisms of heterostructured Ti6Al4V-Cu alloys produced by laser powder bed fusion. *Acta Mater.* **2023**, *256*, 119112. [CrossRef]

14. Xiao, Y.; Song, C.; Liu, Z.; Liu, L.; Zhou, H.; Wang, D.; Yang, Y. In-situ additive manufacturing of high strength yet ductility titanium composites with gradient layered structure using N2. *Int. J. Extreme Manuf.* **2024**, *6*, 35001. [CrossRef]
15. Yin, J.; Zhang, W.; Ke, L.; Wei, H.; Wang, D.; Yang, L.; Zhu, H.; Dong, P.; Wang, G.; Zeng, X. Vaporization of alloying elements and explosion behavior during laser powder bed fusion of Cu–10Zn alloy. *Int. J. Mach. Tools Manuf.* **2021**, *161*, 103686. [CrossRef]
16. Zhan, J.; Wu, J.; Ma, R.; Li, K.; Huang, T.; Lin, J.; Murr, L.E. Effect of microstructure on the superelasticity of high-relative-density Ni-rich NiTi alloys fabricated by laser powder bed fusion. *J. Mater. Process. Technol.* **2023**, *317*, 117988. [CrossRef]
17. Li, K.; Ma, R.; Qin, Y.; Gong, N.; Wu, J.; Wen, P.; Tan, S.; Zhang, D.Z.; Murr, L.E.; Luo, J. A review of the multi-dimensional application of machine learning to improve the integrated intelligence of laser powder bed fusion. *J. Mater. Process. Technol.* **2023**, *318*, 118032. [CrossRef]
18. Qu, S.; Wang, L.; Ding, J.; Fu, J.; Gao, S.; Ma, Q.; Liu, H.; Fu, M.; Lu, Y.; Song, X. Superelastic NiTi Functional Components by High-Precision Laser Powder Bed Fusion Process: The Critical Roles of Energy Density and Minimal Feature Size. *Micromachines* **2023**, *14*, 1436. [CrossRef] [PubMed]
19. Tao, Y.; Lin, L.; Ren, X.; Wang, X.; Cao, X.; Gu, H.; Ye, Y.; Ren, Y.; Zhang, Z. Four-Dimensional Micro/Nanorobots via Laser Photochemical Synthesis towards the Molecular Scale. *Micromachines* **2023**, *14*, 1656. [CrossRef]
20. Li, X.; Liu, Y. Dynamic Compressive Properties and Failure Mechanism of the Laser Powder Bed Fusion of Submicro-LaB6 Reinforced Ti-Based Composites. *Micromachines* **2023**, *14*, 2237. [CrossRef]
21. Lei, L.; Ke, L.; Xiong, Y.; Liu, S.; Du, L.; Chen, M.; Xiao, M.; Fu, Y.; Yao, F.; Yang, F.; et al. Microstructure, Tensile Properties, and Fracture Toughness of an In Situ Rolling Hybrid with Wire Arc Additive Manufacturing AerMet100 Steel. *Micromachines* **2024**, *15*, 494. [CrossRef] [PubMed]
22. Chen, Q.; Jing, Y.; Yin, J.; Li, Z.; Xiong, W.; Gong, P.; Zhang, L.; Li, S.; Pan, R.; Zhao, X.; et al. High Reflectivity and Thermal Conductivity Ag-Cu Multi-Material Structures Fabricated via Laser Powder Bed Fusion: Formation Mechanisms, Interfacial Characteristics, and Molten Pool Behavior. *Micromachines* **2023**, *14*, 362. [CrossRef] [PubMed]
23. Peng, J.; Liu, S.; Wang, D.; Xu, A.; Huang, X.; Ma, T.; Wang, J.; Li, H. Design and Optimization of Thin-Walled Main Support Structure for Space Camera Based on Additive Manufacturing. *Micromachines* **2024**, *15*, 211. [CrossRef] [PubMed]
24. Song, C.; Zou, Z.; Yan, Z.; Liu, F.; Yang, Y.; Yan, M.; Han, C. Process Optimization and Tailored Mechanical Properties of a Nuclear Zr-4 Alloy Fabricated via Laser Powder Bed Fusion. *Micromachines* **2023**, *14*, 556. [CrossRef] [PubMed]
25. Dong, D.; Wang, J.; Chen, C.; Tang, X.; Ye, Y.; Ren, Z.; Yin, S.; Yuan, Z.; Liu, M.; Zhou, K. Influence of Aging Treatment Regimes on Microstructure and Mechanical Properties of Selective Laser Melted 17-4 PH Steel. *Micromachines* **2023**, *14*, 871. [CrossRef] [PubMed]
26. Liu, J.; Ma, H.; Meng, L.; Yang, H.; Yang, C.; Ruan, S.; Ouyang, D.; Mei, S.; Deng, L.; Chen, J.; et al. Laser Powder Bed Fusion of 316L Stainless Steel: Effect of Laser Polishing on the Surface Morphology and Corrosion Behavior. *Micromachines* **2023**, *14*, 850. [CrossRef] [PubMed]
27. Wang, S.; Yang, X.; Chen, J.; Pan, H.; Zhang, X.; Zhang, C.; Li, C.; Liu, P.; Zhang, X.; Gao, L.; et al. Effects of Building Directions on Microstructure, Impurity Elements and Mechanical Properties of NiTi Alloys Fabricated by Laser Powder Bed Fusion. *Micromachines* **2023**, *14*, 1711. [CrossRef]
28. Wang, X.; Liu, J.; Liu, H.; Zhou, Z.; Qin, Z.; Cao, J. The Influence of Laser Process Parameters on the Adhesion Strength between Electroless Copper and Carbon Fiber Composites Determined Using Response Surface Methodology. *Micromachines* **2023**, *14*, 2168. [CrossRef]

micromachines

MDPI

Article

Superelastic NiTi Functional Components by High-Precision Laser Powder Bed Fusion Process: The Critical Roles of Energy Density and Minimal Feature Size

Shuo Qu [1], Liqiang Wang [2], Junhao Ding [1], Jin Fu [3], Shiming Gao [1], Qingping Ma [1], Hui Liu [1], Mingwang Fu [3], Yang Lu [2,4,*] and Xu Song [1,*]

1 Department of Mechanical and Automation Engineering, Chinese University of Hong Kong, Shatin, Hong Kong, China; qushuo@link.cuhk.edu.hk (S.Q.); jhding@link.cuhk.edu.hk (J.D.); shiminggaozju@gmail.com (S.G.); qpma@cuhk.edu.hk (Q.M.); 1155168493@link.cuhk.edu.hk (H.L.)
2 Department of Mechanical Engineering, City University of Hong Kong, Kowloon, Hong Kong, China; liqiawang2-c@my.cityu.edu.hk
3 Department of Mechanical Engineering, The Hong Kong Polytechnic University, Hung Hom, Kowloon, Hong Kong, China; jin0103.fu@connect.polyu.hk (J.F.); ming.wang.fu@polyu.edu.hk (M.F.)
4 Department of Mechanical Engineering, The University of Hong Kong, Pokfulam, Hong Kong, China
* Correspondence: ylu1@hku.hk (Y.L.); xsong@mae.cuhk.edu.hk (X.S.)

Abstract: Additive manufacturing (AM) was recently developed for building intricate devices in many fields. Especially for laser powder bed fusion (LPBF), its high-precision manufacturing capability and adjustable process parameters are involved in tailoring the performance of functional components. NiTi is well-known as smart material utilized widely in biomedical fields thanks to its unique superelastic and shape-memory performance. However, the properties of NiTi are extremely sensitive to material microstructure, which is mainly determined by process parameters in LPBF. In this work, we choose a unique NiTi intricate component: a robotic cannula tip, in which material superelasticity is a crucial requirement as the optimal object. First, the process window was confirmed by printing thin walls and bulk structures. Then, for optimizing parameters precisely, a Gyroid-type sheet triply periodic minimal-surface (G-TPMS) structure was proposed as the standard test sample. Finally, we verified that when the wall thickness of the G-TPMS structure is smaller than 130 μm, the optimal energy density changes from 167 J/m^3 to 140 J/m^3 owing to the lower cooling rate of thinner walls. To sum up, this work puts forward a novel process optimization methodology and provides the processing guidelines for intricate NiTi components by LPBF.

Keywords: 3D printing; laser powder bed fusion; NiTi alloy; energy density; TPMS lattice; robotic cannula; mechanical testing

Citation: Qu, S.; Wang, L.; Ding, J.; Fu, J.; Gao, S.; Ma, Q.; Liu, H.; Fu, M.; Lu, Y.; Song, X. Superelastic NiTi Functional Components by High-Precision Laser Powder Bed Fusion Process: The Critical Roles of Energy Density and Minimal Feature Size. *Micromachines* **2023**, *14*, 1436. https://doi.org/10.3390/mi14071436

Academic Editors: Yang Liu, Jie Yin, Linda Ke and Kai Guan

Received: 11 June 2023
Revised: 15 July 2023
Accepted: 16 July 2023
Published: 18 July 2023

1. Introduction

NiTi, namely Nitinol, is a unique functional alloy known for its superelasticity and shape-memory property, which result from the reversible phase transformation between austenitic B2 and martensitic B19′ phases [1]. It is widely used in aerospace [2], medical biology [3–5], and other fields [6–8]. However, owing to the low thermal conductivity and poor machinability [9], traditional fabrication methods can only manufacture NiTi into simple geometric shapes [10], which cannot satisfy the requirements of intricate components with complex geometries.

Additive manufacturing (AM) has recently been employed for NiTi alloy fabrication [11–14]. Among all the AM methods, laser powder bed fusion (LPBF) is ideal for achieving intricate components with complex geometry [15,16] and high fidelity [17]. The process parameters of LPBF can greatly affect the properties of the printed NiTi components [15,18]. The scan speed, laser power, and hatch space are often adjusted to control the composition and microstructures of printed components [15,16,19]. Some researchers have also manipulated the energy density to design the phases and performance of NiTi alloys [20,21]. However, these studies focus on the process parameter optimization for bulk materials, and the study on the influence of energy density on intricate components and thin wall structures is lacking. The difficulty is that the different wall thicknesses and strut diameters of those components result in various thermal histories and cooling rates [22], which lead to different material microstructures from the bulk material by the same process parameters. Therefore, conventional test specimens, such as cubes, cylinders and dog bones, are unable to represent the property of NiTi intricate components with thin walls and complex structures accurately. The process parameters, geometrical shapes, and minimal feature sizes should be considered holistically in a comprehensive manner.

Moreover, compared to conventional LPBF, the high-precision laser powder bed fusion (HP-LPBF) method has been recently proposed to fabricate intricate components [17,23,24]. It combines fine beam size (25 μm), fine powders (0–25 μm), and thin layer thickness (10 μm) together and results in different microstructures (e.g., grain sizes, dislocation densities, and molten pool sizes), the faster cooling rate of the molten pool, lower surface roughness, higher resolution, and smaller distortion [25]. Researchers have also used HP-LPBF to achieve high-performance NiTi components. Fu et al. [23] employed HP-LPBF to fabricate NiTi samples to study their thermal and mechanical behaviors. However, the microstructural inhomogeneity of as-printed samples resulted in a suppressed martensitic transformation owing to the unexpected precipitation. It could be found that the NiTi fabrication is extremely sensitive to the manufacturing process, including raw material [26], process parameters [27,28], as well as minimal feature size [22]. Xiong et al. [29] used HP-LPBF to successfully achieve the strut lattices with excellent shape recovery. However, the optimal process parameters for lattices with inclined structures were acquired by printing thin wall and grate structures vertically. It is somehow unreasonable because the thermal diffusion of the molten pool varies with the inclination angles [30], especially for NiTi alloy.

In general, intricate components often suffer from complicated stress conditions [31], such as local compression, tension and bending. For example, when the helical robotic cannula tips bend [31], the outer parts are in tension, and the inner parts are in compression. In addition, the snake-like NiTi actuator [32] and NiTi spring [33] are compressed and stretched simultaneously at various locations in the thermal and mechanical responses. The struts of NiTi stents are also under compression and tension when supporting the bending deformation [34]. Moreover, NiTi alloys have asymmetric superelastic behavior in tension and compression modes of deformation [35], which is ascribed to the quicker martensite evolution rate under tension [36]. Thus, a comprehensive mechanical property evaluation of NiTi is necessary for intricate component fabrication. However, conducting mechanical testing (such as compression or bending) on thin wall structures is hard since it will result in the buckling of structures. Furthermore, it is unrealistic to directly fabricate and test intricate components for parameter optimization due to the high cost and low efficiency.

Therefore, in order to obtain the optimal process parameters for NiTi intricate components, a triply periodic minimal surface (TPMS) sheet structure was proposed in this work as the standard test geometry. It is a class of periodic open-cell shell lattices with large specific areas and smooth surfaces. These sheet structures were created by offsetting the zero-level mid-surface. Thus, the wall thickness could be varied easily by adjusting the level parameters of the implicit mathematical expressions. In contrast to skeleton-based TPMS, sheet-based TPMS shows superior mechanical properties for avoiding local failure and owns smaller minimal feature sizes (such as wall thickness) due to the higher surface-to-volume ratio [37]. In this case, the local failure of superelastic NiTi intricate components means not only material fracture but also the generation of irreversible martensite transition and loss of pseudo-elasticity. Among all sheet TPMS structures, Gyroid-type sheet TPMS (G-TPMS) is the most isotropic lattice structure in mechanical response [38]. Besides, G-TPMS has the least stress fluctuations during compression, which means it possesses a more uniform stress distribution and less chance for local failure among all TPMS lattice structures. Therefore, it can be employed as the standard test sample to predict the material performance in the NiTi intricate components.

In this study, the effects of the process parameters on the printing dimensions, microstructure, roughness, phase transformation behavior, and mechanical properties of NiTi by HP-LPBF were studied. Cylinders and G-TPMS structures with different wall thicknesses (65, 130, and 260 μm) were printed with different energy densities. It is demonstrated that when the components possess small wall thickness, the energy density should be reduced accordingly to balance the effect of decreasing the cooling rate to achieve good superelasticity. Finally, robotic cannula tips were fabricated and achieved full recovery with bends greater than 90 degrees at room temperature, which built the knowledge gained from the G-type TPMS sample fabrication.

2. Materials and Methods

2.1. Materials and Processes

The AM fabrication was performed using a HP-LPBF machine (Han's Laser, M100μ, Shenzhen, China). The protective atmosphere was argon with an oxygen content of less than 500 ppm. The laser source had a wavelength of 1070 nm. The substrate material was NiTi in nearly equal proportions. The beam size was 25 μm, and the layer thickness was 10 μm. NiTi powder was provided by the GKN Hoeganaes Corporation, Cinnaminson, NJ, USA. The particle size of the as-received powder ranged from 5 μm to 25 μm, and the size distribution of the powder is shown in Table 1. D_{10} indicates the diameter at 10% cumulative volume. D_{50} means the median diameter, and D_{90} means the diameter at 90% cumulative volume. The chemical composition of the powder is listed in Table 2. Thin-walled samples were printed using a single laser track with a length of 5 mm. Cube samples with dimensions of 5 mm × 5 mm × 5 mm were used to characterize the relative density (RD) and surface roughness. The size of the tensile samples was described in detail in the previous study [25]. The compression samples had a diameter of 2.6 mm and a height of 3 mm. The hatch distance was 50 μm, and the angle of rotation between each layer was 67° to reduce thermal residual stress [23]. Each layer was filled with laser tracks in a zigzag pattern. Thin-walled samples were fabricated using different powers and scan speeds, as shown in Table 3. The linear energy density (LED) was calculated as P/V, where P is the laser power, and V is the scan speed. The bulk samples fabricated using different parameters were named as shown in Table 4. The volumetric energy density (VED) was calculated as P/Vht, where h is the hatch distance of 50 μm, and t is the layer thickness of 10 μm. The scanning path was generated by Materialise Magics®, Leuven, Belgium. Figure 1 shows the models and scanning path of thin wall bulk with a hatch distance of 50 μm, which presents the LED/VED distribution with process parameters. After fabrication, the as-printed samples were cut from the substrate using wire electrical discharge machining (EDM).

Table 1. Particle size distribution of NiTi powder.

Property	Size	D10	D50	D90
Value	5–25 μm	10.0 μm	19.8 μm	28.5 μm

Table 2. Chemical composition of NiTi powder.

Element	Ni	C	O	N	Fe	H	Ti
Value/wt%	55.2	0.01	0.07	0.002	0.01	0.001	Balance

Table 3. Process parameters for thin-walled sample fabrication.

Sample Name	Laser Power (W)	Scan Speed (mm/s)	Linear Energy Density (J/m)
L50	50	1000	50
L60	60	1000	60
L63	50	800	63
L70	70	1000	70
L75	60	800	75
L83	50	600	83
L88	70	800	88
L100	60	600	100
L117	70	600	117

Table 4. Process parameters for bulk sample fabrication.

Sample Name	Laser Power (W)	Scan Speed (mm/s)	Volumetric Energy Density (J/m^3)
P5V6	50	1000	100
P6V6	60	1000	120
P7V6	50	800	125
P8V6	70	1000	140
P9V6	60	800	150
P10V6	50	600	167
P5V8	70	800	175
P6V8	60	600	200
P7V8	70	600	233
P5V10	80	600	267
P6V10	90	600	300
P7V10	100	600	333

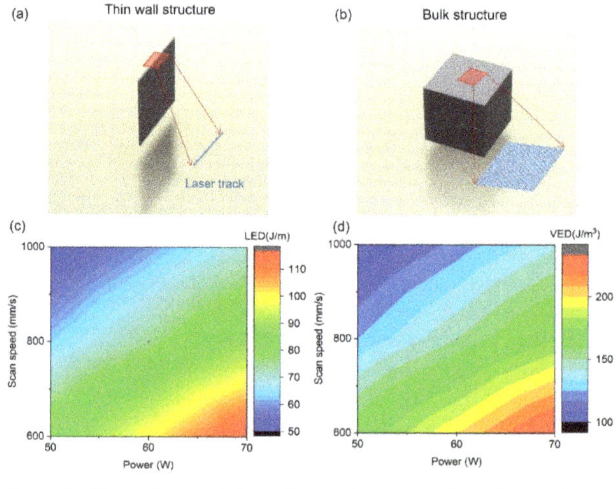

Figure 1. Designing models and scanning path of (**a**) thin wall structure, (**b**) bulk structure, and (**c**,**d**) LED and VED distribution with process parameters.

2.2. Characterization

Scanning electron microscopy (SEM) observations were conducted using a JCM-6000Plus instrument (JEOL Ltd., Tokyo, Japan). Optical microscopy (OM) observations and roughness measurements were conducted using an RH-2000 high-resolution 3D microscope (Hirox–USA, Inc., Hackensack, NJ, USA). The RD values of the samples were measured using the Archimedes method. Three NiTi samples per parameter set were weighed both in the air and in an ethanol medium using an analytical balance with an accuracy of 0.01 mg. To observe the morphology of the molten pools, the side surfaces of the samples were ground, polished, and etched using a solution containing 5% HF, 15% HNO_3, and 80% H_2O by volume fraction. X-ray diffraction (XRD) patterns, obtained using a high-resolution Rigaku SmartLab X-ray diffractometer, were used to determine the phase compositions of the different samples. The 2θ angle range was from 20° to 100°. A Mettler Toledo differential scanning calorimetry (DSC) was used to measure the phase transformation temperatures (TTs) of the as-printed NiTi samples and powder at heating and cooling rates of 10 °C/min. Tensile tests were conducted under quasi-static speed at a strain rate of 10^{-3} s^{-1}. The compression test was conducted at a strain rate of 10^{-3} s^{-1} during strain control loading and at a rate of 100 N/s during stress control loading. Mechanical tests were conducted on the three samples fabricated with each parameter set.

2.3. Geometric Modeling of Component Structures

A robotic cannula designed by Yan et al. [31] for a minimally invasive intracerebral hemorrhage (ICH) evacuation is the focus of the current study, which includes an actuation platform, pre-curved cannula body and flexible cannula tip. As one of the design criteria, the cannula tip, as shown in Figure 2a, requires at least a 90° bending capability in any direction for aspirating the peripheral ICH for clinical decompression [39], for which good superelasticity of NiTi material is required.

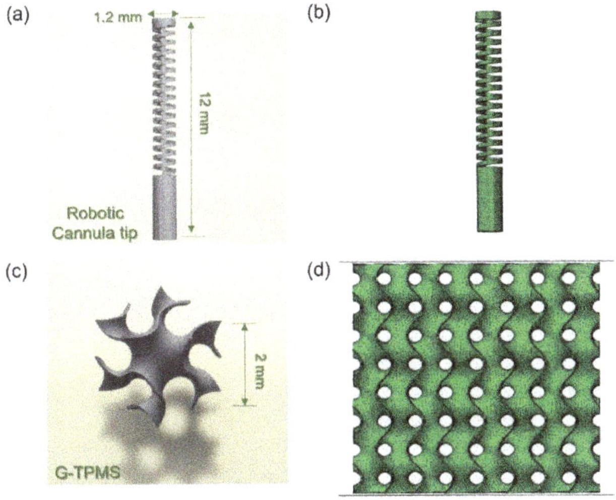

Figure 2. Models of NiTi components: (**a**) robotic cannula tip, (**b**) assembly of tip FE model, (**c**) G-TPMS lattice (unit cell), and (**d**) assembly of G-TPMS FE model.

The models with the TPMS structure, generated using MATLAB software R2020a, were subjected to compression testing. The G-TPMS cube was filled with four unit cells of a G-TPMS, which can be considered as an infinite array in the mechanical behaviors [40,41], as illustrated in Figure 2c. Three groups of G-TPMS cubes with different relative densities (10%, 20% and 40%) are presented in Table 5. The shell thickness was set as 65, 130, and 260 μm, respectively, and the unit cell is 2 mm. The G-TPMS lattice was defined by the following function [42]:

$$\phi_G(x, y, z) = \sin(wx)\cos(wy) + \sin(wz)\cos(wx) + \sin(wy)\cos(wz) = C \tag{1}$$

where $w = \frac{2\pi}{L}$, L indicates the periodic length of TPMS, and C controls the TPMS interface. Different thicknesses were obtained by varying C. The geometric model was rescaled for the fabrication of LPBF into the target unit cell size.

Table 5. Model parameters of TPMS samples.

Sample Name	Cell Size (mm)	Thickness (μm)	RD (%)
G10	2	65	10
G20	2	130	20
G40	2	260	40

To reveal local stresses, simulations on the cannula tip and TPMS shell lattices were performed using finite element (FE) modeling. The mechanical response of G-type TPMS under compression was analyzed using the finite element package Abaqus/Explicit 2017-1. Figure 2b,d show the FE model setup of TPMS and tip. TPMS structure is implemented by modeling and connecting rigid plates to its top and bottom nodes to provide the same condition of experimental tests. All degrees of freedom of the bottom plate are fixed, and the strain rate of the top plate is 10^{-3} s $^{-1}$. An average cell size of 0.2 mm forms the mesh for TPMS shells made from linear triangular prism elements (C3D6) [38]. The cannula tip is meshed by an average cell of 0.1 mm with four-node tetrahedral elements (C3D4). All degrees of freedom of the bottom of the tip are fixed. All points of the tip's top surface are coupled on the center point of the bottom surface and rotated at an angle of 90 degrees. Semi-automatic mass scaling is configured as a minimum target time increment of 5×10^{-4} s for providing a low ratio of kinetic energy to internal energy (<10%) during compression. The tensile property of NiTi produced using P7V6 for FE is shown in Figure S4. The other details can be referred from the previous report [38].

3. Results and Discussion

3.1. Effect of Process Parameters on Thin-Wall Structure Fabrication

The morphology of the powder was observed using SEM (Figure 3a,b). The phase composition of the powder was determined by XRD, as shown in Figure 3c. The phase-TTs were measured using DSC, as shown in Figure 3d. The R-phase was detected before the martensite phase during the heating and cooling process. The R-phase transformation peak is caused by the thermoelastic transition between the austenite and R-phases. The formation of the R-phase has been attributed to crystallographic defects. Thin-walled structures were first printed to evaluate the printing resolution of HP-LPBF and the influence of parameters on molten pool size and microstructure. Figure 4 shows the characterization of thin-walled structure (TWS) samples fabricated using different parameters in Table 3. Figure 4a illustrates the top cross-sectional morphologies of samples. All TWS samples were printed layer-by-layer using a single laser track. Therefore, the wall thickness was the smallest printing size for the corresponding process parameter. The wall thickness was measured, as shown in Figure 4a. The TWS wall thickness results are shown in Figure 4b. The highest resolution of 62 μm was achieved in L50, using a laser power of 50 W and a scan speed of 1000 mm/s.

Moreover, the wall thickness increased with increasing LED (Table 3). Figure 4c shows the XRD diffraction patterns of the groups of low- (L83–L117) and high-speed samples (L50–L70). Similar diffraction patterns are observed for these samples, which indicates that parameter variation has a limited effect on phase composition. The B19′ NiTi phase was the dominant phase. The Ni_3Ti phase, which consumes the Ni content of the matrix, could also be discerned. The secondary Ni-rich phase resulted from precipitation during the layer-by-layer fabrication process. In addition, peaks of a Ti-rich phase ($Ti4Ni_2Ox/Ti_2Ni$) and Ni_4Ti_3 were also detected, which resulted from microstructural inhomogeneity [43]. Figure 4d illustrates the thermally induced phase-transformation behavior of L50–L117, fabricated with different LEDs, respectively, as analyzed using DSC. The austenitic and martensitic transformation peak temperatures (Ap, Mp) of samples were increased by the increase of LED due to the evaporation of Ni element. The TTs of the TWS samples are similar to those of the original powder shown in Figure 3. Therefore, it can be concluded that the loss of Ni content is low during TWSs fabrication. Generally, in TWS fabrication, it is demonstrated that 62 µm of resolution and controllable microstructure of TWS can be realized by HP-LPBF.

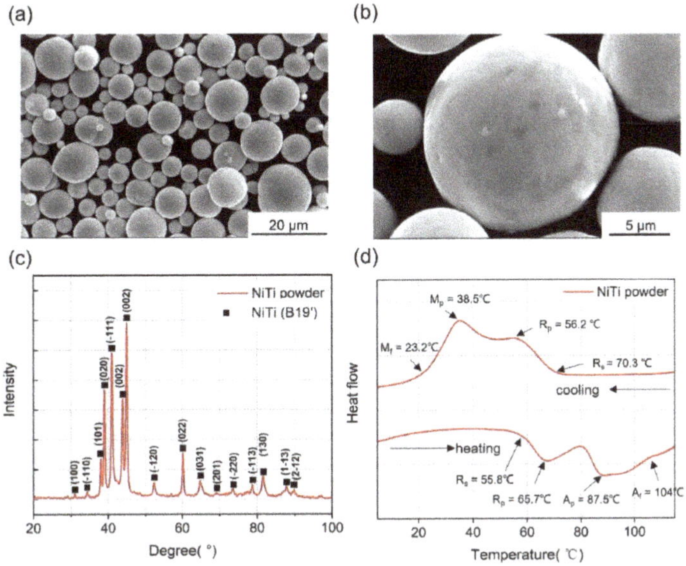

Figure 3. Characterization of NiTi original powder. (**a**,**b**) SEM observations at different magnifications, (**c**) XRD patterns, and (**d**) DSC curves.

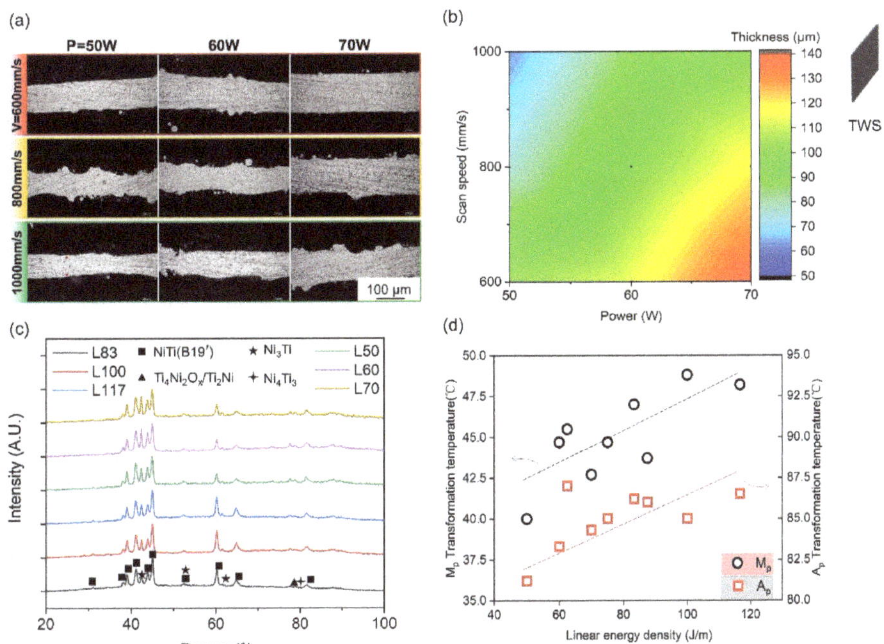

Figure 4. TWS characterizations: (**a**) Cross-sectional morphologies, (**b**) wall thicknesses of TWS samples, (**c**) XRD patterns, and (**d**) DSC analyses of phase transformation behaviors.

3.2. Effect of Process Parameters on Bulk Fabrication

To verify the differences between bulk and thin wall, the cubes were printed to demonstrate the variety of microstructures and phases by VED. Cubes were fabricated using different process parameters given in Table 4. The bulk densification optimization is shown in Figure S1. When the VED is less than 160 J/mm^3, the NiTi alloy cube can be printed with RD > 99.5%, which is considered highly dense [23].

Figure 5 presents the microstructural characterizations of the cube samples. Figure 5a,b show the depths and half-widths of molten pools in the cubes printed with laser powers of 50–70 W and scan speeds of 600–1000 mm/s. Obviously, molten pool depth and width decreased with increasing scan speed and decreasing laser power. The measurement method is shown in the insets. In Figure 5d,i, the polished and etched side surfaces of the cubes show the molten pool morphologies. Vertical cracks occurred in the middles of the molten pools; pores appeared at the bottoms. These pores could be attributed to the keyhole fluctuation in the fabrication process [44,45]. Figure 5c shows the molten pool depth-to-width ratio (D/W) obtained by different printing parameters. Based on the side-surface morphologies, the samples containing pores and cracks were marked with red circles, while dense samples were marked with blue circles. It can be concluded that when D/W exceeded 1.8, the melting condition became unstable [44], and pores and cracks appeared. Figure S2 shows the relationship between D/W and RD. It can be concluded that the RD decreased sharply when D/W was larger than 1.8. In terms of surface roughness, Figure S3 displays the results of the top and side surface roughness of the as-printed cubes. Top surface Ra values were from 0.4 μm to 1.45 μm, as presented in Figure S3c. Side surface Ra values for side roughness were between 6.48 and 10 μm, as presented in Figure S3d. It demonstrated that the process parameter used here can build dense bulk samples with stable molten pools, and this is because no spatter was observed on the top surface. In regard to spatters on the side surface, they were formed due to the interaction between powders and high-temperature metal liquid and thus had an average size of about half of the powder size [46].

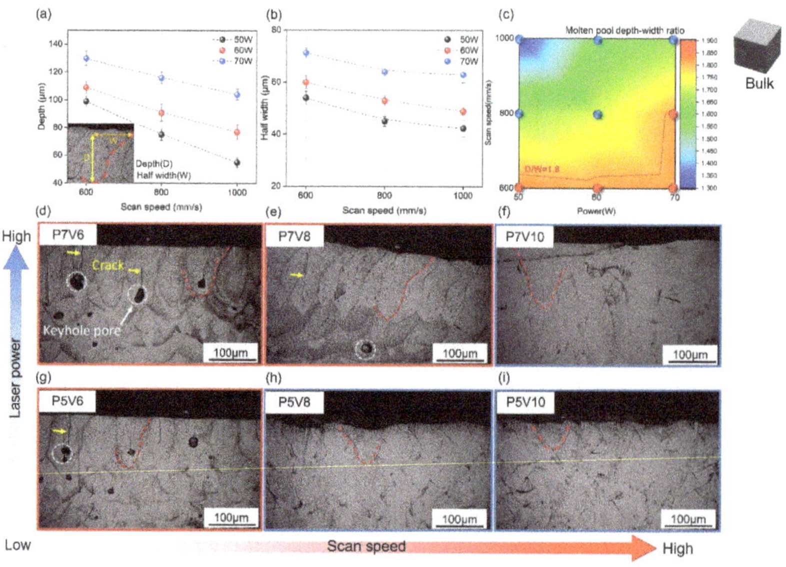

Figure 5. Microstructural characterization of cubes fabricated with different parameters. (**a**) Depth of molten pools. (**b**) Half-width of molten pools. (**c**) Depth-to-width ratios of molten pools. (**d–i**) Side surface morphologies of cubes.

Figure 6a illustrates the XRD patterns of the as-printed NiTi samples obtained using different parameters. The B2 and B19′ phases were detected in all samples. The main secondary phases changed with increases in the VED. Ni-rich phases of Ni_4Ti_3 and Ni_3Ti were observed in cubes fabricated with low energy densities. This is attributed to the lower temperature of the molten pool and faster cooling rate enabled by energy densities of <150 J/m^3, which lead to less Ni loss [47]. When the VED was greater than 150 J/m^3, the Ti_4Ni_2Ox/Ti_2Ni phase appeared. When the VED was greater than 200 J/m^3, peaks of the Ni-rich phases of Ni_4Ti_3 and Ni_3Ti could not be detected because of the large loss of Ni in the high-energy-density LPBF process. Figure 6b illustrates the phase transformation behaviors of the as-printed NiTi samples fabricated using different parameters. The phase TTs of all cubes (Ap = 85–95 °C; Mp = 52–60 °C) were higher than those of the TWS samples and original powder owing to their greater loss of Ni, which resulted from the overlapped molten pools in the printed bulk, as shown in Figure 5. Moreover, the austenitic and martensitic peak temperatures increased with increasing VED, as shown in Figure 6c. The relationship between temperature and VED was determined by linear fitting. The corresponding equations of linear fit between Ap or Mp and VED are given in Figure 6c.

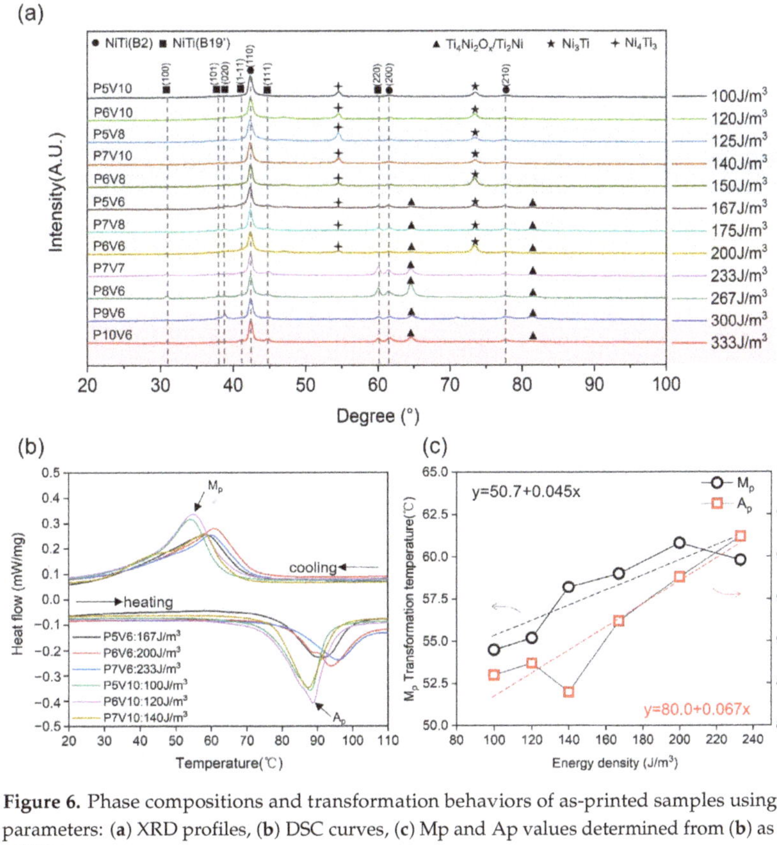

Figure 6. Phase compositions and transformation behaviors of as-printed samples using different parameters: (**a**) XRD profiles, (**b**) DSC curves, (**c**) Mp and Ap values determined from (**b**) as functions of VED.

Overall, phase compositions are varied with energy densities, which do not occur in TWS. It shows that the overlapping of molten pools in bulk printing and different thermal field makes the relationship of bulk between parameters and properties different from that of TWS. Hence, when the minimal feature size decreases down to several molten pool sizes, as shown in Figure 2, the relationship between parameters and properties is different for TWS and bulk structure. In addition, it is hard to conduct mechanical experiments on TWS, such as compression tests. Buckling is prone to occur in the bending/compression testing. Moreover, the tolerance of defects (cracks, pores, and unmelted powders) is also low for TWS mechanical testing. Hence, it is necessary to use a new structure to evaluate the influence of parameters on the NiTi's intricate components.

3.3. Superelastic Property of NiTi

3.3.1. Effect of Process Parameters on Bulk Samples

Figure 7 depicts the loading-unloading compression results of the P5V6, P5V10, P7V6, and P7V10 samples. The compression test was conducted by two types of modes (same stress and same strain). Figure 7a,c shows the representative stress-strain curves of four parameters. Figure 7b,d show irreversible/recovered strain and recovery ratio of different samples arranged from low to high VED. Irreversible strain means the residual strain after the unloading process. Recovered strain indicates the resilient strain during the unloading process. The recovery ratio is defined as the ratio of recovered strain to total strain (recovery strain + irreversible strain). As a result, P5V6 (167 J/m^3) and P7V10 (140 J/m^3) samples exhibited better elastic recovery properties than P5V10 (100 J/m^3) and P7V6 (233 J/m^3) samples in two types of modes (same stress and same strain). P5V6 and P7V10 samples showed a recovery ratio higher than 0.35, and P5V10 and P7V6 samples depicted a recovery ratio lower than 0.30 under the same compression stress. In terms of the same compression strain experiment, P5V6 and P7V10 samples showed a recovery ratio higher than 0.40, and P5V10 and P7V6 samples showed a recovery ratio lower than 0.35. Overall, P5V6 samples possess the best shape recovery property. From Figure 6, it could be found that there are Ni_4Ti_3 and Ti_2Ni phases in P5V6 cube samples. It is well known that the presence of Ni_4Ti_3 is beneficial to the superelastic property of NiTi components [48]. A lack of energy input leads to the heating temperature and duration not being sufficient to allow the Ni_4Ti_3 precipitation to fully form. Thus, the Ni_4Ti_3 precipitation declines in the samples by P5V10 [49]. The P7V10 and P5V6 samples were printed by higher energy densities with higher content of Ni_4Ti_3 precipitation. However, regarding the P7V6 samples, the Ti_2Ni phase precipitated excessively, and the Ni element evaporated, which is harmful to superelasticity. Generally, good control of VED for the balance of Ni_4Ti_3 segregation and Ni evaporation was the key to the deformation recovery property.

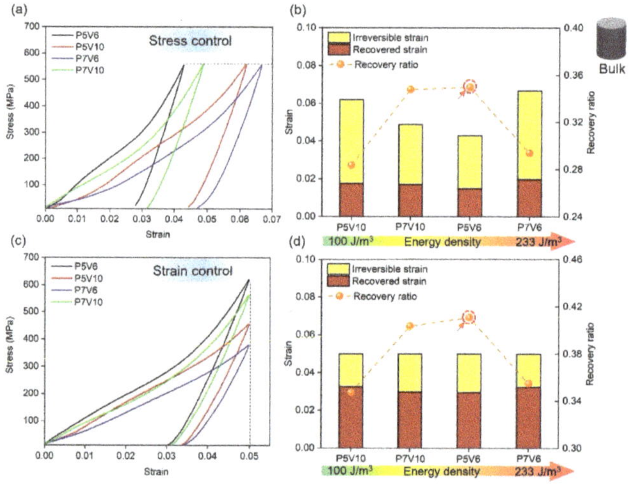

Figure 7. Shape recovery behavior of NiTi samples of P5V6/P5V10/P7V6/P7V10 (**a**) same stress compression, (**c**) same strain compression and (**b**,**d**) irreversible/ recovered strain and recovery ratio of corresponding samples.

3.3.2. Effect of Process Parameters on TPMS Samples

FE simulation was conducted first to estimate the stress distribution of G-TPMS and cannula tip. Firstly, the tensile properties of NiTi fabricated with different parameters are shown in Figure S4, which were used for FE simulation. The phase transformation and work hardening stages are similar for the variety of power shown in Figure S4a. Moreover, Figure S4b presents that higher scan speed leads to a lower work hardening rate owing to fewer precipitations in lower VED. P7V6, which exhibits the highest strength and elongation, is used for simulation. Figure 8a–c shows the stress distribution of G-type TPMS lattice under 6% compressed strain. The tensile stress (red and yellow) and compressive stress (blue and green) are distributed in a regular periodic pattern on the surfaces and internal structures of the lattice (Figure 8a,b). Typically, the regions with larger inclination angles possess larger maximum principal stress and therefore play a more significant role in bearing external loads than those with smaller inclination angles. Moreover, the lattice tends to exhibit contrary tensile and compressive stress states on the two sides of the surface, which indicates an obvious bending behavior (Figure 8c). Furthermore, Figure 8d–f present the stress distribution of the cannula tip with a bend. A similar stress distribution on the surface was also observed in the bending tip. Figure S5 shows the tension and compression stress conditions of all elements extracted from the FE simulation of TPMS and cannula tip. As a result, the overall stress distribution for TPMS and tip is composed of similar tension and compression stress status. Thus, the TPMS structures under compression can well represent the stress condition of cannula tips with a bend.

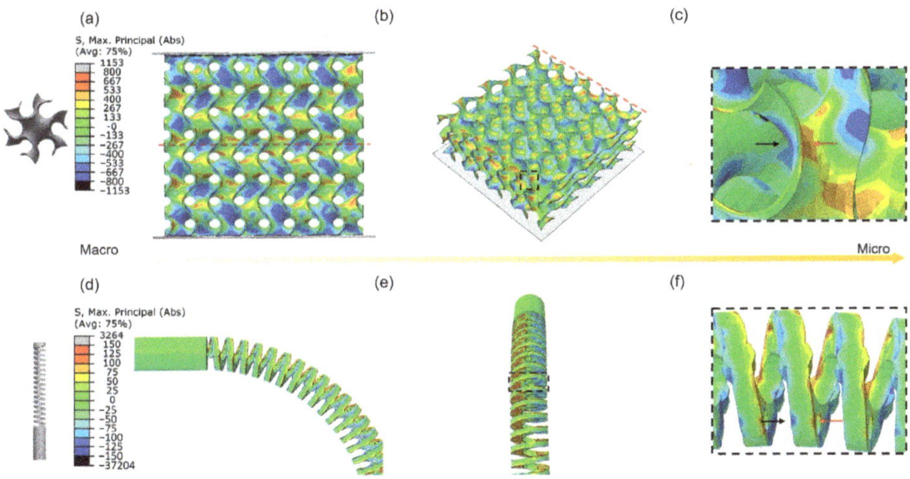

Figure 8. FE results of stress distribution of (**a**–**c**) G-type TPMS lattice under 6% compressed strain (**a**) front view, (**b**) cross-section view, (**c**) local surface, (**d**–**f**) cannula tip with a bend, (**d**) front view, (**b**) side view, and (**c**) local surface.

Figure 9 shows the as-printed TPMS lattice structures of NiTi with RD of 10%/20%/40%. To evaluate the effects of parameters on samples with different RD, all the lattice structures were fabricated using P5V6/P5V10/P7V10. Figure 9d–f illustrates the top surface of the three structures. The wall thickness also varied with cell size, at 65, 130, and 260 μm characterized by SEM. Additionally, the G-TPMS structures failed to be printed using P7V6 due to its excessive energy input during fabrication, which led to the distortion of thin-walled structures. It verified that the optimal process parameters of bulk printing cannot be directly used for intricate component printing.

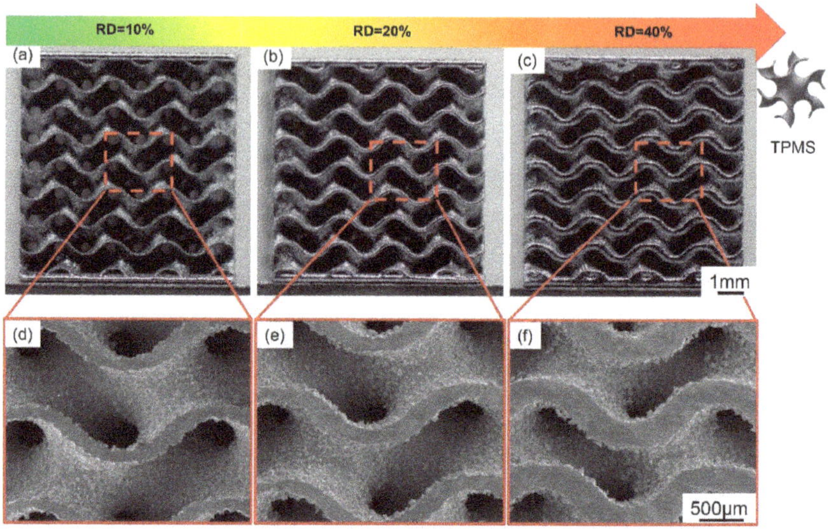

Figure 9. (**a–c**) Images of as-fabricated NiTi G-TPMS lattice structures with different relative densities of 10%/20%/40%, and (**d–f**) SEM images of corresponding TPMS lattices.

Figure 10 depicts the shape recovery properties of TPMS with different RDs using stress or strain control. It can be found that the recovery property of P5V10 is the worst one in the same stress or strain compression experiments due to less precipitation of Ni_4Ti_3 [49]. The deformation recovery properties of P7V10 and P5V6 are approximately equal in performance, and P5V6 samples exhibit a better recovery ratio in line with the cylinder compression results shown in Figure 7. However, P7V10 is better for TPMS with RD of 10%. This is because the heat-affected zone and thermal gradient, which are affected by the thermal diffusion of molten pools, will severely influence the precipitation of the second phase (Ni_4Ti_3, Ti_2Ni).

It is well known that the thermal diffusion coefficient of powder is much lower than as-printed solid parts [24]. Therefore, the cooling rate will be lower when the wall thickness is smaller. For TPMS sheet structures, this equals the condition of a lower RD because thin walls have more surfaces in contact with the powder and fewer surfaces in contact with the solid. Hence, the P5V6 parameter will result in more Ni content evaporation and Ti_2Ni precipitation in 10% RD than 40% RD. Furthermore, P7V10 (140 J/m^3) owns less VED than P5V6 (167 J/m^3), which will generate a similar thermal field and precipitation condition for thin wall printing to that of P5V6 for bulk printing. Additionally, P5V6 and P7V10 result in identical recovery properties for TPMS with RD of 20% (130 μm). Overall, when the structure wall thickness is less than 130 μm, P7V10 is clearly better than P5V6 for its shape recovery property.

3.4. Superelastic Performance of Robot Cannula Tips: A Case Study

Robotic cannulas are used for minimally invasive intracerebral hemorrhage (ICH) evacuation [31]; such structures require high resolution, low roughness, and excellent superelasticity to achieve at least 90-degree bending capability in all directions for dexterous tip motion [39]. From the results of the process parameter optimization, it can be concluded that the VED influences the shape recovery properties of the as-printed parts. The cannula tip is the structure spiral type with a 120 μm strut diameter designed for ICH. Thus, P7V10 is more suitable to print this component to obtain better superelasticity. The morphology of the cannula tip was characterized using OM and SEM, as shown in Figure 11. It can be seen that the holes for the actuation cable manipulation and helix structure of the tip were successfully fabricated at high resolution. Figure 12a–d illustrate that the cannula tip printed with P7V10 parameters exhibited excellent superelastic properties (>90° of bending) at room temperature (see Supplementary Video S1). Although the Mp of the part printed using P7V10 parameters was higher than room temperature, as shown in Figure 6, the matrix also exhibited an austenitic phase in the XRD patterns, and the Ms was close to room temperature. Figure 12e–h show that the cannula tip fabricated with P5V6 parameters remains an unrecovered deformation with a degree of 10.2° after bending (see Supplementary Video S2). It is attributed that the optimal parameter for the minimal feature size (120 μm) of a tip is P7V10, shown in Figure 10. Hence, cannula tips with optimal parameters manifest excellent superelastic performance. HP-LPBF is, therefore, a promising method for fabricating high-performance and high-precision NiTi intricate components.

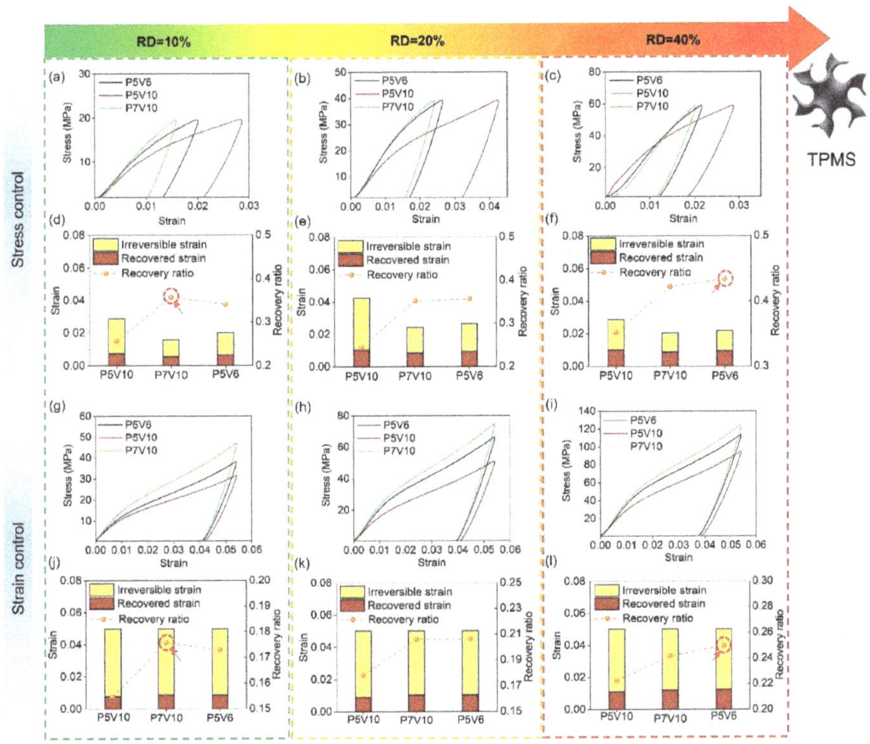

Figure 10. Shape recovery properties of G-TPMS structures, loading-unloading responses of different RD (10%, 20%, 40%) (**a–c**) at stress control, (**g–i**) at strain control, irreversible/recovered strain and recovered ratio of TPMS with different RD (**d–f**) at stress control, and (**j–l**) at strain control.

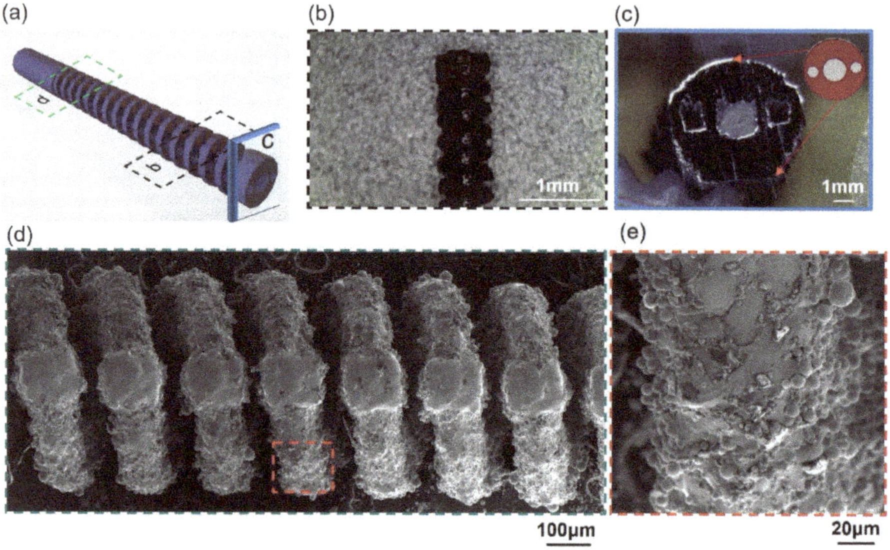

Figure 11. As-printed robotic cannula tip using P7V10 parameters: (**a**) Model, (**b**) top position by OM, (**c**) top cross-section by OM, (**d**) middle position by SEM, and (**e**) high-magnification SEM image of local position.

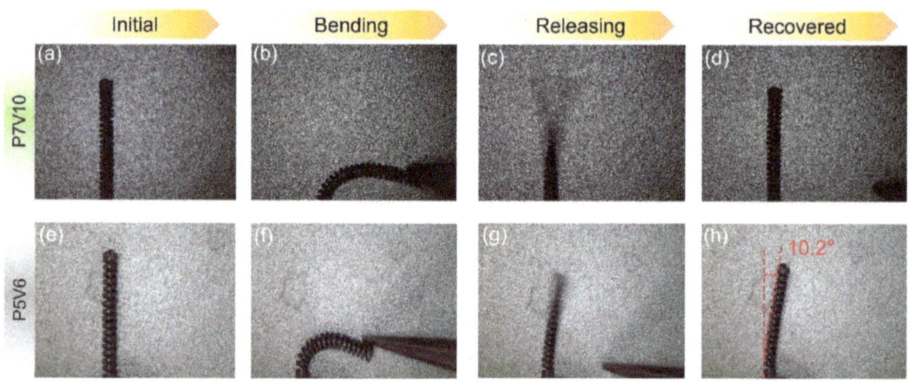

Figure 12. Superelastic properties of NiTi robotic cannula tip using P7V10 and P5V6 parameters at room temperature.

4. Conclusions

Superelastic NiTi intricate components were successfully fabricated by HP-LPBF in this study. Extensive parameter studies were conducted to reveal their influences on the microstructure and mechanical properties of as-printed components with different minimal feature sizes (thin wall structure, bulk samples, TPMS structures, etc.). TPMS structures were proposed and verified to be a good candidate as the standard test for parameter study of printing NiTi intricate components and investigation of NiTi thin wall structure mechanical property. Finally, a robotic cannula tip with fine minimal feature size and high superelasticity was fabricated using HP-LPBF as a case study. The main conclusions of this study are as follows:

(1) Thin-walled structures were fabricated using parameters that yielded different LEDs. The LED had a significant influence on the wall thickness. The thinnest single track of 62 μm was achieved with a laser power of 50 W and a scan speed of 1000 mm/s.
(2) Bulk components with an RD higher than 99.5% could be printed with a VED lower than 160 J/mm^3. Pores and cracks were observed in the microstructure when the D/W ratio of the molten pools was larger than 1.8. The precipitated phase changed from Ni_4Ti_3 and Ni_3Ti to $Ti_4Ni_2O_x$/Ti_2Ni with the increase in VED, which also caused an increase in the TTs.
(3) Cylinders by VED of 140~167 J/m^3 had better shape recovery properties than other parameters (100 J/m^3 or 233 J/m^3) due to a good controlling of VED for the balance of Ni_4Ti_3 segregation and Ni evaporation.
(4) When the wall thickness of the G-TPMS lattice is smaller than 130 μm, the optimal VED for superelasticity is changed from 167 J/m^3 to 140 J/m^3 owing to the lower cooling rate of thinner walls.
(5) The robotic cannula tips with a 120 μm strut diameter fabricated with 140 J/m^3 VED exhibited excellent superelasticity properties thanks to the tailored phase compositions. It reveals the coupling effects of parameters, microstructure and feature size in LPBF, and proposes a new approach to processing high-performance intricate NiTi components.

Supplementary Materials: The following supporting information can be downloaded at: https://www.mdpi.com/article/10.3390/mi14071436/s1, Figure S1. RDs of as-printed cubes; Figure S2. The relationship between D/W and RD of NiTi samples; Figure S3. Surface roughness characterization of cubes fabricated using different parameters: (a) top surface morphology, (b) side surface morphology, (c) Ra of top surfaces obtained using different parameters, (d) Ra of side surfaces obtained using different parameters. Figure S4. Representative tensile curves of samples fabricated with varied (a) laser power and (b) scan speed, (c) UTS and elongation values of NiTi samples with different parameters, (d) UTS and elongation values of four typical samples versus energy density. Figure S5. Tension and compression stress conditions of elements extracted from FE simulation of the G-TPMS and cannula tip. Video S1: Superelastic properties of NiTi robotic cannula tip using P7V10. Video S2: Superelastic properties of NiTi robotic cannula tip using P5V6.

Author Contributions: Conceptualization, X.S. and S.Q.; methodology, S.Q., L.W. and J.D.; software, S.Q.; validation, S.Q., L.W., J.D. and J.F.; formal analysis, S.Q., J.D. and S.G.; investigation, S.Q., L.W. and J.D.; resources, X.S.; data curation, X.S.; writing—original draft preparation, S.Q.; writing—review and editing, J.F., S.G., Q.M., H.L. and X.S.; visualization, S.Q.; supervision, X.S.; project administration, X.S.; funding acquisition, X.S., M.F. and Y.L. All authors have read and agreed to the published version of the manuscript.

Funding: This research is supported by the Innovation and Technology Fund of the Government of the Hong Kong Special Administrative Region (Project No. ITS/008/19), the Shun Hing Institute of Advanced Engineering, The Chinese University of Hong Kong (Project No. RNE-p2-21), the collaborative research fund of the research grants council of Hong Kong (C4074-22GF) and the Changsha Municipal Science and Technology Bureau under Project No. kh2201035 and the City University of Hong Kong under Project No. 7020008.

Data Availability Statement: The data presented in this study are available on request from the corresponding author. The data are not publicly available due to project requirements.

Conflicts of Interest: The authors declare no conflict of interest.

References

1. Otsuka, K.; Ren, X. Physical metallurgy of Ti–Ni-based shape memory alloys. *Prog. Mater. Sci.* **2005**, *50*, 511–678. [CrossRef]
2. Chau, E.T.F.; Friend, C.M.; Allen, D.M.; Hora, J.; Webster, J.R. A technical and economic appraisal of shape memory alloys for aerospace applications. *Mater. Sci. Eng. A* **2006**, *438–440*, 589–592. [CrossRef]
3. Zhang, B.; Chen, J.; Coddet, C. Microstructure and Transformation Behavior of in-situ Shape Memory Alloys by Selective Laser Melting Ti–Ni Mixed Powder. *J. Mater. Sci. Technol.* **2013**, *29*, 863–867. [CrossRef]

4. Mohd Jani, J.; Leary, M.; Subic, A.; Gibson, M.A. A review of shape memory alloy research, applications and opportunities. *Mater. Des.* **2014**, *56*, 1078–1113. [CrossRef]
5. Sandoval, L.; Haskins, J.B.; Lawson, J.W. Stability, structure, and suppression of the martensitic transition temperature by B19′ compound twins in NiTi: Ab initio and classical simulations. *Acta Mater.* **2018**, *154*, 182–189. [CrossRef]
6. Krause, D.; Thomas, B.; Leinenbach, C.; Eifler, D.; Minay, E.J.; Boccaccini, A.R. The electrophoretic deposition of Bioglass® particles on stainless steel and Nitinol substrates. *Surf. Coat. Technol.* **2006**, *200*, 4835–4845. [CrossRef]
7. Dadbakhsh, S.; Speirs, M.; Van Humbeeck, J.; Kruth, J.-P. Laser additive manufacturing of bulk and porous shape-memory NiTi alloys: From processes to potential biomedical applications. *MRS Bull.* **2016**, *41*, 765–774. [CrossRef]
8. Guo, W.; Kato, H. Development of a high-damping NiTi shape-memory-alloy-based composite. *Mater. Lett.* **2015**, *158*, 1–4. [CrossRef]
9. Ma, C.; Wu, M.; Dai, D.; Xia, M. Stress-induced heterogeneous transformation and recoverable behavior of laser powder bed fused Ni-rich Ni50.6Ti49.4 alloys without post treatment. *J. Alloys Compd.* **2022**, *905*, 164212. [CrossRef]
10. Elahinia, M.; Shayesteh Moghaddam, N.; Amerinatanzi, A.; Saedi, S.; Toker, G.P.; Karaca, H.; Bigelow, G.S.; Benafan, O. Additive manufacturing of NiTiHf high temperature shape memory alloy. *Scr. Mater.* **2018**, *145*, 90–94. [CrossRef]
11. Hou, H.; Simsek, E.; Ma, T.; Johnson, N.S.; Qian, S.; Cissé, C.; Stasak, D.; Al Hasan, N.; Zhou, L.; Hwang, Y.; et al. Fatigue-resistant high-performance elastocaloric materials made by additive manufacturing. *Science* **2019**, *366*, 1116–1121. [CrossRef] [PubMed]
12. Speirs, M.; Van Hooreweder, B.; Van Humbeeck, J.; Kruth, J.P. Fatigue behaviour of NiTi shape memory alloy scaffolds produced by SLM, a unit cell design comparison. *J. Mech. Behav. Biomed. Mater.* **2017**, *70*, 53–59. [CrossRef] [PubMed]
13. Dadbakhsh, S.; Speirs, M.; Kruth, J.-P.; Van Humbeeck, J. Influence of SLM on shape memory and compression behaviour of NiTi scaffolds. *CIRP Ann.* **2015**, *64*, 209–212. [CrossRef]
14. Elahinia, M.; Shayesteh Moghaddam, N.; Taheri Andani, M.; Amerinatanzi, A.; Bimber, B.A.; Hamilton, R.F. Fabrication of NiTi through additive manufacturing: A review. *Prog. Mater. Sci.* **2016**, *83*, 630–663. [CrossRef]
15. Guo, W.; Feng, B.; Yang, Y.; Ren, Y.; Liu, Y.; Yang, H.; Yang, Q.; Cui, L.; Tong, X.; Hao, S. Effect of laser scanning speed on the microstructure, phase transformation and mechanical property of NiTi alloys fabricated by LPBF. *Mater. Des.* **2022**, *215*, 110460. [CrossRef]
16. Saedi, S.; Shayesteh Moghaddam, N.; Amerinatanzi, A.; Elahinia, M.; Karaca, H.E. On the effects of selective laser melting process parameters on microstructure and thermomechanical response of Ni-rich NiTi. *Acta Mater.* **2018**, *144*, 552–560. [CrossRef]
17. Qu, S.; Ding, J.; Fu, J.; Fu, M.; Zhang, B.; Song, X. High-precision laser powder bed fusion processing of pure copper. *Addit. Manuf.* **2021**, *48*, 102417. [CrossRef]
18. Shi, G.; Li, L.; Yu, Z.; Liu, R.; Sha, P.; Xu, Z.; Guo, Y.; Xi, R.; Liu, J.; Xin, R.; et al. The interaction effect of process parameters on the phase transformation behavior and tensile properties in additive manufacturing of Ni-rich NiTi alloy. *J. Manuf. Process.* **2022**, *77*, 539–550. [CrossRef]
19. Chen, W.; Yang, Q.; Huang, S.; Huang, S.; Kruzic, J.J.; Li, X. Laser power modulated microstructure evolution, phase transformation and mechanical properties in NiTi fabricated by laser powder bed fusion. *J. Alloys Compd.* **2021**, *861*, 157959. [CrossRef]
20. Ye, D.; Li, S.F.; Misra, R.D.K.; Zheng, R.; Yang, Y.F. Ni-loss compensation and thermomechanical property recovery of 3D printed NiTi alloys by pre-coating Ni on NiTi powder. *Addit. Manuf.* **2021**, *47*, 102344. [CrossRef]
21. Khanlari, K.; Shi, Q.; Li, K.; Hu, K.; Cao, P.; Liu, X. Effects of printing volumetric energy densities and post-processing treatments on the microstructural properties, phase transformation temperatures and hardness of near-equiatomic NiTinol parts fabricated by a laser powder bed fusion technique. *Intermetallics* **2021**, *131*, 107088. [CrossRef]
22. Guo, F.; Guo, Y.; Kong, X.; Xiong, Z.; Hao, S. The Effect of Wall Thickness and Scanning Speed on the Martensitic Transformation and Tensile Properties of Selective Laser Melted NiTi Thin-Wall Structures. *Metals* **2022**, *12*, 519. [CrossRef]
23. Fu, J.; Hu, Z.; Song, X.; Zhai, W.; Long, Y.; Li, H.; Fu, M. Micro selective laser melting of NiTi shape memory alloy: Defects, microstructures and thermal/mechanical properties. *Opt. Laser Technol.* **2020**, *131*, 106374. [CrossRef]
24. Qu, S.; Ding, J.; Song, X. Achieving Triply Periodic Minimal Surface Thin-Walled Structures by Micro Laser Powder Bed Fusion Process. *Micromachines* **2021**, *12*, 705. [CrossRef]
25. Fu, J.; Qu, S.; Ding, J.; Song, X.; Fu, M.W. Comparison of the microstructure, mechanical properties and distortion of stainless steel 316 L fabricated by micro and conventional laser powder bed fusion. *Addit. Manuf.* **2021**, *44*, 102067. [CrossRef]
26. Saedi, S.; Turabi, A.S.; Andani, M.T.; Moghaddam, N.S.; Elahinia, M.; Karaca, H.E. Texture, aging, and superelasticity of selective laser melting fabricated Ni-rich NiTi alloys. *Mater. Sci. Eng. A* **2017**, *686*, 1–10. [CrossRef]
27. Yu, Z.; Xu, Z.; Guo, Y.; Sha, P.; Liu, R.; Xin, R.; Li, L.; Chen, L.; Wang, X.; Zhang, Z.; et al. Analysis of microstructure, mechanical properties, wear characteristics and corrosion behavior of SLM-NiTi under different process parameters. *J. Manuf. Process.* **2022**, *75*, 637–650. [CrossRef]
28. Feng, B.; Wang, C.; Zhang, Q.; Ren, Y.; Cui, L.; Yang, Q.; Hao, S. Effect of laser hatch spacing on the pore defects, phase transformation and properties of selective laser melting fabricated NiTi shape memory alloys. *Mater. Sci. Eng. A* **2022**, *840*, 142965. [CrossRef]
29. Xiong, Z.; Li, H.; Yang, H.; Yang, Y.; Liu, Y.; Cui, L.; Li, X.; Masseling, L.; Shen, L.; Hao, S. Micro laser powder bed fusion of NiTi alloys with superior mechanical property and shape recovery function. *Addit. Manuf.* **2022**, *57*, 102960. [CrossRef]
30. Ding, J.; Qu, S.; Zhang, L.; Wang, M.Y.; Song, X. Geometric deviation and compensation for thin-walled shell lattice structures fabricated by high precision laser powder bed fusion. *Addit. Manuf.* **2022**, *58*, 103061. [CrossRef]

31. Yan, J.; Chen, J.; Chen, J.; Yan, W.; Ding, Q.; Yan, K.; Du, J.; Lam, C.P.; Wong, G.K.C.; Cheng, S.S. A Continuum Robotic Cannula with Tip Following Capability and Distal Dexterity for Intracerebral Hemorrhage Evacuation. *IEEE Trans. Biomed. Eng.* **2022**, *69*, 2958–2969. [CrossRef] [PubMed]
32. Nespoli, A.; Biffi, C.A.; Previtali, B.; Villa, E.; Tuissi, A. Laser and Surface Processes of NiTi Shape Memory Elements for Micro-actuation. *Metall. Mater. Trans. A* **2014**, *45*, 2242–2249. [CrossRef]
33. AbuZaiter, A.; Hikmat, O.F.; Nafea, M.; Ali, M.S.M. Design and fabrication of a novel XYθz monolithic micro-positioning stage driven by NiTi shape-memory-alloy actuators. *Smart. Mater. Struct.* **2016**, *25*, 105004. [CrossRef]
34. Chun, Y.J.; Levi, D.S.; Mohanchandra, K.P.; Fishbein, M.C.; Carman, G.P. Novel micro-patterning processes for thin film NiTi vascular devices. *Smart. Mater. Struct.* **2010**, *19*, 105021. [CrossRef]
35. Safdel, A.; Zarei-Hanzaki, A.; Abedi, H.R.; Pourbabak, S.; Schryvers, D.; Basu, R. Asymmetrical superelastic behavior of thermomechanically processed semi-equiatomic NiTi alloy in tensile and compressive modes of deformation. *J. Alloys Compd.* **2021**, *878*, 160443. [CrossRef]
36. Gall, K.; Sehitoglu, H. The role of texture in tension–compression asymmetry in polycrystalline NiTi. *Int. J. Plast.* **1999**, *15*, 69–92. [CrossRef]
37. Al-Ketan, O.; Rowshan, R.; Abu Al-Rub, R.K. Topology-mechanical property relationship of 3D printed strut, skeletal, and sheet based periodic metallic cellular materials. *Addit. Manuf.* **2018**, *19*, 167–183. [CrossRef]
38. Fu, J.; Ding, J.; Qu, S.; Zhang, L.; Wang, M.Y.; Fu, M.W.; Song, X. Improved light-weighting potential of SS316L triply periodic minimal surface shell lattices by micro laser powder bed fusion. *Mater. Des.* **2022**, *222*, 111018. [CrossRef]
39. Simard, J.M.; Thakor, N.V.; Fukuda, T.; Hogan, N.; Carmena, J.M.; Arkin, R. Plenary lectures: Problems in neurosurgery—A rich environment for engineer. In Proceedings of the 2012 4th IEEE RAS & EMBS International Conference on Biomedical Robotics and Biomechatronics (BioRob), Rome, Italy, 24–27 June 2012; pp. l–lv.
40. Guo, X.; Ding, J.; Li, X.; Qu, S.; Song, X.; Fuh, J.Y.H.; Lu, W.F.; Zhai, W. Enhancement in the mechanical behaviour of a Schwarz Primitive periodic minimal surface lattice structure design. *Int. J. Mech. Sci.* **2022**, *216*, 106977. [CrossRef]
41. Yin, H.; Zheng, X.; Wen, G.; Zhang, C.; Wu, Z. Design optimization of a novel bio-inspired 3D porous structure for crashworthiness. *Compos. Struct.* **2021**, *255*, 112897. [CrossRef]
42. Ding, J.; Zou, Q.; Qu, S.; Bartolo, P.; Song, X.; Wang, C.C.L. STL-free design and manufacturing paradigm for high-precision powder bed fusion. *CIRP Ann.* **2021**, *70*, 167–170. [CrossRef]
43. Wan, X.; Feng, Y.; Lin, X.; Tan, H. Large superelastic recovery and elastocaloric effect in as-deposited additive manufactured Ni50.8Ti49.2 alloy. *Appl. Phys. Lett.* **2019**, *114*, 221903. [CrossRef]
44. Zhao, C.; Parab, N.D.; Li, X.; Fezzaa, K.; Tan, W.; Rollett, A.D.; Sun, T. Critical instability at moving keyhole tip generates porosity in laser melting. *Science* **2020**, *370*, 1080–1086. [CrossRef]
45. Huang, Y.; Fleming, T.G.; Clark, S.J.; Marussi, S.; Fezzaa, K.; Thiyagalingam, J.; Leung, C.L.A.; Lee, P.D. Keyhole fluctuation and pore formation mechanisms during laser powder bed fusion additive manufacturing. *Nat. Commun.* **2022**, *13*, 1170. [CrossRef] [PubMed]
46. Li, Z.; Li, H.; Yin, J.; Li, Y.; Nie, Z.; Li, X.; You, D.; Guan, K.; Duan, W.; Cao, L.; et al. A Review of Spatter in Laser Powder Bed Fusion Additive Manufacturing: In Situ Detection, Generation, Effects, and Countermeasures. *Micromachines* **2022**, *13*, 1366. [CrossRef] [PubMed]
47. Michael, A.; Zhou, Y.N.; Yavuz, M.; Khan, M.I. Modelling the alloy element composition change in NiTi achieved through laser induced vaporization. *Mater. Chem. Phys.* **2019**, *231*, 87–94. [CrossRef]
48. Shi, G.; Li, L.; Yu, Z.; Sha, P.; Cao, Q.; Xu, Z.; Liu, Y.; Guo, Y.; Si, J.; Liu, J. Effect of crystallographic anisotropy on phase transformation and tribological properties of Ni-rich NiTi shape memory alloy fabricated by LPBF. *Opt. Laser Technol.* **2023**, *157*, 108731. [CrossRef]
49. Ma, J.; Franco, B.; Tapia, G.; Karayagiz, K.; Johnson, L.; Liu, J.; Arroyave, R.; Karaman, I.; Elwany, A. Spatial Control of Functional Response in 4D-Printed Active Metallic Structures. *Sci. Rep.* **2017**, *7*, 46707. [CrossRef]

micromachines

MDPI

Review

Four-Dimensional Micro/Nanorobots via Laser Photochemical Synthesis towards the Molecular Scale

Yufeng Tao [1,2,3,*], Liansheng Lin [1], Xudong Ren [1], Xuejiao Wang [1], Xia Cao [4], Heng Gu [1], Yunxia Ye [1], Yunpeng Ren [1] and Zhiming Zhang [3]

1 Institute of Micro-Nano Optoelectronics and Terahertz Technology, Jiangsu University, Zhenjiang 212013, China
2 Wuhan National Laboratory for Optoelectronics, Huazhong University of Science and Technology, Wuhan 430074, China
3 Postdoctoral Workstation, Zhejiang Chuangge Technology Co., Ltd., Zhuji 311899, China
4 School of Pharmacy, Jiangsu University, Zhenjiang 212013, China
* Correspondence: taoyufeng@ujs.edu.cn

Abstract: Miniaturized four-dimensional (4D) micro/nanorobots denote a forerunning technique associated with interdisciplinary applications, such as in embeddable labs-on-chip, metamaterials, tissue engineering, cell manipulation, and tiny robotics. With emerging smart interactive materials, static micro/nanoscale architectures have upgraded to the fourth dimension, evincing time-dependent shape/property mutation. Molecular-level 4D robotics promises complex sensing, self-adaption, transformation, and responsiveness to stimuli for highly valued functionalities. To precisely control 4D behaviors, current-laser-induced photochemical additive manufacturing, such as digital light projection, stereolithography, and two-photon polymerization, is pursuing high-freeform shape-reconfigurable capacities and high-resolution spatiotemporal programming strategies, which challenge multi-field sciences while offering new opportunities. Herein, this review summarizes the recent development of micro/nano 4D laser photochemical manufacturing, incorporating active materials and shape-programming strategies to provide an envisioning of these miniaturized 4D micro/nanorobots. A comparison with other chemical/physical fabricated micro/nanorobots further explains the advantages and potential usage of laser-synthesized micro/nanorobots.

Keywords: additive manufacturing; micro/nano 4D fabrication; smart materials; laser photochemistry

Citation: Tao, Y.; Lin, L.; Ren, X.; Wang, X.; Cao, X.; Gu, H.; Ye, Y.; Ren, Y.; Zhang, Z. Four-Dimensional Micro/Nanorobots via Laser Photochemical Synthesis towards the Molecular Scale. *Micromachines* **2023**, *14*, 1656. https://doi.org/10.3390/mi14091656

Academic Editors: Yang Liu, Jie Yin, Linda Ke and Kai Guan

Received: 30 June 2023
Revised: 11 August 2023
Accepted: 19 August 2023
Published: 24 August 2023

1. Introduction

There has been growing interest in the development of dynamic shape-transforming 3D-printed structures, which is considered the next big breakthrough in additive manufacturing techniques. This innovative research aims to create 3D-printed structures that can actively change their shape over time. The term "4D printing" was first introduced by Tibbets et al. in 2013 to describe this concept of creating objects with the ability to transform their shape dynamically. Since then, researchers and engineers have been exploring various materials and design strategies to realize the potential of 4D printing and its applications across different industries. The recently emerged stimuli-responsive 4D micro/nanostructured robot (4DM/NR) is a non-negligible high-tech field at this early stage [1] and has already exhibited remarkable application prospects in aspects of micromachines [2], micro/nano-mechanics [3], embeddable robots [4,5], sensors, and bio-science [5]. These smart 4DM/NRs have become the center of many multi-disciplinary studies and are envisioned to work in various micro-environments. As pioneering studies have reported [6,7], 4DM/NRs' small size and precise shape programming [8] offer great potential in bionics, neurorrhaphy [9], micro-optics [10], drug delivery, and highly directional locomotion [11] that are far beyond their macroscopic counterparts. Despite the tempting already achieved benefits, the high-freedom shape reconfiguration [12] of 4DM/NRs is restricted inside the laboratories due to the major challenges involving the complexity of practical motion control [13], dynamics [14], and tedious fabrication steps out of the massive application. The basic motion of 4DM/NRs, such as simple bending or expansion, is becoming common [1–14], while complex 4D programming, such as out-of-plane reversible gripping, selecting stimuli type, dynamic information encoding/transmission, and self-healing ability from physical damage, remains rare and pose a significant challenge for researchers.

To break through the technological/material limitations of implantable 4DM/NR applications, much effort has been devoted to shape-controlling strategies [15] to incorporate advanced light-triggered photochemistry [16,17] and soft-responsive materials [18]. In this respect, we have summarized and analyzed our previous work and peer research about laser-direct synthesized 4DM/NRs (see schematic in Figure 1) for review. To begin, the photochemical mechanism, which mostly relies on the photon-induced crosslinking process [19] (a photochemical reaction known as single-photon [20], two-photon [12], and multi-photon [21] polymerization) was introduced with an optical setup. To scale the robots' volume down to micro/nano-environments [22] with improvements to intelligence, the fabrication resolution of laser synthesis has to be confined to nano-size [23] at a flexible scanning path. In parameter-tunable laser photochemical fabrication [24], the spatiotemporal programming depends on the method of modulating photons [18]. The obtained stimuli-responsive micro/nanostructure is an energy converter that reversely transforms input energy to other forms, promising geometrical reconfiguration and leading to optical, bio-electrocatalytic [25], optical, electrical, or biological responsiveness [26]. Therefore, how to evaluate and optimize the micro/nanoscale interaction with various stimuli is the second challenge, which determines the behaviors of biomimetic 4DM/NRs.

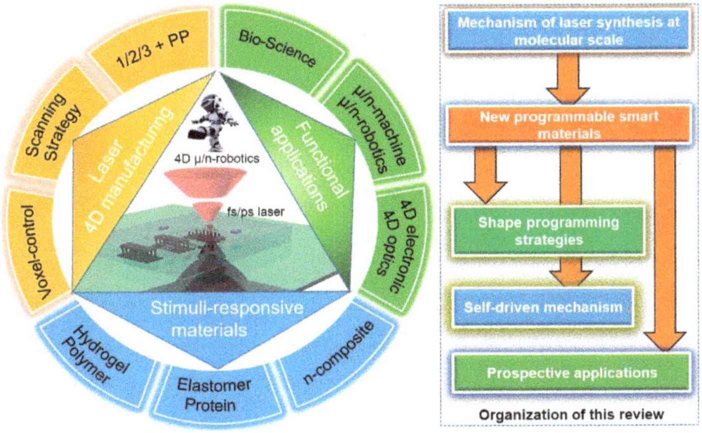

Figure 1. Schematic illustration of this review, covering three main sections: 1. laser-associated 4D manufacturing approaches and programming strategies around shape programming, which mostly operate on high-resolution two-photon polymerization and differ from macroscopic extrusion, material jetting, or other 3D printing methods; 2. stimuli-responsive materials, especially functional hydrogels/polymers, inorganic nonmetallic materials, and their nanocomposites; 3. future prospective interdisciplinary usage via 4D micro/nanofabrication is potentially applicable to micro/nanomechanics, adaptive optics, and bio-sciences. The organization logic throughout this manuscript follows this order: laser synthesis $\rightarrow$ smart materials $\rightarrow$ the programming strategies of 4D micro/nanorobotics $\rightarrow$ self-driven mechanism $\rightarrow$ application and comparison.

The key characteristic of active 4D materials is their stimuli-responsiveness. Currently used 4D materials include functional hydrogels [24], liquid crystal elastomers [27], responsive organic polymers [28], proteins, and their nanocomposite materials [29]. For this ultrafast photochemical polymeric reaction, its underlying molecular dynamics and material science deserve specific investigation, which intrinsically determines the fabrication resolution [12,18] and programming method. At the end of this review, we compare laser-synthesized mini-robots with robots obtained via other current chemical/biological/physical methods [30] in various fields of use to highlight the advantages of the reviewed digital photosynthesis methods.

2. Micro/Nanoscale Laser 4D Additive Manufacturing with Smart Materials

Accelerated by the coming era of "Industry 4.0", the digital 3D production processes became a revolutionary development in efficiency, resolution, geometric capacity, and functionalization, incorporating new smart materials. The geometry, volume, and resolution optimization of laser photochemistry [31,32] has developed beyond the traditional aim, contributing to new highly valued 4D products; its fabrication process can be explained as a series of photochemical reactions starting from photon absorption (as seen in Figure 2a). The photon energy excites the ground-state atoms of photoinitiators or monomers [33] to a higher energy level (known as photon absorption), making the photoinitiators or monomers unleash unstable free radicals, the free radicals trigger chemical reactions to link the molecular chain of monomers into a matrix (known as crosslinking or polymerization) or reduce the metallic ions into crystals (known as reduction). The short-wavelength photons possess higher light energy than long-wavelength ones, therefore, the short-wavelength laser beam is prone to incur single-photon [34] absorption, while the photoinitiator absorbs two [35,36] or several [21,33,37] photons simultaneously to incur two/multi-photon absorption. The subsequent reactions transform the liquid precursors into a solidification state at a micro/nano resolution.

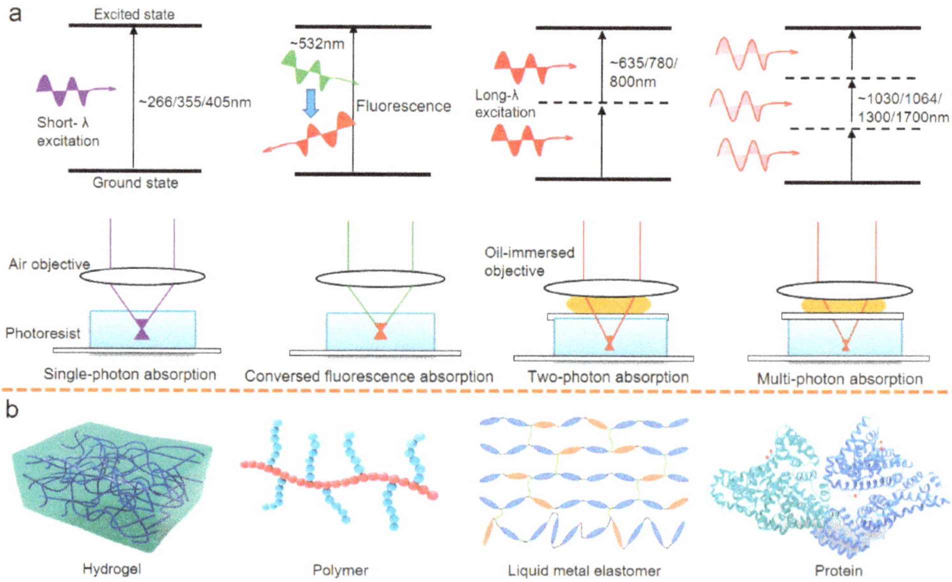

Figure 2. (**a**) Single–photon absorption at relatively short-wavelength laser projection, green-light-triggered fluorescence absorption, typical two-photon absorption, multi-photon absorption at infrared wavelength laser projection. (**b**) There are four representative smart materials used in laser photochemistry for robotic applications: stimuli-responsive hydrogel, shape memory polymer, liquid metal elastomer, and protein materials.

Many studies have successfully proved several stimuli-responsive smart materials for micro/nano 4D printing [25,33,37,38] (Figure 2b). Hydrogel [39–43] is a typical kind of functional material with a crosslinking three-dimensional network [18], and it is hydrophilic and insoluble in solvents. Especially, our recent work report an amphiphilic hydrogel containing both hydrophilic and hydrophobic groups [12], making the micro/nano 4D products respond to various stimuli rather than simple humidity. The hydrogels/polymers are also ideal soft carriers of magnetic nanoparticles [35,44–46], mechanics-reinforced carbon nanotubes [24], photon-sensitive fillers, drug particles, or electrically conductive dopants. Using these advantages, the cross-linked network can generate responsiveness under small loading of environmental stimuli, such as temperature, magnetic field, electric field, or multiple stimuli. The intelligent hydrogel networks usually present a mesoporous or fibrous network with unfoldable space to absorb or desorb the solvent and produce swelling or shrinkage, the recent work shows the swelling/shrinkage volume change ratio easily exceeds 200% [37–42]. Due to hydrogel's desirable biocompatibility [47] and mechanical properties, micro/nano 4D hydrogel can be further combined with in vitro cell culture, vascular dredging [48], or be used to emulate the cargo/reorganization of organisms [49].

Besides hydrogel, shape memory polymer [14,15,50], another special material, has attracted widespread attention, as it takes the glass transition temperature as the critical temperature for phase transferring. When the ambient temperature is lower than the glass transition temperature, the shape memory polymer is in a fixed phase, with a memory structure shape having a large modulus and hardness [51]. When the temperature is higher than the glass transition temperature, its physical properties undergo significant changes, with the modulus and hardness gradually decreasing and softening, which is a reversible phase that ensures shape changes by temperature variation [52–55].

As has been previously reported, liquid crystal [56] is another stimuli-responsive substance between crystals and liquids, it possesses the fluidity of liquids and the orderliness of crystals. Micro/nanocrystals are connected to the polymer skeleton and moderately crosslinked to obtain liquid crystal elastomer [57]. The increase in environmental temperature or under the irradiation of light causes liquid crystal molecules to change from a liquid crystal state [58] to an isotropic state, resulting in thermally induced deformation. The change of orderliness of the crystals also induces tunable optical function. In the current literature, liquid crystal transforms into elastomer via polymerization, and the elastomer is both soft, rigid, and widely used in robotics and optics.

Some other prior research has demonstrated protein [59] as another bio-compatible smart substance, as its certain spatial structure is formed by the winding and folding of peptide chains composed of amino acids through dehydration and condensation [60]. In laser-induced synthesis, some functional proteins form micro/nanostructures through amino acid bonds, the amino acids carry diverse functional groups that endow proteins with the ability to respond to external stimuli as biological actuators [61]. For example, bovine serum protein has carboxyl and amino groups on its molecular chain; it can cause morphological changes in protein molecules by pH variation.

These above-mentioned smart materials have garnered significant attention for their impressive deformability and scalability for micro/nano 4D printing. However, what we know about them is largely based on observation and laboratory experiments. Their imperfection still deserves in-depth studies for practical use. For example, devices based on hydrogel materials often suffer from relatively poor mechanical properties, limiting their practical applications. Liquid crystal elastomers tend to exhibit smaller degrees of deformation compared to other materials, and their response speeds are generally slower. The deformation of shape-memory polymers is primarily induced by changes in temperature, its simplicity in actuation mode comes at the cost of slower response speeds. In a nutshell, the innovative approaches enhancing mechanical properties, shape-morphing controllability, capabilities, and ability, need to be further explored.

After the introduction of used smart materials, we summarize the laser systems (Figure 3a$_1$,a$_2$) for photochemistry. The beginning is the coherent laser source; previous researches tell us that the nonlinear laser crystals (neodymium-doped yttrium aluminum garnet or yttrium aluminate, titanium-doped sapphire, potassium dihydrogen phosphate, and so on) work with advanced modulation techniques to irradiate laser beams covering ultraviolet-to-infrared wavelengths. Optical parametric oscillators [62] hold the responsibility to tune light wavelengths based on the nonlinear four-wave mixing effect [63], the wavelength-tuning is realized by physically changing the angle between the incident pump light and the laser crystal axis [64].

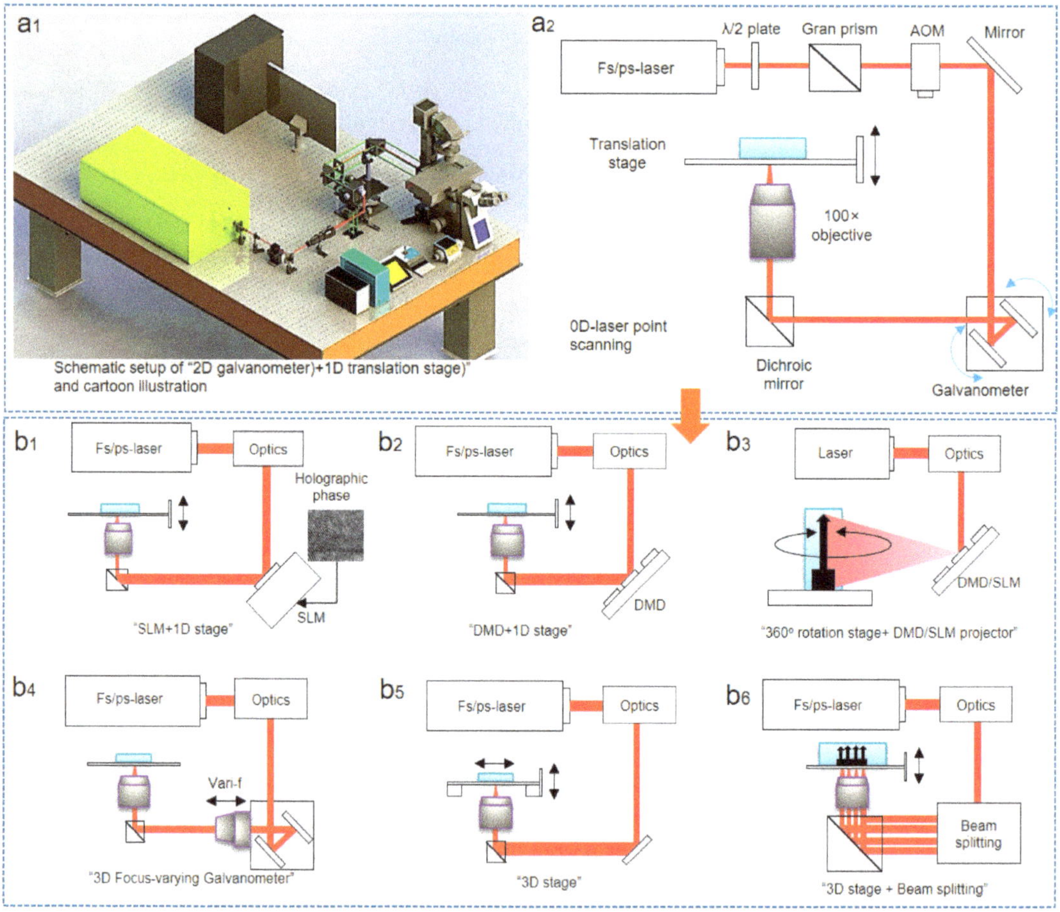

Figure 3. (**a$_1$**) Illustration of laser direct writing system and (**a$_2$**) its optical principle; (**b$_1$–b$_6$**) The current optical principle for high-freedom formation has developed to new categories: 2D spatial modulator (SLM) + 1D stage; the high-speed digital mirror (DMD) + 1D stage; the rotational substrate + projector; the newly-emerged 3D galvanometer with dynamically tunable focus lens; integrated 3D stage; 3D stage + beam splitting for paralleled fabrication.

To fabricate delicate 4D shape-reconfigurable micro/nanostructures using these lasers, a digital optical platform (Figure 3) consisting of the above-discussed light sources and modulation setup is required. As previous works reported that a green micro/nano 4D manufacturing optical platform [65] developed from the digital 3D nano printing system [66] (Figure 3a$_1$,a$_2$) should simultaneously possess the interactive functions of complex model analysis and slicing, spatially laser beam control and modulation ability, scanning path programming, high compatibility with multi-materials, and so on. For this aim, the optical setup of conventional micrometer-resolution lithography or projection evolves to a 4D printing nano-accuracy structure pre-designed for shape reconfiguration.

Studies [67] show that the stepsize of shape-programming is a single layer in 3D lithography or projection; it is the minimum programmable unit that needs to be designed and unevenly patterned, limiting the shape-programming resolution at its area size [68] at the micrometer scale. For standard micro/nanoscale 4D printing (as seen in Figure 3), when using a femtosecond laser direct writing setup [18], the basic programmable unit is the laser-focused voxel [69]; therefore, the shape-morphing resolution reaches the <100 nm level. In the setup with the stimulated emission depletion (STED) [70], even a resolution of <50 nm [71] becomes achievable. However, STED setup strongly depends on negative photoresists [72], limiting the functions of smart materials.

The literature review [22,25,33–37,62–70] also showcases the development of opto-mechatronics. The laser direct-writing setup adopts an orthodox "2 + 1" configuration, 2D scanning galvanometer/micromirror [73] plus 1D translation stage, advanced 3D galvanometer [74] with focus-adjusting ability, or inserting spatial phase modulators to form a "2 + 1" configuration. The recently reported rotating strategy uses the rotation freedom of the substrate to print 3D structures, also showing desirable feature sizes. By modulating a single laser beam to a point array or specific pattern, the fabrication efficiency or volume is ameliorated over one magnitude. In summary, the laser photochemical synthesis utilizes an optical setup of the digitalized laser additive manufacturing system [75–78] with precise optical components (focusing optical path, radio-frequency acoustic–optic modulator, 3D scanning system, large N.A. objective lens) to confine the photochemical reaction at the nanometric resolution, and is being developed by researchers from several fields.

3. Shape Programming Strategies by Ultrafine Nanostructures

Numerous current works emphasize the programming strategies of 4DM/NR biomimetic function [12–52], allowing the artificially made 4DM/NRs to realize high-level motions beyond nature. To spatiotemporally program reverse 4D behaviors, the shape-programming strategy is crucially important and not inferior to structural design or smart materials. The shape-programming strategies of previous studies can be categorized into three levels: the layer (Figure 4a), the microscale meta structures (biomimetic cilia with meta microstructures in Figure 4b–d), and the single nanowire (Figure 5, linewidth controlled by single laser focus). The 2D single-layer pattern at different layouts generates different swelling or shrinkage degrees and is used in light-projection fabrication. By designing the layout of each layer, the shape reconfiguration can be controlled by the light projection fabrication. In dual/multi-layer design [79,80], shape-morphing happens by using different materials, forming a soft–rigid structure, where the rigid layer works as a skeleton, and the soft layer works as a muscle [81] to generate direction bending. However, according to existing analyses [32–41,79–110], the mechanical mismatch between multiple layers is a pitfall for repeated shape-morphing in actual usage.

The latest literature indicates that these fundamental shape-programming strategies have evolved by introducing meta-mechanical structures into the multi-layers. For instance, our peer researchers built small 3D micro cage-like micropillar meta structures [82] to replace the planar soft layer along with the rigid layer, creating the unique meta-mechanical micropillar-triggered shape configuration of cilia using two-photon polymerization (Figure 4b), where the meta-mechanical micropillar cells present anisotropic shape-morphing and strong adherence to rigid layers, ensuring high-order direction in reverse photothermal shape reconfiguration, and the spatial resolution of shape-programming depends on the volume of a single micropillar at the microscale.

In the work of [82], one remarkable advantage of smart hydrogel is its ability to undergo precise and programmable light-induced transformations with exceptional spatial resolution. Researchers focus a laser beam (approximately 1 μm^2) on either the face or edge of a cubic woodpile structure measuring $50 \times 50 \times 50$ μm^3 (length $\times$ width $\times$ height,

Figure 4c). Upon exposure to the laser, the woodpile exhibits a fascinating response—it shrank around the laser focus area and rapidly assumed various stable 3D morphologies. This newly found dynamic behavior demonstrates the versatility and controllability of smart hydrogels in achieving complex shape transformations at microscopic scales. As seen in Figure 4d, researchers also report a 3D microclamp consisting of a circular array of six micropillars, and its functionality is demonstrated through optical micrographs taken under different light-stimulation powers in an aqueous environment. This meta-mechanical micropillar provides a much larger surface/volume ratio than the traditional multi-layer structure, making the stimuli-responsive behavior violent. Predictably, by devising meta-mechanical cells and implanting them into the soft-to-grid design, high-performance biomimetic 4DM/NRs could be realized.

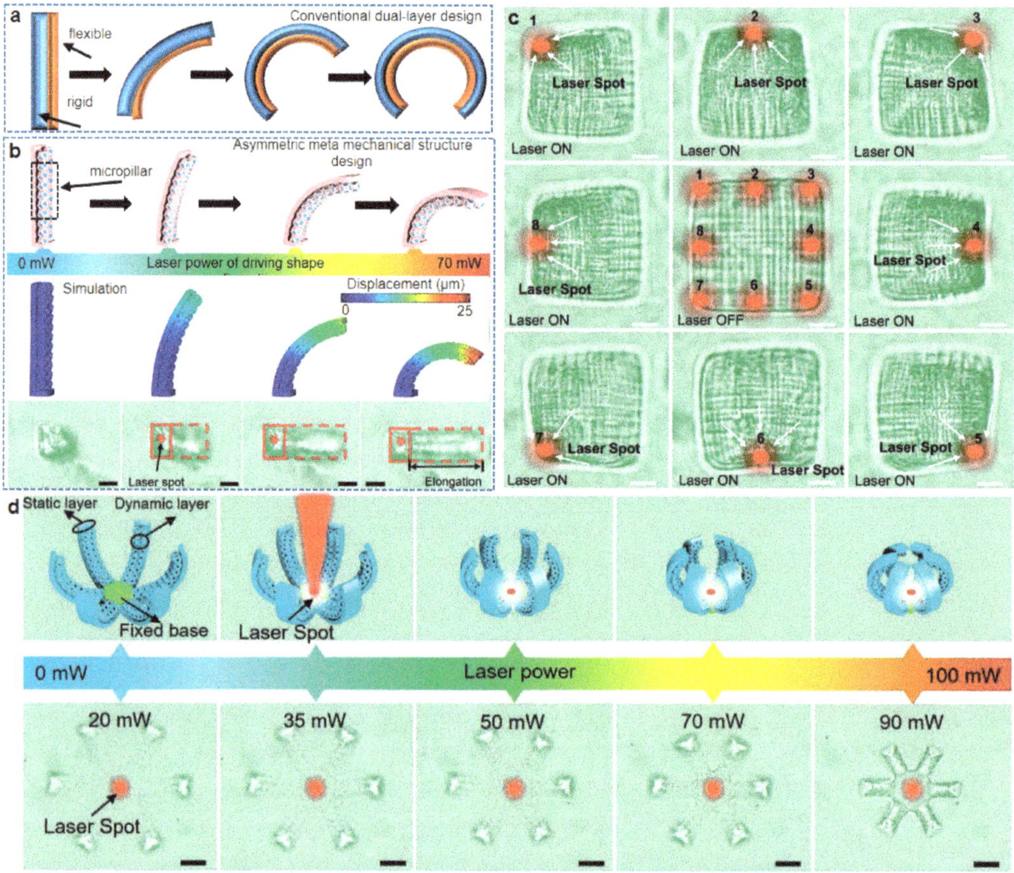

Figure 4. (**a**) The classical soft–rigid dual layer for shape bending. (**b**) The comprehensively enhanced shape reconfiguration of meta mechanical structure (biomimetic cilia). The simulation using finite element analysis matches well with experiment. (**c**) Optical micrographs of different shape transition states of woodpile photonic crystal microstructures regulated by the laser stimulation. There are eight positions of laser spot as indicated by the arrows. (**d**) Meta-structure-enhanced flower-shaped microclamp using a laser beam as photon stimuli. These figures are reproduced from Ref. [82] with copyright permission. The scale bar is 10 μm. The scale bars of (**b**–**d**) are 10 μm.

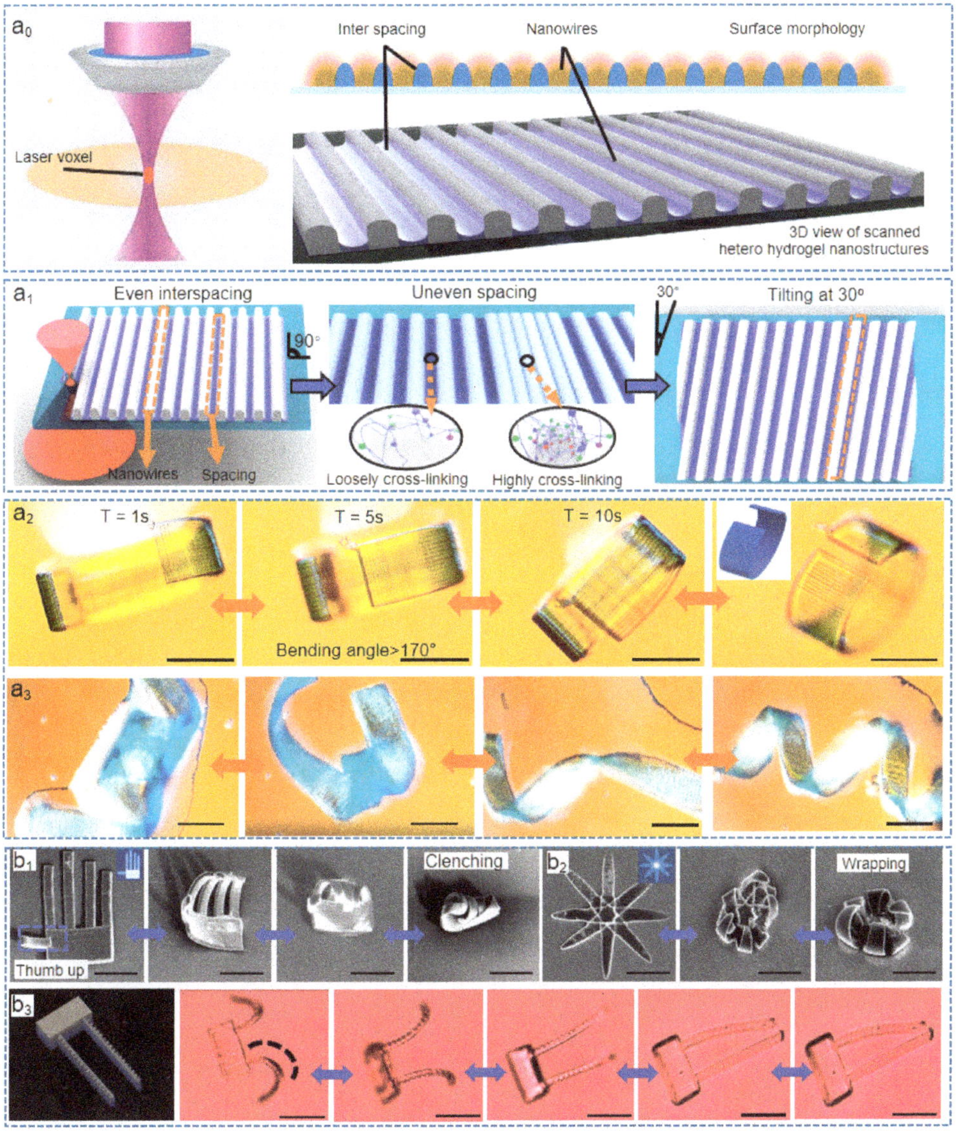

Figure 5. ($\mathbf{a_0}$) Illustration of laser focus voxel during the laser synthesis process. The red ellipse denotes the focused laser point inside the precursor material. The other figure is the section view of the laser-scanned interlaced nanowire/interspacing nanostructure reproduced from Ref. [18], which is known as a heterojunction nanostructure in Ref. [12]. ($\mathbf{a_1}$) The nanowire-based heterojunction scanning strategies in monolayer. ($\mathbf{a_2}$,$\mathbf{a_3}$) The accumulated intermolecular force realized "bracelet" and chiral torsion, respectively. ($\mathbf{b_1}$–$\mathbf{b_3}$) The out-of-plane bent biomimetic hand, flower, and grippers. These figures are reproduced from Ref. [12] with copyright permission from the authors. The scale bars of ($\mathbf{a_2}$,$\mathbf{a_3}$) are 50 μm, and the scale bars of ($\mathbf{b_1}$–$\mathbf{b_3}$) are all 50 μm.

In Figure 5a$_0$, it can be seen that the volume of the laser focus (known as "voxel") directly affects the scanned nanowires; the photon polymerization reaction happens only inside the "voxel". Previous investigations [18,62–70] conclude that the voxel shape focused by a Gauss laser beam is an eclipse at the nanoscale and that the volume of formed nanowires (its height and linewidth) depends on the laser voxel during scanning. Accordingly, the spatial resolution of the laser-scanning-induced nanowires, patterns, and structures is determined by the specific volume of the laser voxel. To push the shape-programming resolution and capability to their limits, in our latest work, heterojunction scanning [18] is experimentally confirmed as an effective strategy, in which laser scanning maintains single-voxel resolution and introduces uneven crosslinking degrees into the monolayer (a 2D metastructure containing both hard and soft layers, Figure 5a_0–a_3) with different densities, directions, and positions, making the shape reconfiguration anisotropic (expansion happened at pointed direction, Figure 5b_1–b_3) instead of the traditional slow isotropic swelling.

Our reported experiment results have guessed that the 4DM/NRs of hands and grippers in Figure 5a_0–a_3 act based on the non-covalent bonding effect [12,18]. Non-covalent bonding refers to the interactions between molecules or particles that do not involve the sharing or transfer of electrons. These interactions include hydrogen bonding, van der Waals forces, electrostatic interactions, and hydrophobic interactions. To precisely harness intermolecular forces, heterojunction scanning is proposed to manipulate more complex 4D behaviors in the monolayer than a tedious multi-material multi-layer reciprocating scanning strategy, which also eliminates the mechanical mismatch during multi-layer interfacing [51,80].

In heterojunction scanning, the direction of shape bending/expansion is perpendicular to the laser-scanned nanowires, and the bending direction can be controlled by adjusting the laser scanning direction to realize chiral torsion, anisotropic deformation, and site-specific mutation, in which the hydrogel structures are selectively modified or altered at specific locations. Moreover, the hydrogel nanowires demonstrated spontaneous self-repairing capabilities, allowing them to recover their original shape or functionality after being damaged or deformed. This reusability makes them promising candidates for micro/nanorobotics. Beneficially, the dimension of the 1D nanowire is traceable to the 0D voxel (Figure 5a_0–a_3) of laser focus [81–83]; therefore, the monolayer or formed 2D metastructures realize nano-accuracy programming based on the spacing and linewidth of the nanowire without losing shape reconfigurable capacity, promising future molecule-level robots.

In the above-demonstrated programming strategies, digitalized programming contributes to the tempting advantages of precise spatial controllability, outperforming traditional hydrothermal, sonic, and electrochemical methods. Moreover, the micro/nanoscale volume after shape programming decreases material cost and offers a larger surficial area than those bulky ones, and the inside mesoporous [83,84] or fibrous [12,18] network naturally offers numerous diffusion channels for transporting stimuli (solvent, ions, heating, and so on), making the responsiveness violent and fast when compared to slowly swelled, bulky ones [85]. Anyway, shape programming has a huge developing space and allows for huge geometric flexibility compared to conventional chemistry for better complexity in motion control, feature size, and function integration, and there exists a huge developing space.

4. Self-Driven Mechanism by Various Stimuli

The driven mechanism of the 4DM/NRs has been an extensively studied field, as it is crucially important to performance controlling and improvement of 4DM/NRs. Among the existing works, the noncontact magnetic field drive (as seen in Figure 6a_1–a_3) is often used for the remote control of microrobots due to its strong penetration ability, long driving distance, and high-order direction [45,86–89]. Because photocurable precursor materials are doped with magnetic nanoparticles (Fe, Ni, NdFeB, etc.) [44], by rotating the outside-applied magnetic field, the micro/nanostructured filter rotates at desired angles and frequencies accordingly for the filtering modes.

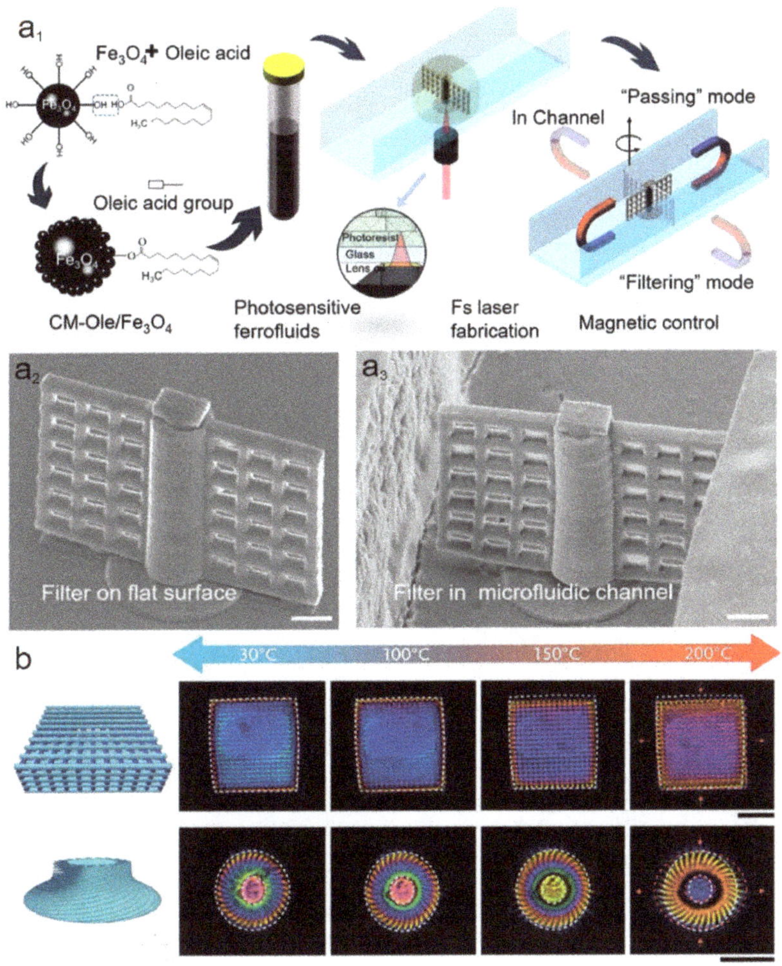

Figure 6. (**a$_1$**) Fabrication procedures for the magnetically driven rotary microfilter reproduced from Ref. [86] with permission from © The Optical Society. (**a$_2$**,**a$_3$**) SEM images of the two-photon polymerization fabricated filter on a flat surface and embedded inside a fluidic channel. The scale bars are 20 μm. (**b**) Example of temperature response: a woodpile and spiral disk thermally change their dimension and lead to coloration change. Reproduced from Ref. [90]. The scale bars represent 20 μm.

This innovative microfilter enables switching between two distinct modes: filtering and passing (Figure 6a$_2$,a$_3$). One notable advantage of this magnetically driven rotary microfilter is its ability to perform multi-mode filtering functions. For example, it can effectively capture particles with a size of 8 μm while allowing particles measuring 2.5 μm to pass through. It also can accommodate situations where both types of particles need to be passed through without filtration. The responsive characteristic of the microfilter, enabled by magnetic control, significantly enhances its reusability.

Numerous works have proved that temperature-dependent response [90,91] is another widely driven mechanism (Figure 6b). As known, the polymer, hydrogel, liquid crystal elastomers, or proteins generally shrink under heating conditions. Based on this, the reported woodpile photonic crystal/spiral disk shrinks to change its dimension after heating. The second driven force is the intermolecular-effect-induced dynamic force by responsive stimuli of various solvents [18]. Due to the presence of molecular interactions between the material matrix and solvent, the micro/nano 4D product exhibits significantly different swelling or shrinkage abilities with different solvents. pH stimulation responsiveness [83,91] has received widespread attention in micro/nano 4D printing due to high response speed [92] in milliseconds and excellent actuation amplitude. pH changes induce structural deformation, as demonstrated in Figure 7, the carboxyl groups in polymers/proteins undergo deionization in acidic solutions or ionization in alkaline solutions [93,94], causing expansion or contraction due to the electrostatic forces in the polymer or protein molecular chain [95–97].

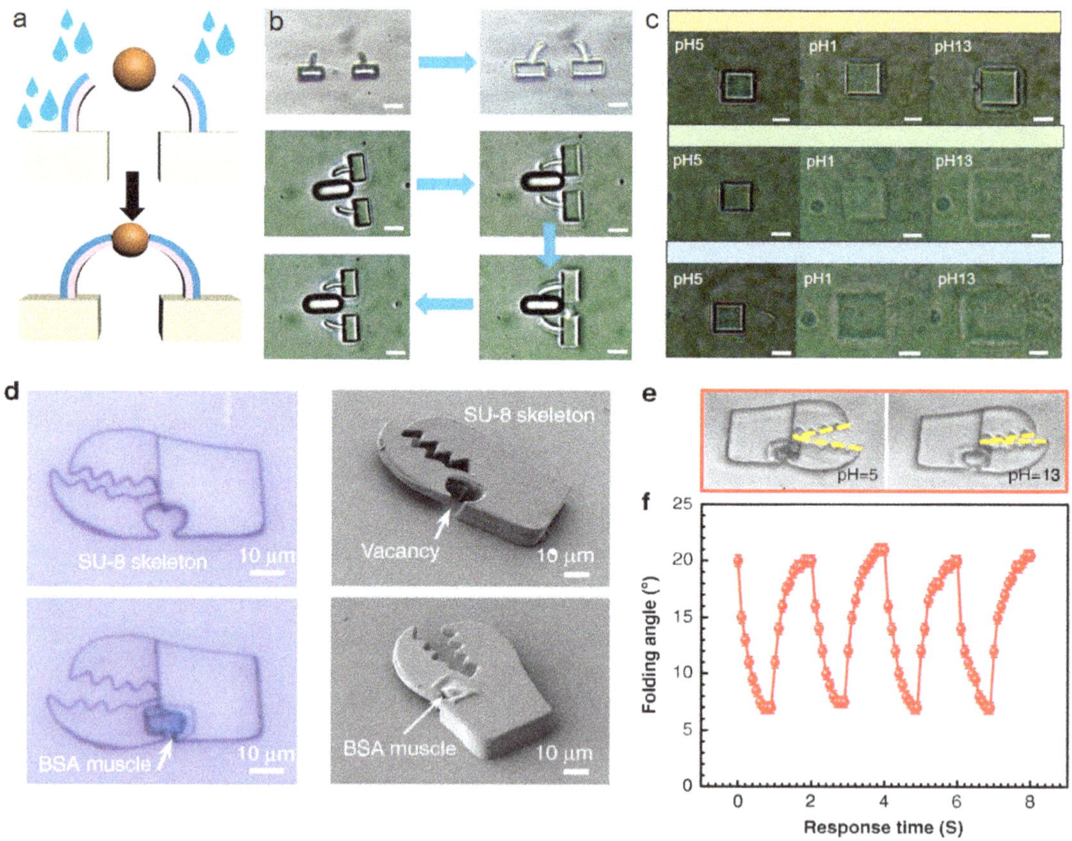

Figure 7. (**a**) Illustration of dual-layer structure shape changing by pH variation. Reproduced from Ref. [96]. (**b**) The reverse gripping action from a pair of designed protein cantilevers, where the scale bar is 5 µm. (**c**) The swelling/shrinkage of a cube shape made of proteins by two-photon polymerization. Reproduced from Ref. [96]. The scale bar is 10 µm. (**d**–**f**) The authors of Ref. [83] demonstrate a composite-material micro claw robot responsive to pH variation, where SU–8 photoresist works as the rigid part and BSA is bovine serum protein, which is polymerized on the claw to form the flexible muscle.

As illustrated by Figure 7, pH responsiveness is a novel-driven mechanism for fabricating micromechanical devices with a nonuniform internal lattice density. The technique employed in this research is the femtosecond laser two-photon polymerization, which allows for the flexible design and high-precision production of 3D micro/nano devices. The increase in scanning step length from 100 nm to 200 nm led to a higher swelling ratio of BSA micro/nano blocks [96]. This can be attributed to a decrease in the cross-linking density of the processed structures. In a pH 1 solution, the swelling ratio was approximately 1.9 when the step length was 200 nm. On the other hand, in a pH 13 solution, the swelling ratio reached 265. When using a step length of 100 nm in an alkaline solution, the swelling ratio was found to be 76.7% of that observed in structures fabricated with a step length of 200 nm.

According to the current literature, light [98] is another remote noncontact drive for 4DM/NRs and is used to precisely stimulate other physicochemical effects to induce structural actuation, such as the photothermal effect, photochemical effect, and optical force. Photothermal conversion [99,100] is an effective form of energy conversion [101,102] that uses the photothermal effect of materials to induce local temperature gradients to induce structural shape changes. Some designs are implemented through the transformation of azobenzene moieties in a liquid crystalline elastomer matrix by laser direct writing to exhibit motion triggered by ultraviolet irradiation and near-infrared responsive mechanical deformation [102]. Photochemical effects are based on specific photoresponsive chemical materials, such as azobenzene or spiropyran [103] derivatives to produce photoisomerization under light stimulation, changing the polymer chain length and thus deforming the device structure. Several envisioning robotic devices with functionality have been produced from a single azobenzene-functionalized cholesteric liquid crystal elastomer. The other driven mechanism is pure optical force, which realizes the motion of an object by using the gradient force of the light field itself or the momentum of photons.

In the above-listed driven methods, each driving method comes with its own set of challenges. Magnetic responses are limited by their driving distance capabilities and raise concerns about the safety of high-intensity magnetic fields in biomedical applications. Achieving accurate control of pH and solvent responses can be challenging. Temperature responses tend to be slow. Light responses suffer from low energy conversion rates and issues with tissue transparency. To achieve widespread application scenarios in the future, an ideal actuation method for micro/nano 4D printing needs to fulfill several requirements simultaneously and should be capable of noncontact stimulation over long distances to accommodate complex working environments. Additionally, it should offer precise control and agile responses to meet the demands of intelligent integrated devices. Among previously reported driven methods [82–103], light responses hold great promise as an ideal driving and control method. This is due to their rich wavelength selectivity, precise spatial targeting capabilities, and multi-dimensional regulatory abilities. Light responses offer the potential for achieving complex movements and deformations in micro/nano structures with high precision and efficiency.

5. Prospective Future and Applications of 4DM/NRs

After the introduction of versatile shape-programming strategies and the driven mechanisms of the 4DM/NRs, determining a method to predict and apply 4DM/NRs has grasped our attention [104–120]. To enhance the control over shape transformations while minimizing response time, researchers have been exploring nanofabrication approaches with sophisticated designs. A groundbreaking four-dimensional microprinting strategy is firstly introduced for constructing 3D-to-3D shape-morphing micromachines [104] in a single-material–single-step mode. By utilizing direct laser writing, the researchers can spatially distribute heterogeneous stimulus-responsive hydrogels into arbitrary 3D shapes with sub-micrometer features. The material properties of the hydrogels, including crosslinking densities, stiffnesses, and swelling/shrinking degrees, can be precisely modulated by programming the exposure dosage of femtosecond laser pulses.

Taking use of proper structural design and numerical prediction, more complex biomimetic shape reorganization (machine-level mutation [104], untethered jumping [105], directional swimming [106,107], and so on) becomes realizable, and response time is greatly shortened with improved performance. These micromachines exhibit rapid, precise, and reversible 3D-to-3D shape transformations in response to multiple external stimuli. To predict the shape reconfiguration trend for complex structures, researchers have increasingly employed finite element analysis (FEA) simulation [18,104]. This computational method allows them to simulate, predict, and analyze the behavior of 4DM/NRs during shape transformation. By combining FEA with advanced manufacturing techniques like four-dimensional direct writing (4DLW), researchers can accurately program the higher-order motions of hollow 3D microstructures. For example, uniaxial/biaxial contraction and articulated folding (Figure 8) are among the motions that match well with the numerical predictions made by FEA. This convergence between simulation results and actual performance enhances our understanding of how these innovative micro/nanorobots behave during shape reconfiguration.

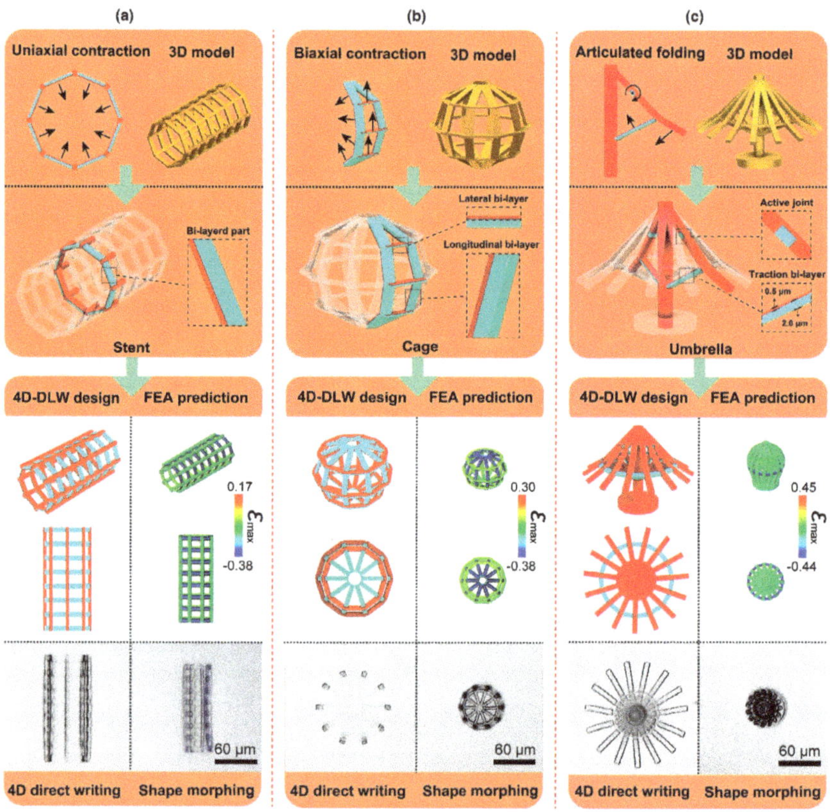

Figure 8. The authors of Ref. [104] present demonstrations of innovative technologies that enable the creation of complex 3D reconfigurable micro-architectures. These architectures are made possible through the use of advanced micromachines, such as microstents, microcages, and micro-umbrellas (**a**–**c**). These micromachines are capable of achieving various forms of reconfiguration, including rapid and precise uniaxial contraction, biaxial contraction, and articulated-lever folding. This level of flexibility allows for a wide range of applications in fields such as robotics and biomedical engineering. The process used to create these structures is known as 4DLW (four-dimensional direct laser writing). It involves several steps, including the embedding of deformation-amplifying mechanisms, the design of the 4DLW system, FEA prediction to optimize performance, the actual 4D direct writing process using lasers or other techniques, and finally, shape-morphing to achieve the desired configuration. The pictures are reproduced from Ref. [104].

The time-dependent shape changing [108] belongs to the fourth dimension. However, the shape-mutation-induced functional changing brings the fifth-dimensional 5D ability of optical functions into the 4DM/NRs. The 4DM/NRs possess 5D biomimetic functions of sensory function [109,110], chameleon-like coloration [111–114], dynamic optical devices [115–119] with variable focus, information encoding/transmission [120], wettability tailoring [121], and even electronic skin [122]. The typical 5D optical devices can be represented by color-shifted photonic crystals [18] or smart biomimetic compound eyes [118]. Dynamic micro/nanoscale 4D optical devices inherently have better integration ability and tunable imaging mechanisms, leading to optical performance improvement.

For example, compound eyes have long been recognized for their potential in modern optics, but both natural and artificial compound eyes suffer from a limitation—they lack the ability to achieve variable-focus imaging due to their fixed focal lengths within individual ommatidia. The researchers of Ref. [118] sought inspiration from the tunable crystalline lens found in human eyes to overcome this limitation. They successfully fabricated smart stimuli-responsive compound eyes using bovine serum albumin (BSA) protein through the innovative technique of femtosecond laser fabrication (Figure 9). By leveraging the unique swelling and shrinking properties of BSA under different pH conditions, the research team created compound eyes with a tunable field of view ranging from 35° to 80°, as well as variable focal lengths within each ommatidium. This composite compound eye exhibited nearly 400% focal length tuning while maintaining a fixed tunable field of view.

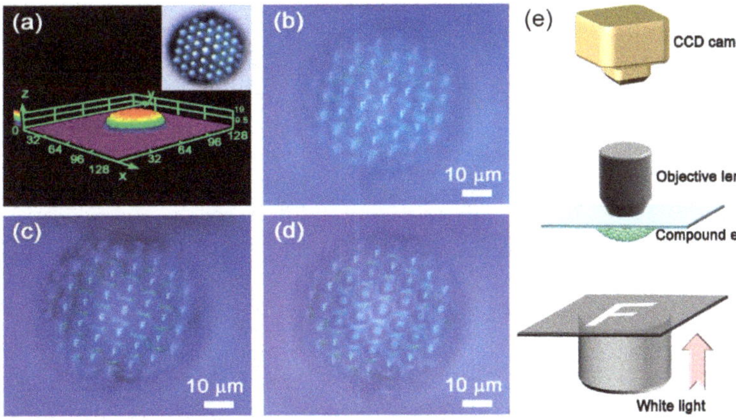

Figure 9. The focusing and imaging performances of a batch of protein-based compound eyes reproduced from Ref. [118] with the author's permission. The compound eye under investigation is composed of ommatidia, each measuring 10 μm in diameter and 3 μm in height. (**a**) Displays a 3D laser confocal microscopy image capturing the detailed structure of the compound eye. To assess its imaging capabilities, the researchers analyzed the performance from the inner part to the outer part of the compound eye, as depicted in (**b–d**). (**e**) The schematic of the optical setup used to measure the imaging performance of compound eyes.

To evaluate its imaging capabilities, the researchers of Ref. [118] conducted tests using the letter "F" as a test object. They captured images of the letter "F" using the inner, middle, and outer parts of the compound eye separately. They specifically examined the inner, middle, and outer ommatidia individually to assess their respective focusing results (Figure 9a–d). The quality of focus achieved by these ommatidia and the grayscale intensity distribution of focal spots is extracted from six symmetrical ommatidia belonging to each part (inner, middle, and outer). By studying these distributions, researchers gained insights into how well each part of the compound eye focused on specific objects.

In addition to optical applications, another potential trend of 4DM/NRs is the interdisciplinary application in bio-science [123–137]. To enable more intricate micro/nano cargo manipulation, such as encapsulation and release, in biological settings, it is crucial to equip microrobots with the ability to adapt and morph their shapes in dynamic environments (Figure 10). By combining magnetic propulsion with shape-morphing capabilities, Ref. [123] creates a remarkable shape-morphing micro crab capable of performing targeted microparticle delivery (Figure 10c–f).

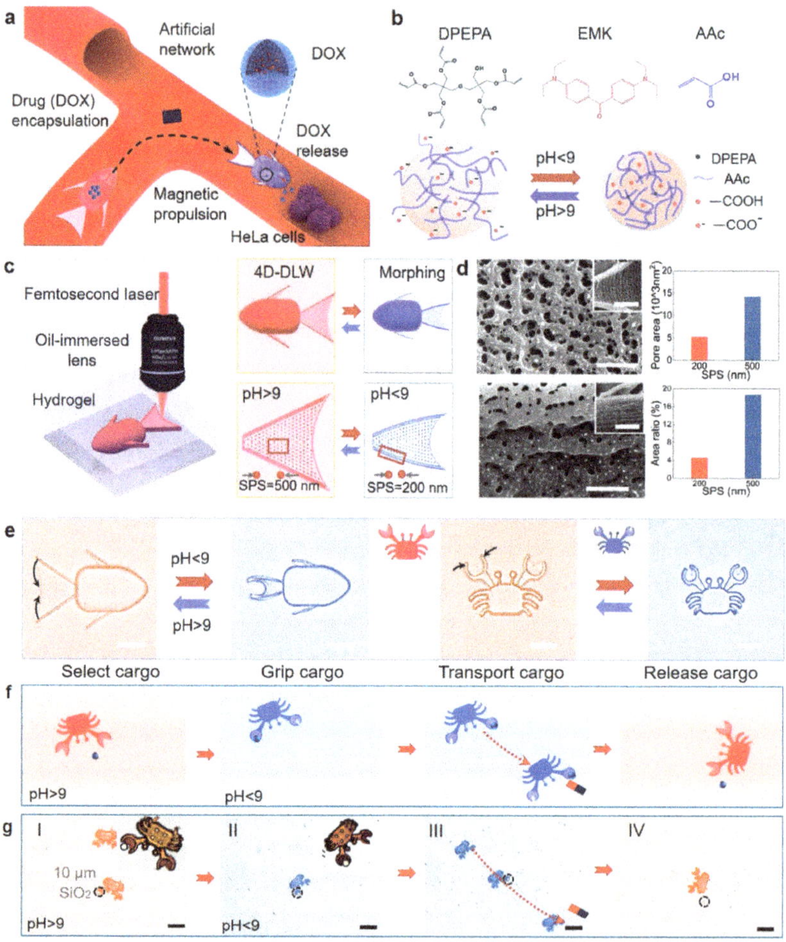

Figure 10. The researchers of Ref. [123] achieved one-step 4D printing of shape-morphing microrobots for HeLa cell treatment. The fabricated magnetic microrobots (in the shapes of crabs and fishes) can undergo shape-morphing to achieve targeted drug release for treating cancer cells. (**a**) Schematic of the magnetic shape-morphing microfish (SMMF) for targeted doxorubicin (DOX) release to treat cancer cells by shape-morphing. (**b**) Main compositions and schematic of pH-responsive hydrogels, where AAc is acrylic acid, DPEPA is the cross-linker thdipentaerythritol penta acrylate, and EMK is the photoinitiator 4,4′–bis–(diethylamino)benzophenone. (**c**) Four-dimensional printing with a designable point density to encode shape-morphing. (**d**) Scanning electron microscopy (SEM) images of the gel pores in the tail and body (inset images). (**e**) Optical images of the 4D fish and crab, illustrating the opening and closure of the fins and claws, respectively. (**f**,**g**) Schematic procedures and time-lapse images of I. selecting, II. tightly gripping, III. transporting, and IV. releasing targeted cargo by the 4D crab. Scale bars: (**d**) 1 μm, and inset images, 5 μm; (**e**) 25 μm; (**f**,**g**) 50 μm. All pictures here are reproduced from Ref. [123].

As a proof-of-concept demonstration, they have also designed a shape-morphing micro fish that can encapsulate a drug (doxorubicin (DOX)) by closing its mouth in phosphate-buffered saline with a pH of around 7.4. Moreover, 4DM/NRs successfully achieved localized HeLa cell treatment within an artificial vascular network by utilizing the "opening–closing" motion. Another example of bio-application via 4DM/NRs may be the robotic devices and the light-driven artificial organs demonstrated by researchers of Ref. [82], they developed a miniature aortic valve that could potentially revolutionize medical treatments and advance biomedical research. The fabricated robotic aortic valve is controlled by applied light to open or close (Figure 11), and the specific opening amplitude of the aortic valve under different light stimulation powers is tested. Potentially, these novel organs can be implanted into real human/animal organs as cardiac pacemakers or for other adjuvant therapy.

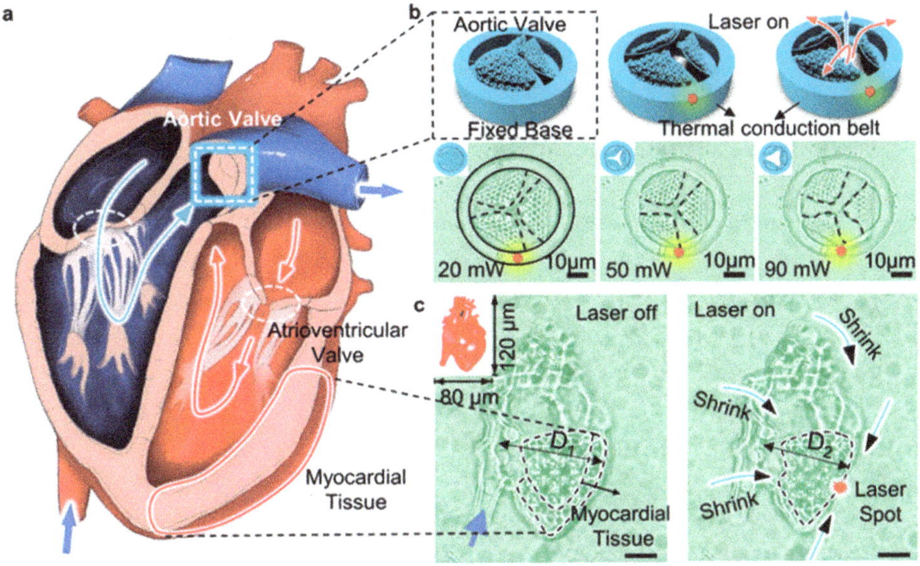

Figure 11. The researchers of Ref. [82] have successfully printed and demonstrated a functioning microscale artificial heart. Their schematic of the heart structure is depicted in (**a**). To assess its functionality, the opening amplitude of the aortic valve is examined under different light stimulation powers, as shown in (**b**), and optical micrographs captured the changes in valve opening as a result of varying light stimulation. Furthermore, optical micrographs were taken to compare the appearance of the microheart with and without light stimulation, as seen in (**c**). It is worth noting that this remarkable piece of technology has incredibly small dimensions, measuring merely $80 \times 120 \times 60\ \mu m^3$, as indicated by the schematic illustration inset, in which the scale bar denotes 20 μm.

In recent years, an increasing number of bio-applications have emerged that make use of Four-Dimensional Micro/NanoRobots (4DM/NRs). These tiny robots hold tremendous potential for various biomedical applications. The continuous advancements in size optimization, motion control, and imaging technology have significantly contributed to unlocking the capabilities of these 4DM/NRs. Researchers are constantly working towards optimizing their size to ensure compatibility with biological systems. This allows for minimally invasive procedures and precise manipulation at the microscale. Moreover, improvements in motion control mechanisms enable these robots to navigate complex environments with accuracy and precision. They can be potentially guided through intricate pathways, such as blood vessels or tissues, facilitating targeted delivery and microcargo operations. Furthermore, the integration of imaging technology provides real-time monitoring and feedback during the operation of these 4DM/NRs. Overall, 4DM/NRs hold great promise as ideal platforms for complex micro cargo operations and on-demand drug release. Their ability to navigate within biological systems coupled with their precise manipulation capabilities makes them valuable tools in biomedical research and healthcare applications. Continued advancements in this field will further expand their potential impact on improving medical treatments and interventions.

6. Advantages of Current Miniaturized 4D Robots, Challenges, and Future

In the era of micro/nano 4D products, there simultaneously exist other competitive 4D products (Figure 12). The first one may be the microscale self-propelled bio-robots that use microorganisms [138] from the animal or plant world. As we know, there exist animal robots, such as trained dogs, bees, wasps, termites, or ants, being used as robots. On the micro/nanoscale, the available tiny biological robots can be microorganisms, cells [139], bacteria, eggs of insects, and even sperm. Peer researchers employ fish sperm cells as micromotors to interact with the surrounding environment to trigger a swimming speed, generating a snake-like moving path. The tiny microorganism moves and self-navigates to execute specific tasks. For example, aqua sperm [140], whose size is similar to that of bacteria, is reported to destroy biofilms that colonize medical and laboratory tubing. In addition, the other plant tissue robots are transformed by humans using modified cultivation engineering in the presence of functional nanoparticles, which happen through the cell growth media, and these particles are taken up into the plant tissue cells for remote magnet-driven motion ability.

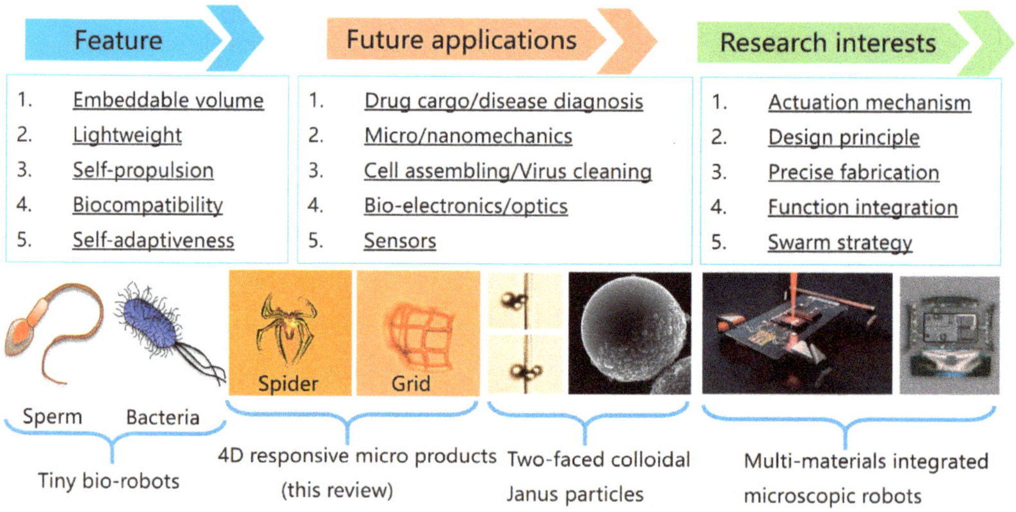

Figure 12. The other existing 4DM/NR types reported, including their features, prospective applications, and research interests.

The second competitor is two-aced colloidal Janus nanoparticles [141]. These anisotropic nanoparticles are synthesized by stepwise chemical methods, where one face of a Janus nanoparticle is colloidal (polymers, plastics, SiO_2, and so on), and the other face is metal-deposited, such as iron, titanium, copper, or platinum for magnetic force or photocatalysis ability [142,143]. Janus nanoparticles generally show desirable dispersion characteristics in organic aqueous solutions, like surfactants, amd by rotating the magnetic field's direction, the Janus particles move in the controlled direction. Another driving force is gas propulsion from chemical catalysis reactions to the photothermal effect [144]. The deposited metal decomposes organic solvent into gaseous substances under light projection to put themselves forward by the generated gas.

The third competitor is the modern integrated micro-robots using multi-material stepwise semiconductor fabrication [145], and digital control [146] for the swarm character, high intelligence [147], autonomy, and even wireless communication. The above-studied four main 4DM/NRs types are reshaping our modern bio-science, robotics, micro-electro-mechanical systems, and the Internet of Things by playing ever-increasingly important roles. However, their similarity, differences, advantages, and disadvantages vary a lot in terms of volumes, fabrication methods, application ranges, flexibility and self-recovery, and feature size, which can be found in the comparative table (Figure 13). In conclusion, laser synthesis is a green fabrication method [148] requiring no tedious semiconductor process, nor the cultivation of microorganisms, while maintaining bio-friendliness [149,150] and the high-order regulation of wanted functions.

4DM/NR types	Biomimetic μ/n robots 4D-printing products	Sperm Bacteria Microorganism	Colloidal Janus particles Two-faced particles	M/NEMS robotics Integrated μ-robots
Embeddable volume/light weight	☑	☑	☑	☑
4D applications (gripping/optical/electrical...)	☑	☑	☑	☑
All-in-one rapid fabrication	☑			
Bio-electronics/optics Sensors	☑			☑
Self-adaptiveness Self recovery	☑	☑		
Self-propulsion/Self power supply	☑	☑	☑	
Flexible geometric design/high resolution	☑			☑

Figure 13. Comparative analysis of four different currently studied 4DM/NRs in terms of applications, fabrication, unique advantages, design capability, and feature size, where laser photochemically synthesized 4D robots possess the most comprehensive advantage, slowing the application range covering micro/nano optics, electronics, and robotics with flexible geometric design outperforming the human-cultured microorganisms or nanoparticle-shaped Janus robotics. Moreover, the laser synthesis allows multi-function integration in tiny volumes requires only one-step fabrication, which is much more efficient and beneficial to shorten the design circles.

7. Conclusions

Conclusively, the laser photochemically synthesized 4DM/NRs have compliant body structures made of smart materials, or their composites can be printed directly using the current digital computer-assisted laser system. The modern digitalized ultrafast laser system [148] promises high compatibility with compound materials and high complexity in structure/pattern design, allowing for the implantation of shape-reconfiguration memory during fabrication. It is worth affirming that the digital laser system provides high flexibility in nanostructured heterojunction or meta-mechanical structures to responsive materials for performance improvement.

The reviewed 4D laser-synthesized monolithic soft robots are capable of linear and turning locomotion at the micro/nanoscale for breaking the limitations of volume and controllability by traditional chemical/physical methods. By doping functional nanoparticles, the synthesized 4DM/NRs possess potential far beyond the currently reported robotic applications. Thereby, we strongly recommend the photochemistry method for customizing next-generation 4DM/NRs for envisioned applications. Using this method, the driving modes/programming strategies/spatial resolution and volume are versatile and applicable to various extreme conditions.

Author Contributions: Conceptualization, Y.T.; validation, L.L. and X.R.; investigation, X.W.; resources, X.R. and X.C.; writing—original draft preparation, H.G. and Y.R.; writing—review and editing, Y.Y. and Z.Z. All authors have read and agreed to the published version of the manuscript.

Funding: This research was financially supported by the National Natural Science Youth Fund of China (61805094, 62205132), the Natural Science Foundation of Jiangsu Province (BK20221363, BK20210746), and the National Key R&D Program of China (SQ2018YFB110138).

Data Availability Statement: Data is contained within this article.

Acknowledgments: The authors are grateful to Xiong Wei from the Huazhong University of Science and Technology for his instruction on laser direct writing techniques.

Conflicts of Interest: The authors declare no conflict of interest.

References

1. Kim, H.; Ahn, S.; Mackie, D.; Kwon, J.; Kim, S.; Choi, C.; Moon, Y.; Lee, H.B.; Ko, S. Shape morphing smart 3D actuator materials for micro soft robot. *Mater. Today* **2020**, *41*, 243–269. [CrossRef]
2. Daryadela, S.; Behroozfarb, A.; Minary-Jolandan, M. A microscale additive manufacturing approach for in situ nanomechanics. *Mater. Sci. Eng. C* **2019**, *767*, 138441. [CrossRef]
3. Magdanz, V.; Sánchez, S.; Schmidt, O.G. Development of a Sperm-Flagella Driven Micro-Bio-Robot. *Adv. Mater.* **2013**, *25*, 6581–6588. [CrossRef] [PubMed]
4. Panda, S.; Hajra, S.; Rajaitha, P.M.; Kim, H.J. Stimuli-responsive polymer-based bioinspired soft robots. *Micro. Nano. Syst. Lett.* **2023**, *11*, 2. [CrossRef]
5. Wu, S.; Hong, Y.; Zhao, Y.; Yin, J.; Zhu, Y. Caterpillar-inspired soft crawling robot with distributed programmable thermal actuation. *Sci. Adv.* **2023**, *9*, eadf8014. [CrossRef] [PubMed]
6. Yu, X.T.; Deng, Z.T.; Li, H.; Ma, Y.C.; Ma, X.B.; Zheng, Q. Anisotropic hydrogel fabricated by controlled diffusion as a bio-scaffold for the regeneration of cartilage injury. *RSC Adv.* **2022**, *12*, 28254. [CrossRef]
7. Li, G.; Yang, S.; Wu, W.; Chen, F.; Li, X.; Tian, Q.; Wang, K.; Zhao, Y.; Wang, J.; Liu, Q.; et al. Biomimetic 4D Printing Catapult: From Biological Prototype to Practical Implementation. *Adv. Func. Mater.* **2023**, *33*, 2301286. [CrossRef]
8. Apsite, I.; Biswas, A.; Li, Y.Q.; Ionov, L. Microfabrication Using Shape-Transforming Soft Materials. *Adv. Func. Mater.* **2020**, *30*, 1908028. [CrossRef]
9. Joshi, A.; Choudhury, S.; Baghel, V.S.; Ghosh, S.; Gupta, S.; Lahiri, D.; Ananthasuresh, G.K.; Chatterjee, K. 4D Printed Programmable Shape-Morphing Hydrogels as Intraoperative Self-Folding Nerve Conduits for Sutureless Neurorrhaphy. *Adv. Healthc. Mater.* **2023**, *1*, e2300701. [CrossRef]
10. Gonzalez-Hernandez, D.; Sanchez-Padilla, B.; Gailevicius, D.; Thodika, S.C.; Juodkazis, S.; Brasselet, E.; Malinauskas, M. Single-Step 3D Printing of Micro-Optics with Adjustable Refractive Index by Ultrafast Laser Nanolithography. *Adv. Opt. Mater.* **2023**, *11*, 2300258. [CrossRef]
11. Lee, Y.; Choi, J.; Choi, Y.; Park, S.M.; Yoon, C.K. Four-Dimensional Printing of Stimuli-Responsive Hydrogel-Based Soft Robots. *J. Vis. Exp.* **2023**, 191. [CrossRef]
12. Tao, Y.F.; Lu, C.C.F.; Wang, X.J.; Xin, Z.D.; Cao, X.; Ren, Y.P. Monolayer heterojunction interactive hydrogels for high-freedom 4D shape reconfiguration by two-photon polymerization. *Int. J. Bioprinting* **2023**, *9*, 678. [CrossRef]
13. Wang, L.L.; Zhang, F.H.; Liu, Y.J.; Du, S.; Leng, J. Photosensitive Composite Inks for Digital Light Processing Four-Dimensional Printing of Shape Memory Capture Devices. *ACS Appl. Mater. Interfaces* **2021**, *13*, 18110–18119. [CrossRef] [PubMed]
14. Spiegel, C.A.; Hackner, M.; Bothe, V.P.; Spatz, J.; Blasco, E. 4D Printing of Shape Memory Polymers: From Macro to Micro. *Adv. Fuct. Mater.* **2022**, *32*, 2110580. [CrossRef]
15. Gastaldi, M.; Spiegel, C.A.; Vazquez-Martel, C.; Barolo, C.; Roppolo, I.; Blasco, E. 4D Printing of Light Activated Shape Memory Polymers with Organic Dyes. *Mol. Syst. Des. Eng.* **2022**, *8*, 323–329. [CrossRef]
16. Zhang, B.; Li, H.G.; Cheng, J.X.; Ye, H.T.; Sakhaei, A.H.; Yuan, C.; Rao, P.; Zhang, Y.F.; Chen, Z.; Wang, R.; et al. Mechanically Robust and UV-Curable Shape-Memory Polymers for Digital Light Processing Based 4D Printing. *Adv. Mater.* **2021**, *33*, 2101298. [CrossRef]
17. Zhang, Y.; Huang, L.M.; Song, H.J.; Ni, C.J.; Wu, J.J.; Zhao, Q.; Xie, T. 4D Printing of Digital Shape Memory Polymer with Tunable High Performance. *ACS Appl. Mater. Interfaces* **2019**, *11*, 32408–32413. [CrossRef]
18. Tao, Y.F.; Lu, C.C.F.; Deng, C.S.; Long, J.; Ren, Y.P.; Dai, Z.J.; Tong, Z.P.; Wang, X.J.; Meng, S.; Zhang, W.G.; et al. Four-Dimensional Stimuli-Responsive Hydrogels Micro-Structured via Femtosecond Laser Additive Manufacturing. *Micromachine* **2022**, *13*, 32. [CrossRef]

19. Li, S.G.; Liu, X.L.; Zhang, S.; Zhou, Y.W.; Wan, X.; Li, N.; Li, J.; Zhang, L. D–π–A–π–D Initiators Based on Benzophenone Conjugate Extension for Two-Photon Polymerization Additive Manufacturing. *Photonics* **2022**, *9*, 183. [CrossRef]
20. Preuß, J.; Gehring, H.; Schmidt, R.; Jin, L.; Wendland, D.; Kern, J.; Pernice, W.; Vasconcellos, S.M.; Bratschitsch, R. Low-Divergence hBN Single-Photon Source with a 3D-Printed Low-Fluorescence Elliptical Polymer Microlens. *Nano Lett.* **2023**, *23*, 407–413. [CrossRef]
21. Gonzalez-Hernandez, D.; Varapnickas, S.; Bertoncini, A.; Liberale, C.; Malinauskas, M. Micro-Optics 3D Printed via Multi-Photon Laser Lithography. *Adv. Opt. Mater.* **2023**, *11*, 2201701. [CrossRef]
22. Bernardeschi, I.; Ilyas, M.; Beccai, L. A Review on Active 3D Microstructures via Direct Laser Lithography. *Adv. Intell. Syst.* **2021**, *3*, 2100051. [CrossRef]
23. Xiong, W.; Zhou, Y.S.; Hou, W.J.; Jiang, L.J.; Mahjouri-Samani, M.; Park, J.; He, X.N.; Gao, Y.; Fan, L.S.; Baldacchini, T.; et al. Laser-based micro/nanofabrication in one, two and three dimensions. *Front. Optoelectron.* **2015**, *8*, 351–378. [CrossRef]
24. Tao, Y.; Wei, C.; Liu, J.; Deng, C.; Cai, S.; Xiong, W. Nanostructured electrically conductive hydrogels via ultrafast laser processingand self-assembly. *Nanoscale* **2019**, *11*, 9176–9184. [CrossRef] [PubMed]
25. Jodeiri, K.; Foerster, A.; Trindade, G.; Im, J.; Diego Carballares, R.; Fernández-Lafuente, P.M.; Lacey, A.; Parmenter, C.; Tuck, C. Additively Manufactured 3D Micro-bioelectrodes for Enhanced Bioelectrocatalytic Operation. *ACS Appl. Mater. Interfaces* **2023**, *15*, 14914–14924. [CrossRef]
26. Chen, X.; Han, S.Y.; Wu, W.H.; Wu, Z.H.; Yuan, Y.; Wu, J.; Liu, C.S. Harnessing 4D Printing Bioscaffolds for Advanced Orthopedics. *Small* **2022**, *18*, e2106824. [CrossRef]
27. Zhang, J.; Guo, Y.; Hu, W.; Soon, R.H.; Davidson, Z.S.; Sitti, M. Liquid Crystal Elastomer-Based Magnetic Composite Films for Reconfigurable Shape-Morphing Soft Miniature Machines. *Adv. Mater.* **2021**, *33*, e2006191. [CrossRef]
28. Pugliese, R.; Regondi, S. Artificial Intelligence-Empowered 3D and 4D Printing Technologies toward Smarter Biomedical Materials and Approaches. *Polymers* **2022**, *14*, 2794. [CrossRef]
29. Zhao, C.Y.; Wang, Y.; Gao, L.; Xu, Y.Q.; Fan, Z.Y.; Liu, X.J.; Ni, Y.; Xuan, S.H.; Deng, H.X.; Gong, X.L. High-Performance Liquid Metal/Polyborosiloxane Elastomer toward Thermally Conductive Applications. *ACS Appl. Mater. Interfaces* **2022**, *14*, 21564–21576. [CrossRef]
30. Gardi, G.; Ceron, S.; Wang, W.; Petersen, K.H.; Sitti, M. Micro robot collectives with reconfigurable morphologies, behaviors, and functions. *Nat. Commun.* **2022**, *13*, 2239. [CrossRef]
31. Li, H.Y.; Gong, X.; Ni, H.; Lu, P.; Luo, X.; Wen, J.; Yang, Y.J.; Qian, X.; Sun, Z.R.; Wu, J. Light-Induced Ultrafast Molecular Dynamics: From Photochemistry to Optochemistry. *J. Phys. Chem. Lett.* **2022**, *13*, 5881–5893. [CrossRef] [PubMed]
32. Tao, Y.F.; He, K.; Zhang, E.J.; Tan, J.; Hao, H.; Ren, X.D. Cellulose-Derived Wearable Carbon Nanoflake Sensors Customized by Semiconductor Laser Photochemistry. *Adv. Sens. Res.* **2023**, *2*, 2200020. [CrossRef]
33. Muenchinger, A.; Hahn, V.; Beutel, D.; Woska, S.; Monti, J.; Rockstuhl, C.; Blasco, E.; Wegener, M. Multi-Photon 4D Printing of Complex Liquid Crystalline Microstructures by In Situ Alignment Using Electric Fields. *Adv. Mater. Technol.* **2022**, *7*, 2100944. [CrossRef]
34. Falandt, M.; Bernal, P.N.; Dudaryeva, O.; Florczak, S.; Grossbacher, G.; Schweiger, M.; Longoni, A.; Greant, C.; Assuncao, M.; Nijssen, O.; et al. Spatial-Selective Volumetric 4D Printing and Single-Photon Grafting of Biomolecules within Centimeter-Scale Hydrogels via Tomographic Manufacturing. *Adv. Mater. Technol.* **2023**, *8*, 2300026. [CrossRef]
35. Suter, M.; Zhang, L.; Siringil, E.C.; Peters, C.; Luehmann, T.; Ergeneman, O.; Peyer, K.E.; Nelson, B.J.; Hierold, C. Superparamagnetic microrobots: Fabrication by two-photon polymerization and biocompatibility. *Biomed. Microdevices* **2013**, *15*, 997–1003. [CrossRef]
36. Ritacco, T.; Aceti, D.M.; De Domenico, G.; Giocondo, M.; Mazzulla, A.; Cipparrone, G.; Pagliusi, P. Tuning Cholesteric Selective Reflection In Situ Upon Two-Photon Polymerization Enables Structural Multicolor 4D Microfabrication. *Adv. Opt. Mater.* **2022**, *10*, 2101526. [CrossRef]
37. Vinciguerra, M.R.; Tavakoli, M.; Majidi, C.; Yao, L.; Patel, D.K.; Zu, W. Multimaterial Printing of Liquid Crystal Elastomers with Integrated Stretchable Electronics. *ACS Appl. Mater. Interfaces* **2023**, *15*, 24777–24787. [CrossRef]
38. Chen, M.; Gao, M.; Bai, L.; Zheng, H.; Qi, H.J.; Zhou, K. Recent Advances in 4D Printing of Liquid Crystal Elastomers. *Adv. Mater.* **2023**, *35*, e2209566. [CrossRef]
39. Tan, L.; Lee, H.; Fang, L.; Cappelleri, D.J.J. A Power Compensation Strategy for Achieving Homogeneous Microstructures for 4D Printing Shape-Adaptive PNIPAM Hydrogels: Out-of-Plane Variations. *Gels* **2022**, *8*, 828. [CrossRef]
40. Li, K.; Zhao, J.; Zhussupbekova, A.; Shuck, C.E.; Hughes, L.; Dong, Y.; Barwich, S.; Vaesen, S.; Shvets, I.V.; Mobius, M.; et al. 4D printing of MXene hydrogels for high-efficiency pseudo capacitive energy storage. *Nat. Commum.* **2022**, *13*, 6884. [CrossRef]
41. Ding, A.; Jeon, O.; Cleveland, D.; Gasvoda, K.L.; Wells, D.; Lee, S.J.; Alsberg, E. Jammed Micro-Flake Hydrogel for Four-Dimensional Living Cell Bioprinting. *Adv. Mater.* **2022**, *34*, e2109394. [CrossRef] [PubMed]
42. Hu, Y.; Wang, Z.; Jin, D.; Zhang, C.; Sun, R.; Li, Z.; Hu, K.; Ni, J.; Cai, Z.; Pan, D.; et al. Botanical-Inspired 4D Printing of Hydrogel at the Microscale. *Adv. Funct. Mater.* **2020**, *30*, 1907377. [CrossRef]
43. Tao, F.; Deng, C.; Long, J.; Liu, J.; Wang, X.; Song, X.; Lu, C.; Yang, J.; Hao, H.; Wang, C.; et al. Multiprocess Laser Lifting-Off for Nanostructured Semiconductive Hydrogels. *Adv. Mater. Interfaces* **2021**, *9*, 2101250. [CrossRef]

44. Xu, Z.; Wang, X.; Chen, F.; Chen, K. 4D printing of hard magnetic soft materials based on NdFeB particles. *J. Intell. Mater. Syst. Struct.* **2023**. [CrossRef]
45. Hu, X.; Yasa, I.C.; Ren, Z.; Goudu, S.R.; Ceylan, H.; Hu, W.; Sitti, M. Magnetic soft micromachines made of linked microactuator networks. *Sci. Adv.* **2021**, *7*, eabe8436. [CrossRef]
46. Yang, Z.; Li, Z. Magnetic Actuation Systems for Miniature Robots: A Review. *Adv. Intell. Syst.* **2020**, *2*, 2000082. [CrossRef]
47. Liu, C.; Lou, Y.; Sun, Z.; Ma, H.; Sun, M.; Li, S.; You, D.; Wu, J.; Ying, B.; Ding, W.; et al. 4D Printing of Personalized-Tunable Biomimetic Periosteum with Anisotropic Microstructure for Accelerated Vascularization and Bone Healing. *Adv. Healthc. Mater.* **2023**, *2*, e2202868. [CrossRef]
48. Wang, X.; Zhang, Y.; Shen, P.; Cheng, Z.; Chu, C.; Xue, F.; Bai, J. Preparation of 4D printed peripheral vascular stent and its degradation behavior under fluid shear stress after deployment. *Biomater. Sci.* **2022**, *10*, 2302–2314. [CrossRef] [PubMed]
49. Pan, Y.; Lee, L.H.; Yang, Z.; Ul Hassan, S.; Shum, H.C. Co-doping optimized hydrogel-elastomer micro-actuators for versatile biomimetic motions. *Nanoscale* **2021**, *13*, 18967–18976. [CrossRef]
50. Liu, S.; Dong, X.; Wang, Y.; Xiong, J.; Guo, R.; Xiao, J.; Sun, C.; Zhai, F.; Wang, X. 4D Printing of Shape Memory Epoxy for Adaptive Dynamic Components. *Adv. Mater. Technol.* **2023**, *8*, 2202004. [CrossRef]
51. Huang, S.; Zhang, H.; Sheng, J.; Agyenim-Boateng, E.; Wang, C.; Yang, H.; Jiang, G.; Zhou, J.; Lu, J.; Zhang, J. Digital light processing 4D printing multilayer polymers with tunable mechanical properties and shape memory behavior. *Chem. Eng. J.* **2023**, *465*, 142830. [CrossRef]
52. Gu, T.; Bi, H.; Sun, H.; Tang, J.; Ren, Z.; Zhou, X.; Xu, M. Design and development of 4D-printed cellulose nanofibers reinforced shape memory polymer composites: Application for self-deforming plant bionic soft grippers. *Addit. Manuf.* **2023**, *70*, 103544. [CrossRef]
53. Chowdhury, J.; Anirudh, P.V.; Karunakaran, C.; Rajmohan, V.; Mathew, A.T.; Koziol, K.; Alsanie, W.F.; Kannan, C.; Balan, A.S.S.; Thakur, V.K. 4D Printing of Smart Polymer Nanocomposites: Integrating Graphene and Acrylate Based Shape Memory Polymers. *Polymers* **2021**, *13*, 3660. [CrossRef] [PubMed]
54. Zhao, W.; Li, N.N.; Liu, L.W.; Leng, J.; Liu, Y.J. Mechanical behaviors and applications of shape memory polymer and its composites. *Appl. Phys. Rev.* **2023**, *10*, 011306. [CrossRef]
55. Ge, Q.; Sakhaei, A.H.; Lee, H.; Dunn, C.K.; Fang, N.X.; Dunn, M.L. Multimaterial 4D Printing with Tailorable Shape Memory Polymers. *Sci. Rep.* **2016**, *6*, 31110. [CrossRef]
56. Ceamanos, L.; Mulder, D.J.; Kahveci, Z.; Lopez-Valdeolivas, M.; Schenning, A.P.H.J.; Sanchez-Somolinos, C. Photomechanical response under physiological conditions of azobenzene-containing 4D-printed liquid crystal elastomer actuators. *J. Mater. Chem. B* **2023**, *11*, 4083–4094. [CrossRef]
57. Lv, P.; Yang, X.; Bisoyi, H.K.; Zeng, H.; Zhang, X.; Chen, Y.; Xue, P.; Shi, S.; Priimagi, A.; Wang, L.; et al. Stimulus-driven liquid metal and liquid crystal network actuators for programmable soft robotics. *Mater. Horiz.* **2021**, *8*, 2475–2484. [CrossRef]
58. Zhang, Y.; Yang, G.; Hayat, U.; Yu, J.; Liu, C.; Wang, H.J.; Wang, J.Y. Water-responsive 4D printing based on self-assembly of hydrophobic protein "Zein" for the control of degradation rate and drug release. *Bioact. Mater.* **2023**, *23*, 343–352. [CrossRef]
59. Lin, Q.; Hu, Y.; Qiu, C.; Li, X.; Sang, S.; McClements, D.J.; Chen, L.; Long, J.; Xu, X.; Wang, J.; et al. Peanut protein-polysaccharide hydrogels based on semi-interpenetrating networks used for 3D/4D printing. *Food Hydrocoll.* **2023**, *137*, 108332. [CrossRef]
60. Narupai, B.; Smith, P.T.; Nelson, A. 4D Printing of Multi-Stimuli Responsive Protein-Based Hydrogels for Autonomous Shape Transformations. *Adv. Funct. Mater.* **2021**, *31*, 2011012. [CrossRef]
61. Li, S.; Bai, H.D.; Liu, Z.; Zhang, X.Y.; Huang, C.Q.; Wiesner, E.N.R.; Silberstein, M.R.D.; Shepherd, R.O.E. Digital light processing of liquid crystal elastomers for self-sensing artificial muscles. *Sci. Adv.* **2021**, *7*, eabg3677. [CrossRef]
62. Zhu, X.; He, J.; Su, Y.; Chen, Y.; Hu, X. Theoretical Investigation of Backward Optical Parametric Oscillator Pumped by Vortex Beams. *Front. Phys.* **2022**, *10*, 886962. [CrossRef]
63. Pirzio, F.; Negri, J.R.; Agnesi, A. Femtosecond optical parametric oscillator with 3D-printed polymeric parts. *Opt. Laser Technol.* **2022**, *147*, 107657. [CrossRef]
64. Barros, R.F.; Alves, G.B.; Khoury, A.Z. Gouy-phase effects in the frequency combs of an optical parametric oscillator. *Phys. Rev. A* **2021**, *103*, 023511. [CrossRef]
65. Wu, X.; Gross, B.; Leuschel, B.; Mougin, K.; Dominici, S.; Gree, S.; Belqat, M.; Tkachenko, V.; Cabannes-Boue, B.; Chemtob, A.; et al. On-Demand Editing of Surface Properties of Microstructures Made by 3D Direct Laser Writing via Photo-Mediated RAFT Polymerization. *Adv. Funct. Mater.* **2022**, *32*, 2109446. [CrossRef]
66. Tao, Y.; Ren, Y.; Wang, X.; Zhao, R.; Liu, J.; Deng, C.; Wang, C.; Zhang, W.; Hao, H. A femtosecond laser-assembled SnO2 microbridge on interdigitated Au electrodes for gas sensing. *Mater. Lett.* **2022**, *308*, 131120. [CrossRef]
67. Li, W.; Cheng, X.; Wang, Y.; Wang, S. Projection printing of scaffolds with shape recovery capacity and simultaneously improved stiffness and toughness using an ultra-fast-curing poly(propylene fumarate)/hyperbranched additive resin. *Addit. Manuf.* **2021**, *48*, 102446. [CrossRef]
68. Ge, Q.; Li, Z.; Wang, Z.; Kowsari, K.; Zhang, W.; He, X.; Zhou, J.; Fang, N.X. Projection micro stereolithography based 3D printing and its applications. *Int. J. Extreme Manuf.* **2020**, *2*, 022004. [CrossRef]
69. Wang, J.Y.; Jin, F.; Dong, X.Z.; Liu, J.; Zheng, M.L. Flytrap Inspired pH-Driven 3D Hydrogel Actuator by Femtosecond Laser Microfabrication. *Adv. Mater. Technol.* **2022**, *7*, 2200276. [CrossRef]

70. Yu, W.; Ji, Z.; Dong, D.; Yang, X.; Xiao, Y.; Gong, Q.; Xi, P.; Shi, K. Super-resolution deep imaging with hollow Bessel beam STED microscopy. *Laser Photonics Rev.* **2016**, *10*, 147–152. [CrossRef]
71. He, X.; Li, T.; Zhang, J.; Wang, Z. STED Direct Laser Writing of 45 nm Width Nanowire. *Micromachines* **2019**, *10*, 726. [CrossRef] [PubMed]
72. Wollhofen, R.; Buchegger, B.; Eder, C.; Jacak, J.; Kreutzer, J.; Klar, T.A. Functional photoresists for sub-diffraction stimulated emission depletion lithography. *Opt. Mater. Exp.* **2017**, *7*, 2538–2559. [CrossRef]
73. Yang, W.; Wang, Z.; Wang, X.; Yu, T.; Xie, S.; Ge, Z. 3D printing of bioinspired hydrogel microstructures with programmable and complex shape deformations based on a digital micro-mirror device. *Opt. Laser Technol.* **2023**, *157*, 108759. [CrossRef]
74. Li, X.; Liu, B.; Mei, X.; Wang, W.; Wang, X.; Li, X. Development of an In-Situ Laser Machining System Using a Three-Dimensional Galvanometer Scanner. *Engineering* **2020**, *6*, 68–76. [CrossRef]
75. Chen, Q.; Jing, Y.; Yin, J.; Li, Z.; Xiong, W.; Gong, P.; Zhang, L.; Li, S.; Pan, R.; Zhao, X.; et al. High Reflectivity and Thermal Conductivity Ag-Cu Multi-Material Structures Fabricated via Laser Powder Bed Fusion: Formation Mechanisms, Interfacial Characteristics, and Molten Pool Behavior. *Micromachines* **2023**, *14*, 362. [CrossRef]
76. Li, Z.; Li, H.; Yin, J.; Li, Y.; Nie, Z.; Li, X.; You, D.; Guan, K.; Duan, W.; Cao, L.; et al. A Review of Spatter in Laser Powder Bed Fusion Additive Manufacturing: In Situ Detection, Generation, Effects, and Countermeasures. *Micromachines* **2022**, *13*, 1366. [CrossRef] [PubMed]
77. Yin, J.; Zhang, W.; Ke, L.; Wei, H.; Wang, D.; Yang, L.; Zhu, H.; Dong, P.; Wang, G.; Zeng, X. Vaporization of alloying elements and explosion behavior during laser powder bed fusion of Cu-10Zn alloy. *Int. J. Mach. Tools Manuf.* **2021**, *161*, 103686. [CrossRef]
78. Yin, J.; Wang, D.; Wei, H.; Yang, L.; Ke, L.; Hu, M.; Xiong, W.; Wang, G.; Zhu, H.; Zeng, X. Dual-beam laser-matter interaction at overlap region during multi-laser powder bed fusion manufacturing. *Addit. Manuf.* **2021**, *46*, 102178. [CrossRef]
79. Veiga, J.S.; Carneiro, M.R.; Molter, R.; Vinciguerra, M.; Yao, L.; Majidi, C.; Tavakoli, M. Toward Fully Printed Soft Actuators: UV-Assisted Printing of Liquid Crystal Elastomers and Biphasic Liquid Metal Conductors. *Adv. Mater. Technol.* **2023**, *25*, 2300144. [CrossRef]
80. Tang, Z.; Gao, Z.; Jia, S.; Wang, F.; Wang, Y. Graphene-Based Polymer Bilayers with Superior Light-Driven Properties for Remote Construction of 3D Structures. *Adv. Sci.* **2017**, *4*, 1600437. [CrossRef]
81. Benyahia, K.; Gomes, S.; Andre, J.-C.; Qi, H.J.; Demoly, F. Influence of interlocking blocks assembly on the actuation time, shape change, and reversibility of voxel-based multi-material 4D structures. *Smart. Mater. Struct.* **2023**, *32*, 065011. [CrossRef]
82. Deng, C.S.; Liu, Y.C.; Fan, X.H.; Jiao, B.Z.; Zhang, Z.X.; Zhang, M.D.; Chen, F.Y.; Gao, H.; Deng, L.M.; Xiong, W. Femtosecond Laser 4D Printing of Light-Driven Intelligent Micromachines. *Adv. Funct. Mater.* **2023**, *33*, 2211473. [CrossRef]
83. Ma, Z.C.; Zhang, Y.L.; Han, B.; Hu, X.Y.; Li, C.H.; Chen, Q.D.; Sun, H.B. Femtosecond laser programmed artificial musculoskeletal systems. *Nat. Commun.* **2020**, *11*, 4536. [CrossRef] [PubMed]
84. Liu, J.W.; Fan, X.H.; Tao, Y.F.; Deng, C.S.; Yu, K.W.; Zhang, W.G.; Deng, L.M.; Xiong, W. Two-Step Freezing Polymerization Method for Efficient Synthesis of High-Performance Stimuli-Responsive Hydrogels. *ACS Omega* **2020**, *5*, 5921–5930. [CrossRef]
85. Yao, X.; Chen, H.; Qin, H.; Cong, H.-P. Nanocomposite Hydrogel Actuators with Ordered Structures: From Nanoscale Control to Macroscale Deformations. *Small Methods* **2023**, *5*, e2300414. [CrossRef]
86. Wang, C.; Hu, Z.; Yang, L.; Zhang, C.; Zhang, L.; Ji, S.; Xu, L.; Li, J.; Hu, Y.; Wu, D.; et al. Magnetically driven rotary microfilter fabricated by two-photon polymerization for multimode filtering of particles. *Opt. Lett.* **2021**, *46*, 2968–2971. [CrossRef]
87. Xiong, Z.; Zheng, C.; Jin, F.; Wei, R.; Zhao, Y.; Gao, X.; Xia, Y.; Dong, X.; Zheng, M.; Duan, X. Magnetic-field-driven ultra-small 3D hydrogel microstructures: Preparation of gel photoresist and two-photon polymerization microfabrication. *Sens. Actuat. B-Chem.* **2018**, *274*, 541–550. [CrossRef]
88. Dai, Y.; Jia, L.; Wang, L.; Sun, H.; Ji, Y.; Wang, C.; Song, L.; Liang, S.; Chen, D.; Feng, Y.; et al. Magnetically Actuated Cell-Robot System: Precise Control, Manipulation, and Multimode Conversion. *Small* **2022**, *18*, 2105414. [CrossRef]
89. Cong, Z.; Tang, S.; Xie, L.; Yang, M.; Li, Y.; Lu, D.; Li, J.; Yang, Q.; Chen, Q.; Zhang, Z.; et al. Magnetic-Powered Janus Cell Robots Loaded with Oncolytic Adenovirus for Active and Targeted Virotherapy of Bladder Cancer. *Adv. Mater.* **2022**, *34*, 2201042. [CrossRef]
90. del Pozo, M.; Delaney, C.; da Cunha, M.P.; Debije, M.G.; Florea, L.; Schenning, A. Temperature-Responsive 4D Liquid Crystal Microactuators Fabricated by Direct Laser Writing by Two-Photon Polymerization. *Small Struct.* **2022**, *3*, 2100158. [CrossRef]
91. Cen, S.; Li, Z.; Guo, Z.; Li, H.; Shi, J.; Huang, X.; Zou, X.; Holmes, M. 4D printing of a citrus pectin/beta-CD Pickering emulsion: A study on temperature induced color transformation. *Addit. Manuf.* **2022**, *56*, 102925.
92. Lee, Y.W.; Ceylan, H.; Yasa, I.C.; Kilic, U.; Sitti, M. 3D-Printed Multi-Stimuli-Responsive Mobile Micromachines. *ACS Appl. Mater. Interfaces* **2021**, *13*, 12759–12766. [CrossRef] [PubMed]
93. Pan, H.M.; Goto, A. Topology-Dependent pH-Responsive Actuation and Shape Memory Programming for Biomimetic 4D Printing. *Macromol. Rapid Commun.* **2023**, *44*, e2300074. [CrossRef]
94. Wei, S.; Liu, J.; Zhao, Y.; Zhang, T.; Zheng, M.; Jin, F.; Dong, X.; Xing, J.; Duan, X. Protein-Based 3D Microstructures with Controllable Morphology and pH-Responsive Properties. *ACS Appl. Mater. Interfaces* **2017**, *9*, 42247–42257. [CrossRef]
95. Ghazal, A.F.; Zhang, M.; Bhandari, B.; Chen, H. Investigation on spontaneous 4D changes in color and flavor of healthy 3D printed food materials over time in response to external or internal pH stimulus. *Food. Res. Int.* **2021**, *142*, 110215. [CrossRef]

96. Hu, X.; Ma, Z.; Han, B.; Li, C.; Zhang, Y. Femtosecond Laser Fabrication of Protein-Based Smart Soft Actuators. *Chin. J. Lasers* **2021**, *48*, 1402001.
97. Wang, H.; Di, J.; Sun, Y.; Fu, J.; Wei, Z.; Matsui, H.; Alonso, A.d.C.; Zhou, S. Biocompatible PEG-Chitosan@Carbon Dots Hybrid Nanogels for Two-Photon Fluorescence Imaging, Near-Infrared Light/pH Dual-Responsive Drug Carrier, and Synergistic Therapy. *Adv. Funct. Mater.* **2015**, *25*, 5537–5547. [CrossRef]
98. Sol, J.A.H.P.; Douma, R.F.; Schenning, A.P.H.J.; Debije, M.G. 4D Printed Light-Responsive Patterned Liquid Crystal Elastomer Actuators Using a Single Structural Color Ink. *Adv. Mater. Technol.* **2023**, *8*, 2200970. [CrossRef]
99. Lugger, S.J.D.; Ceamanos, L.; Mulder, D.J.; Sanchez-Somolinos, C.; Schenning, A.P.H.J. 4D Printing of Supramolecular Liquid Crystal Elastomer Actuators Fueled by Light. *Adv. Mater. Technol.* **2023**, *8*, 2201472. [CrossRef]
100. Martella, D.; Antonioli, D.; Nocentini, S.; Wiersma, D.S.; Galli, G.; Laus, M.; Parmeggiani, C. Light activated non-reciprocal motion in liquid crystalline networks by designed microactuator architecture. *Rsc. Adv.* **2017**, *7*, 19940–19947. [CrossRef]
101. Ceamanos, L.; Kahveci, Z.; Lopez-Valdeolivas, M.; Liu, D.; Broer, D.J.; Sanchez-Somolinos, C. Four-Dimensional Printed Liquid Crystalline Elastomer Actuators with Fast Photoinduced Mechanical Response toward Light-Driven Robotic Functions. *ACS Appl. Mater. Interfaces* **2020**, *12*, 44195–44204. [CrossRef] [PubMed]
102. Chen, L.; Dong, Y.; Tang, C.Y.; Zhong, L.; Law, W.C.; Tsui, G.C.P.; Yang, Y.; Xie, X. Development of Direct-Laser-Printable Light-Powered Nanocomposites. *ACS Appl. Mater. Interfaces* **2019**, *11*, 19541–19553. [CrossRef] [PubMed]
103. Li, C.; Xue, Y.; Han, M.; Palmer, L.C.; Rogers, J.A.; Huang, Y.; Stupp, S.I. Synergistic photoactuation of bilayered spiropyran hydrogels for predictable origami-like shape change. *Matter* **2021**, *4*, 1377–1390. [CrossRef]
104. Jin, D.; Chen, Q.; Huang, T.-Y.; Huang, J.; Zhang, L.; Duan, H. Four-dimensional direct laser writing of reconfigurable compound micromachines. *Mater. Today* **2020**, *32*, 19–25. [CrossRef]
105. Li, M.; Wang, X.; Dong, B.; Sitti, M. In-air fast response and high speed jumping and rolling of a light-driven hydrogel actuator. *Nat. Commun.* **2020**, *11*, 3988. [CrossRef]
106. Dong, M.; Wang, X.P.; Chen, X.Z.; Mushtaq, F.; Deng, S.Y.; Zhu, C.H.; Torlakcik, H.; Terzopoulou, A.; Qin, X.H.; Xiao, X.Z.; et al. 3D-Printed Soft Magnetoelectric Microswimmers for Delivery and Differentiation of Neuron-Like Cells. *Adv. Funct. Mater.* **2020**, *30*, 1910323. [CrossRef]
107. Chen, Z.; Song, X.; Mu, X.; Zhang, J.; Cheang, U.K. 2D Magnetic Microswimmers for Targeted Cell Transport and 3D Cell Culture Structure Construction. *ACS Appl. Mater. Interfaces* **2023**, *15*, 8840–8853. [CrossRef]
108. Wang, J.Y.; Jin, F.; Dong, X.Z.; Liu, J.; Zhou, M.X.; Li, T.; Zheng, M.L. Dual-Stimuli Cooperative Responsive Hydrogel Microactuators Via Two-Photon Lithography. *Small* **2023**, *8*, e2303166. [CrossRef]
109. Di, X.; Hou, J.W.; Yang, M.M.; Wu, G.L.; Sun, P.C. A bio-inspired, ultra-tough, high-sensitivity, and anti-swelling conductive hydrogel strain sensor for motion detection and information transmission. *Mater. Horizons* **2022**, *9*, 3057–3069. [CrossRef]
110. Huang, H.; Liao, C.; Zou, M.; Liu, D.; Liu, S.; Wang, Y.; Bai, Z.; Liu, D.; Li, B.; Huang, J.; et al. Four-Dimensional Printing of a Fiber-Tip Multimaterial Microcantilever as a Magnetic Field Sensor. *ACS Photonics* **2023**, *10*, 1916–1924. [CrossRef]
111. Wang, H.; Zhang, W.; Ladika, D.; Yu, H.; Gailevicius, D.; Wang, H.; Pan, C.-F.; Nair, P.N.S.; Ke, Y.; Mori, T.; et al. Two-Photon Polymerization Lithography for Optics and Photonics: Fundamentals, Materials, Technologies, and Applications. *Adv. Funct. Mater.* **2023**, *6*, 2214211. [CrossRef]
112. Liu, B.; Dong, B.; Xin, C.; Chen, C.; Zhang, L.; Wang, D.; Hu, Y.; Li, J.; Zhang, L.; Wu, D.; et al. 4D Direct Laser Writing of Submerged Structural Colors at the Microscale. *Small* **2023**, *19*, e2204630. [CrossRef] [PubMed]
113. Liang, S.Y.; Liu, Y.F.; Ji, Z.K.; Xia, H.; Sun, H. Chameleon-inspired design of dynamic patterns based on femtosecond laser-induced forward transfer. *Chem. Eng. J.* **2023**, *466*, 143121. [CrossRef]
114. Peng, S.; Sun, S.; Zhu, Y.; Qiu, J.; Yang, H. Colourful 3D anti-counterfeiting label using nanoscale additive manufacturing. *Virtual Phys. Prototyp.* **2023**, *18*, e2179929. [CrossRef]
115. Zhang, W.; Wang, H.; Tan, A.T.L.; Ranganath, A.S.; Zhang, B.; Wang, H.; Chan, J.Y.E.; Ruan, Q.; Liu, H.; Ha, S.T.; et al. Stiff Shape Memory Polymers for High-Resolution Reconfigurable Nanophotonics. *Nano Lett.* **2022**, *22*, 8917–8924. [CrossRef]
116. Zhai, Y.Q.; Niu, J.Q.; Liu, Y.S.; Chen, X.; Liu, J.Q.; Yang, B. Biologically Inspired, Optical Waveguide with Isolation Layer Integrated Microlens Array for High-Contrast Imaging. *Adv. Opt. Mater.* **2023**, *11*, 2300020. [CrossRef]
117. Cao, J.J.; Hou, Z.S.; Tian, Z.N.; Hua, J.G.; Zhang, Y.L.; Chen, Q.D. Bioinspired Zoom Compound Eyes Enable Variable-Focus Imaging. *ACS Appl. Mater. Interfaces* **2020**, *12*, 10107–10117. [CrossRef]
118. Ma, Z.C.; Hu, X.Y.; Zhang, Y.L.; Liu, X.Q.; Hou, Z.S.; Niu, L.G.; Zhu, L.; Han, B.; Chen, Q.D.; Sun, H.B. Smart Compound Eyes Enable Tunable Imaging. *Adv. Funct. Mater.* **2019**, *29*, 1903340. [CrossRef]
119. Lei, X.; Peng, S.; Niu, Y.; Sun, S.; Zhu, Y.; Qiu, J. Magnetically driven micro-optical choppers fabricated by two-photon polymerization. *Opt. Lett.* **2023**, *48*, 835–838. [CrossRef]
120. Li, C.Z.; Jiao, Y.L.; Zhang, Y.Y.; Jiang, S.J.; Lv, X.D.; Wu, S.Z.; Li, J.W.; Hu, Y.L.; Ye, J.X.; Liu, K.; et al. Noncontact All-In-Situ Reversible Reconfiguration of Femtosecond Laser-Induced Shape Memory Magnetic Microcones for Multifunctional Liquid Droplet Manipulation and Information Encryption. *Adv. Funct. Mater.* **2021**, *31*, 2100543. [CrossRef]
121. Wang, B.; Handschuh-Wang, S.; Shen, J.; Zhou, X.; Guo, Z.; Liu, W.; Pumera, M.; Zhang, L. Small-Scale Robotics with Tailored Wettability. *Adv. Mater.* **2023**, *35*, e2205732. [CrossRef]

122. Hu, L.; Chee, P.L.; Sugiarto, S.; Yu, Y.; Shi, C.; Yan, R.; Yao, Z.; Shi, X.; Zhi, J.; Kai, D.; et al. Hydrogel-Based Flexible Electronics. *Adv. Mater.* **2023**, *35*, e2205326. [CrossRef]
123. Xin, C.; Jin, D.; Hu, Y.; Yang, L.; Li, R.; Wang, L.; Ren, Z.; Wang, D.; Ji, S.; Hu, K.; et al. Environmentally Adaptive Shape-Morphing Microrobots for Localized Cancer Cell Treatment. *ACS Nano* **2021**, *15*, 18048–18059. [CrossRef]
124. Lao, Z.; Sun, R.; Jin, D.; Ren, Z.; Xin, C.; Zhang, Y.; Jiang, S.; Zhang, Y.; Zhang, L. Encryption/decryption and microtarget capturing by pH-driven Janus microstructures fabricated by the same femtosecond laser printing parameters. *Int. J. Extreme Manuf.* **2021**, *3*, 025001. [CrossRef]
125. Wang, Y.; Cui, H.; Esworthy, T.; Mei, D.; Wang, Y.; Zhang, L.G. Emerging 4D Printing Strategies for Next-Generation Tissue Regeneration and Medical Devices. *Adv. Mater.* **2022**, *34*, 2109198. [CrossRef] [PubMed]
126. Ding, A.; Lee, S.J.; Ayyagari, S.; Tang, R.; Cong Truc, H.; Alsberg, E. 4D biofabrication via instantly generated graded hydrogel scaffolds. *Bioact. Mater.* **2022**, *7*, 324–332. [CrossRef]
127. Yan, Y.; Yao, R.; Zhao, J.; Chen, K.; Duan, L.; Wang, T.; Zhang, S.; Guan, J.; Zheng, Z.; Wang, X.; et al. Implantable nerve guidance conduits: Material combinations, multi-functional strategies and advanced engineering innovations. *Bioact. Mater.* **2022**, *11*, 57–76. [CrossRef] [PubMed]
128. Xu, Y.; Bian, Q.; Wang, R.; Gao, J. Micro/nanorobots for precise drug delivery via targeted transport and triggered release: A review. *Int. J. Pharmaceut.* **2022**, *616*, 121551. [CrossRef] [PubMed]
129. Shakoor, A.; Gao, W.; Zhao, L.; Jiang, Z.; Sun, D. Advanced tools and methods for single-cell surgery. *Microsyst. Nanoeng.* **2022**, *8*, 47. [CrossRef] [PubMed]
130. Torlakcik, H.; Sarica, C.; Bayer, P.; Yamamoto, K.; Iorio-Morin, C.; Hodaie, M.; Kalia, S.K.; Neimat, J.S.; Hernesniemi, J.; Bhatia, A.; et al. Magnetically Guided Catheters, Micro- and Nanorobots for Spinal Cord Stimulation. *Front. Neurorobot.* **2021**, *15*, 749024. [CrossRef]
131. Soto, F.; Wang, J.; Ahmed, R.; Demirci, U. Medical Micro/Nanorobots in Precision Medicine. *Adv. Sci.* **2020**, *7*, 2002203. [CrossRef] [PubMed]
132. Chen, S.; Tan, Z.; Liao, P.; Li, Y.; Qu, Y.; Zhang, Q.; Yang, M.; Chan, K.W.Y.; Zhang, L.; Man, K.; et al. Biodegradable Microrobots for DNA Vaccine Delivery. *Adv. Healthc. Mater.* **2023**, *2*, e2202921. [CrossRef] [PubMed]
133. Wang, X.; Lin, D.; Zhou, Y.; Jiao, N.; Tung, S.; Liu, L. Multistimuli-Responsive Hydroplaning Superhydrophobic Microrobots with Programmable Motion and Multifunctional Applications. *ACS Nano* **2022**, *16*, 14895–14906. [CrossRef] [PubMed]
134. Liu, S.; Gao, C.; Peng, F. Micro/nanomotors in regenerative medicine. *Mater. Today Adv.* **2022**, *16*, 100281. [CrossRef]
135. Xu, Z.; Wu, Z.; Yuan, M.; Chen, Y.; Ge, W.; Xu, Q. Versatile magnetic hydrogel soft capsule microrobots for targeted delivery. *Iscience* **2023**, *26*, 106727. [CrossRef]
136. Chinnakorn, A.; Nuansing, W.; Bodaghi, M.; Rolfe, B.; Zolfagharian, A. Recent progress of 4D printing in cancer therapeutics studies. *SLAS Technol.* **2023**, *28*, 127–141. [CrossRef]
137. Lim, X.Y.; Capinpin, S.M.; Bolem, N.; Foo, A.S.C.; Yip, W.-C.G.; Kumar, A.P.; Teh, D.B.L. Biomimetic nanotherapeutics for targeted drug delivery to glioblastoma multiforme. *Bioeng. Transl. Med.* **2023**, *28*, 127–141. [CrossRef]
138. Ceccarini, M.R.; Ripanti, F.; Raggi, V.; Paciaroni, A.; Petrillo, C.; Comez, L.; Donato, K.; Bertelli, M.; Beccari, T.; Valentini, L. Development of Salmon Sperm DNA/Regenerated Silk Bio-Based Films for Biomedical Studies on Human Keratinocyte HaCaT Cells under Solar Spectrum. *J. Funct. Biomater.* **2023**, *14*, 280. [CrossRef]
139. Xie, L.; Cong, Z.; Tang, S.; Yang, M.; Li, Y.; Ren, C.; Chen, Q.; Lu, D.; Wan, F.; Zhang, X.; et al. Oncolytic adenovirus-loaded magnetic-driven Janus tumor cell robots for active and targeted virotherapy of homologous carcinoma. *Mater. Today Chem.* **2023**, *30*, 101560. [CrossRef]
140. Mayorga-Martinez, C.C.; Zelenka, J.; Grmela, J.; Michalkova, H.; Ruml, T.; Mares, J.; Pumera, M. Swarming Aqua Sperm Micromotors for Active Bacterial Biofilms Removal in Confined Spaces. *Adv. Sci.* **2021**, *8*, 2101301. [CrossRef]
141. Wang, H.C.; Yu, S.M.; Liao, J.J.; Qing, X.D.; Sun, D.X.; Ji, F.T.; Song, W.P.; Wang, L.; Li, T.L. A Robot Platform for Highly Efficient Pollutant Purification. *Front. Bioeng. Biotechnol.* **2022**, *10*, 903219. [CrossRef] [PubMed]
142. Tang, S.; Zhang, F.; Gong, H.; Wei, F.; Zhuang, J.; Karshalev, E.; de Avila, B.E.-F.; Huang, C.; Zhou, Z.; Li, Z.; et al. Enzyme-powered Janus platelet cell robots for active and targeted drug delivery. *Sci. Robot.* **2020**, *5*, eaba6137. [CrossRef] [PubMed]
143. Xiao, Y.Y.; Jiang, Z.C.; Tong, X.; Zhao, Y. Biomimetic Locomotion of Electrically Powered "Janus" Soft Robots Using a Liquid Crystal Polymer. *Adv. Mater.* **2019**, *31*, 1903452. [CrossRef]
144. Peng, X.; Liu, T.Q.; Shang, C.; Jiao, C.; Wang, H.l. Mechanically strong Janus poly(N-isopropylacrylamide)/graphene oxide hydrogels as thermo-responsive soft robots. *Chin. J. Polym. Sci.* **2017**, *35*, 1268–1275. [CrossRef]
145. Han, M.D.; Guo, X.G.; Chen, X.X.; Liang, C.M.; Zhao, H.B.; Zhang, Q.H.; Bai, W.B.; Zhang, F.; Wei, H.M.; Wu, C.S.; et al. Submillimeter-scale multimaterial terrestrial robots. *Sci. Robot.* **2022**, *7*, eabn0602. [CrossRef] [PubMed]
146. Reynolds, M.F.; Cortese, A.J.; Liu, Q.K.; Zheng, Z.Q.; Wang, W.; Norris, S.L.; Lee, S.; Miskin, M.Z.; Molnar, A.C.; Cohen, I.; et al. Microscopic robots with onboard digital control. *Sci. Robot.* **2022**, *7*, eabq2296. [CrossRef]
147. Miskin, M.Z.; Cortese, A.J.; Dorsey, K.; Esposito, E.P.; Reynolds, M.F.; Liu, Q.K.; Cao, M.C.; Muller, D.A.; McEuen, P.L.; Cohen, I. Electronically integrated, mass-manufactured, microscopic robots. *Nature* **2020**, *584*, 557. [CrossRef] [PubMed]
148. Liu, J.W.; Liu, Y.C.; Deng, C.S.; Yu, K.W.; Fan, X.H.; Zhang, W.G.; Tao, Y.F.; Hu, H.C.; Deng, L.M.; Xiong, W. 3D Printing Nano-Architected Semiconductors Based on Versatile and Customizable Metal-Bound Composite Photoresins. *Adv. Mater. Technol.* **2022**, *7*, 2101230. [CrossRef]

149. Do, A.; Worthington, K.S.; Tucker, B.A.; Salem, A.K. Controlled drug delivery from 3D printed two-photon polymerized poly(ethylene glycol) dimethacrylate devices. *Int. J. Pharm.* **2018**, *552*, 217–224. [CrossRef]
150. Feliciano, A.J.; Blitterswijk, C.V.; Moroni, L.; Baker, M.B. Realizing Tissue Integration with Supramolecular Hydrogels. *Acta Biomater.* **2021**, *124*, 1–14. [CrossRef]

micromachines

MDPI

Article

Dynamic Compressive Properties and Failure Mechanism of the Laser Powder Bed Fusion of Submicro-LaB6 Reinforced Ti-Based Composites

Xianghui Li [1,2] and Yang Liu [1,2,3,*]

1 Chaozhou Branch of Chemistry and Chemical Engineering Guangdong Laboratory, Chaozhou 515600, China
2 Faculty of Mechanical Engineering & Mechanics, Ningbo University, Ningbo 315211, China
3 MOE Key Laboratory of Impact and Safety Engineering, Ningbo University, Ningbo 315211, China
* Correspondence: liuyang1@nbu.edu.cn; Tel.: +86-18857400630

Abstract: In this study, lanthanum hexaboride (LaB_6) particle-reinforced titanium matrix composites (PRTMCs, TC4/LaB_6) were successfully manufactured using the laser powder bed fusion (LPBF) process. Thereafter, the effect of the mass fraction of LaB_6 on the microstructure and the dynamic compressive properties was investigated. The results show that the addition of LaB_6 leads to significant grain refinement. Moreover, the general trend of grain size reveals a concave bend as the fraction increases from 0.2% to 1.0%. Furthermore, the texture intensity of prior β grains and α grains was found to be weakened in the composites. It was also observed that the TC4/LaB_6 have higher quasi-static and dynamic compressive strengths but lower fracture strain when compared with the as-built TC4. The sample with 0.5 wt.% LaB_6 was found to have the best strength–toughness synergy among the three groups of composites due to having the smallest grain size. Furthermore, the fracture mode of TC4/LaB_6 was found to change from the fracture under the combined action of brittle and ductility to the cleavage fracture. This study was able to provide a theoretical basis for an in-depth understanding of the compressive properties of additive manufacturing of PRTMCs under high-speed loading conditions.

Keywords: laser powder bed fusion; Ti-based composites; dynamic compressive properties; high strain rate; failure mechanism

Citation: Li, X.; Liu, Y. Dynamic Compressive Properties and Failure Mechanism of the Laser Powder Bed Fusion of Submicro-LaB6 Reinforced Ti-Based Composites. *Micromachines* **2023**, *14*, 2237. https://doi.org/10.3390/mi14122237

Academic Editor: Antonio Ancona

Received: 14 November 2023
Revised: 8 December 2023
Accepted: 12 December 2023
Published: 13 December 2023

Correction Statement: This article has been republished with a minor change. The change does not affect the scientific content of the article and further details are available within the backmatter of the website version of this article.

1. Introduction

Laser powder bed fusion, as one of the most important additive manufacturing (AM) technologies, is able to build digitally designed parts up layer by layer by melting a feedstock using a micro-scale laser beam with a high energy density [1,2]. This unique production method provides more design freedom and flexible manufacturability compared with conventional fabrication methods [3,4]. Titanium alloys, in particular Ti-6Al-4V (TC4), have drawn enormous attention due to their high strength-to-weight ratio and excellent corrosion resistance [5]. Nowadays, the LPBF of TC4 is one of the most important research topics in AM, and it is mainly used in the fabrication of value-added components in the fields of aerospace, weapons, marine engineering, etc. [6,7]. However, the LPBF of TC4 suffers from the formation of hierarchical structures of acicular martensitic α' grains within a large columnar of prior β grains, which results in inferior mechanical properties, especially under high-strain-rate conditions [8,9].

Particle-reinforced metal matrix composites (PRMMCs) are regarded as one of the alternatives to overcoming the performance defects of metals under impacting loads. However, conventional fabrication methods for PRMMCs—including powder metallurgy [10] mechanical alloying [11], self-propagation high-temperature synthesis [12], etc.—often have various shortcomings, such as interface cracks, segregation, etc. Recently, the continuous development of additively manufactured particle-reinforced Ti-based composite to foster their industrial adoption relies on achieving properties superior to the counterparts fabricated using conventional methods [13–17]. Rare earth (RE) or RE compounds are regarded as effective grain refiners, typically acting as heterogeneous nucleation sites for the β and α phases [14]. They induce the formation of new grains during the solidification process in additive manufacturing, leading to refined grain structures and impeding excessive grain growth. Barriobero et al. [15] found that the addition of La into Ti can change the texture orientation relationship between the α and parent β grains, resulting in the generation of equiaxed grains and weakening the anisotropy of LPBF titanium alloy. Bermingham et al. [16,17] found that the addition of LaB_6 nanoparticles into LENS of TC4 reversed the direction of the Marangoni flow in the molten pool and refined the grain size, and the tensile strength was enhanced at a cost of ductility. Liu et al. [14] found that the addition of trace LaB_6 into electron beam melting TC4 can reduce the size of prior β columnar and α grains as well as the texture intensity, thus reducing the anisotropy.

Dynamic impact resistance is one of the most important performance benchmarks of PRMMCs due to its wide application in the fields of aerospace, defense, etc. However, the literature shows that the dynamic mechanical properties of PRMMCs are closely related to the reinforcement and the interface between the matrix and the particles, which are different from the corresponding matrix [18]. Research on the dynamic mechanical response of AM titanium alloys had been reported frequently, encompassing aspects such as strength [19], toughness [20], constitutive behavior [21], dynamic damage [22], etc. However, there is still a notable gap regarding the investigation of the dynamic mechanical response of particle-reinforced titanium matrix composites in additive manufacturing. Therefore, Ti-based composites reinforced using the sub-micro particles were fabricated with LPBF through the mixing of TC4 with various mass fractions of LaB_6. Furthermore, dynamic deformation behaviors were tested using the split Hopkinson pressure bar (SHPB) at various strain rates, and then the effects of the addition of LaB_6 on the microstructure and dynamic compressive properties were revealed. These findings contribute to the understanding of the performances and deformation mechanism of LPBF of TC4/LaB_6, which also provides a reference for designing high-performance particle-reinforced titanium matrix composites.

2. Experimental Details

2.1. Raw Materials

Gas-atomized spherical TC4 powder with an average particle size of 38.5 μm was used as the matrix, and irregular LaB_6 particles (typically large particle sizes: 10–50 μm, average size: approximately 1 μm) were used as the reinforcement, as shown in Figure 1a,b. The two materials were mixed using planetary high-energy ball milling (NanDa Instrument QM-3SP4, Nanjing Nanda Instrument Co., Ltd., Nanjing, China). The mass fractions of the LaB_6 particle were set as 0.2 wt.%, 0.5 wt.%, and 1.0 wt.%, respectively. The parameters of ball milling were set at a rotation speed of 350 rpm, milling time of 4 h, and ball-to-powder ratio of 4:1. From Figure 1c, it is observed that the composite powder maintains good sphericity. Additionally, Figure 1d shows that the nanoparticles adhere tightly to the surface of the Ti particles.

2.2. Specimen Preparation

The LPBF equipment used for the specimen preparation is the Concept Laser Mlab 100R, which consists of a 100 W continuous fiber laser (50 μm spot diameter), a high-accuracy galvo scanning system, a powder delivery system, a protective atmosphere filling device, and a circulation purification device. The forming chamber was filled with argon as a protective gas in order to ensure that the oxygen content in the chamber was below 100 ppm, preventing the metal material from oxidizing during the process. After a series of parameter evaluation experiments, the optimal process parameters were determined using the orthogonal scanning method, with a laser power of 72.5 W, scanning speed of 700 mm/s, powder layer thickness of 30 μm, and scanning hatch spacing of 70 μm. Furthermore, blocks with $5 \times 5 \times 5$ mm^3 were fabricated for the microstructure characterization, and a cylinder with $\phi 5 \times 5$ mm^3 was fabricated for the loading test. As shown in Table 1, the specimens were named TMC0, TMC1, TMC2, and TMC3 in accordance with the mass fractions of LaB_6: 0 wt.%, 0.2 wt.%, 0.5 wt.%, and 1.0 wt.%, respectively.

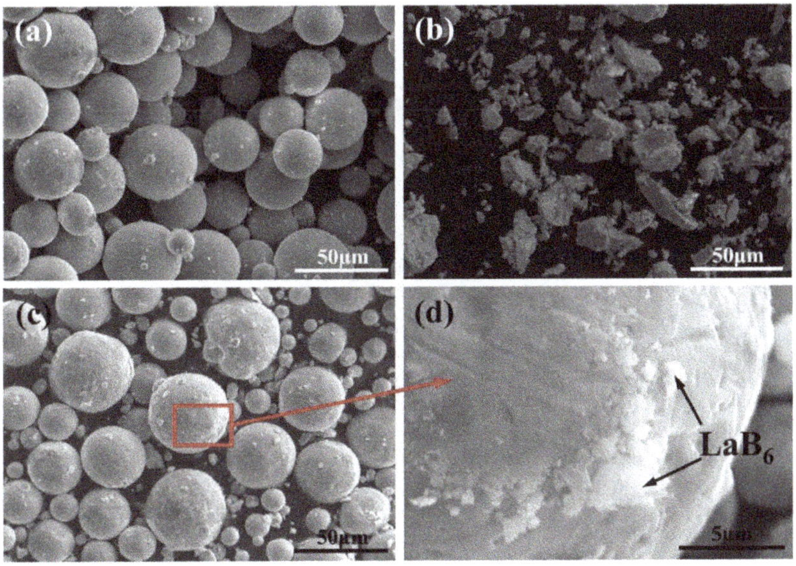

Figure 1. SEM images of raw powders: (**a**) gas-atomized spherical TC4 powder, (**b**) irregular LaB_6 nanoparticle, (**c**) composite powder after ball milling, and (**d**) higher magnification image showing the nanoparticles adhering to the surface of Ti particles.

Table 1. Different names corresponding to various LaB_6 mass fractions.

Name	Mass Fraction (wt.%)
TMC0	0
TMC1	0.2
TMC2	0.5
TMC3	1.0

2.3. Characterization and Analysis

Scanning electron microscopy (SEM, Hitachi SU5000, Hitachi High-Tech Co., Ltd, Tokyo, Japan), electron backscatter diffraction (EBSD, AZTEC 2.0, Oxford Instruments, Tubney Woods, Abingdon, Oxon OX13 5QX, UK), bright-field transmission electron microscopy (TEM, FEI Talos F200X, Beam Energy: 200 kV, Thermo Fisher Scientific, Waltham, MA, USA), and selected area electron diffraction (SAED, Selected Area Aperture: 10 μm) were used in this study in order to systematically characterize the microstructure, phase, and texture of the samples. Kroll reagent was used to etch the specimens. Electrolytic polishing was performed using Perchloric acid and Glacial acetic acid with a ratio of 1:19 at a voltage of 55 V and a current of 1.95 A. Ar ion beam thinning was used in the preparation of samples for TEM. A Materials Testing System (MTS 810) was used to perform the quasi-static compression tests at an initial strain rate of 10^{-3} s^{-1}. A split Hopkinson pressure bar (SHPB) device was used to conduct impacting tests, the methods and test principles are specifically described in Ref. [20]. The impacting velocity was set as 15, 20, and 23 m/s, thus resulting in strain rates of about 1800, 2500, and 3000/s. The mechanical testing was conducted along the direction of sample fabrication, with each test condition repeated three times. Thereafter, the average value was taken into consideration.

3. Results and Discussions

3.1. Initial Microstructure

As shown in Figure 2a, it can be observed that the TMC0 exhibits distinct prior β columnar grain and grain boundaries (marked by white dashed lines in the figure), with a length of several millimeters and a width approximately equal to the hatch spacing. The formation of columnar grains is caused by epitaxial growth during the LPBF process [4,23]. A fine interlaced acicular α'/α phase with an average aspect ratio of 15–25 was formed within these prior β grains, and the grain size and orientation were found to be highly heterogeneous, which is in agreement with the findings of Refs. [14,17]. Although the traces of grain boundaries are still visible, the prior β grain of the TMC1 was found to be significantly narrowed, and a small number of pores formed in the grain. Moreover, the aspect ratios of acicular α'/α do not vary considerably (in the range between 15 and 25), and the direction is still strongly aligned (as shown in Figure 2b). When LaB_6 increases to 0.5 wt.%, the prior β grain boundaries are no longer discernible in the TMC2, and a small number of acicular α'/α with decreasing aspect ratios are chaotically distributed, as shown in Figure 2c. When LaB_6 further increases to 1.0 wt.%, most of the acicular α'/α disappear. This is consistent with Ref. [14], which revealed that an increase in LaB_6 considerably decreased the width of the prior β grains as well as the aspect ratios of the acicular α'/α in the EBM of TC4 alloy. Marangoni convection is the dominant fluid flow that impacts the molten pool [24]. Due to steep temperature gradients in the molten pool, surface tension gradients are initially generated, leading to surface flow from regions of a lower surface tension to those of a higher surface tension [25]. Bermingham et al. [17] demonstrated that the addition of the RE element La can reverse the surface tension temperature coefficient, leading to a reversal of fluid flow direction, thereby influencing the molten pool. This impeded the epitaxial growth process of prior β columnar grains, ultimately leading to a more uniform microstructure.

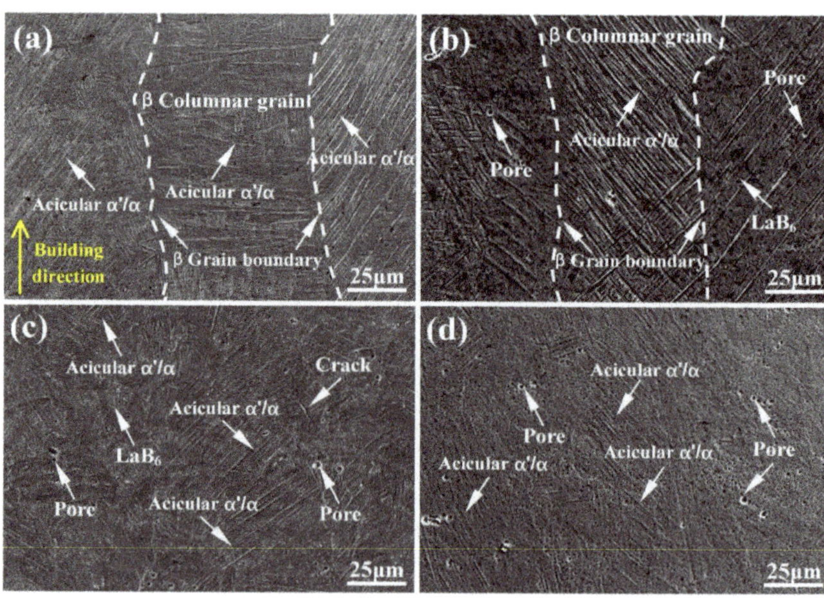

Figure 2. SEM images of microstructure of (**a**) TMC0, (**b**) TMC1, (**c**) TMC2, and (**d**) TMC3.

In order to further identify the phase composition of LPBF TC4 and TC4/LaB_6 composites, the specimens TMC0 and TMC3 were analyzed using bright-field transmission electron microscopy (TEM) and selected area electron diffraction (SAED). In Figure 3a, it is observed that the TMC0 is mainly composed of large grains with well-defined grain boundaries. The grains and boundaries in Figure 3a were analyzed using SAED, and the results show that the grains consist of α-Ti and the boundaries consist of β-Ti. This phenomenon is consistent with the findings of Ref. [26]. In order to determine whether the introduced reinforcing particle LaB_6 can react with the Ti matrix, the TMC3 sample was analyzed using TEM and SAED, as shown in Figure 3d–f. Many nano-sized particles were formed in the β-Ti matrix (Figure 3d). The TEM image of the nano-sized particle at a high magnification (Figure 3e) shows that these particles are well combined in the matrix with clean interfaces and without debonding or cracking. Thereafter, the composition of nano-sized particles was determined to be La_2O_3 and β-Ti using SAED, as illustrated in Figure 3f. This indicated that LaB_6 reacts with the matrix to produce the RE oxide La_2O_3. La_2O_3 serves as a heterogeneous nucleation site for the β phase, promoting the nucleation of the β phase and weakening the grain boundaries of the prior β columnar grains [27].

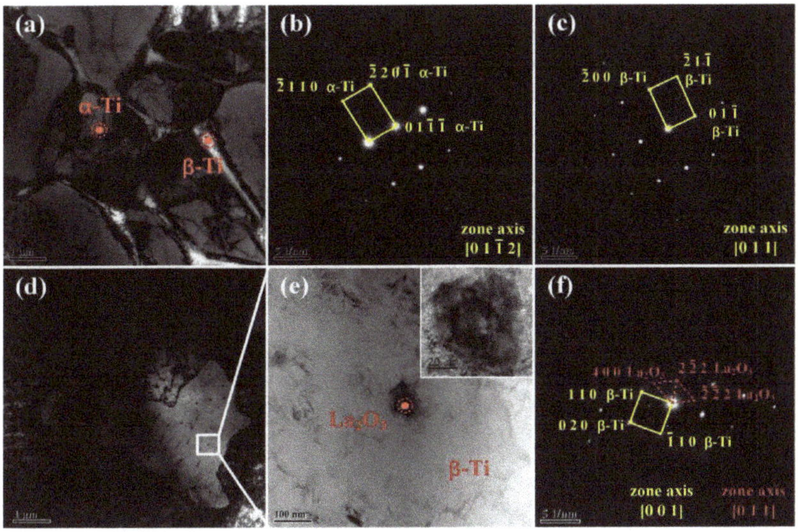

Figure 3. Bright-field TEM images of LPBF samples and corresponding selected area electron diffraction patterns: (**a–c**) TMC0, (**d–f**) TMC3.

High-resolution transmission electron microscopy (HRTEM) was employed for an in-depth investigation of nanoparticles La_2O_3. As shown in Figure 4a, dislocations are observed at the boundaries between nanoparticles and the matrix. As shown in Figure 4b, multiple phases with different plane spacings be easily found in the magnified image of the nanoparticles' boundaries. The lattice–fringe separations of 0.322 nm, 0.351 nm, and 0.366 nm were computed to identify the planes as (222) La_2O_3, (101) β-Ti, and (101) TiB, respectively. Regarding the parallel crystal planes of (222) La_2O_3 and (101) β-Ti, a mismatch occurs between the phase interfaces due to differences in interplanar distances. Consequently, dislocations are present in the phase interface, serving to ease the so-called mismatch [28,29].

The prior β columnar grains and grain boundary contours of the TMC0 are found to be more distinct on the EBSD inverse-pole figures (IPFs), as shown in Figure 5a. The fine acicular α'/α inside the prior β columnar grain are more noticeable. It was clear that most acicular α'/α could partially penetrate the β columnar grain boundary due to their length, demonstrating a well-organized internal structure and clean morphology. Because of the layer-by-layer manner of LPBF, the extremely high cooling rate resulted in distinct temperature difference between the adjacent prior β columnar grains. Then, it was observed that the α phase precipitated in the prior β grains as the temperature decreased, and the α grains were more likely to nucleate on the β grain boundaries, resulting in a lot of fine α equiaxed grain formations at the boundaries [30]. As shown in Figure 5b, the grain boundaries of the prior β columnar grain of the TMC1 are vaguely distinguishable, and the boundary areas of the distributed acicular α'/α start to show a certain degree of disorder. Moreover, many acicular α'/α also form in the intergranular regions of the prior β grains. When LaB_6 increases to 0.5 wt.% and 1.0 wt.%, the prior β grains in the TMC2 and TMC3 are found to be indistinguishable, the acicular α'/α are significantly reduced and disordered, the number of α equiaxed grains rapidly increases, and they become randomly distributed, as illustrated in Figure 5c,d. As shown in Figure 5e–h, the four samples exhibited the highest texture intensity in the <0001> direction, according to the pole figures (PFs). Moreover, the TMC0 and TMC1 that could identify the prior β columnar grains have a higher texture intensity on the <0001> crystal planes than the remaining two samples that could not be used for this identification. The above results show that the addition of LaB_6 weakens or even eliminates the grain boundaries of the prior β grains, refining the grains. This could be because the LaB_6 particles could provide a large number of low-energy heterogeneous nucleation sites for α and β grains during the $L \rightarrow \beta$ and $\beta \rightarrow \alpha + \beta$ phase transitions, resulting in a narrower β grain and eventually the formation of more equiaxed α grains [14,16].

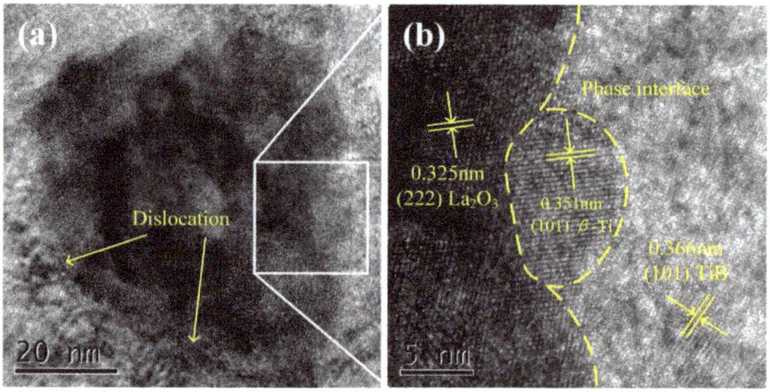

Figure 4. (**a**,**b**) High-resolution transmission electron microscopy (HRTEM) images of the nano-sized particle of TMC3.

Figure 5i exhibits the grain size obtained from the EBSD data, and the general trend of grain size comprises a concave bend when the mass fraction increases from 0 wt.% to 1.0 wt.%. The value of the TMC0 is 4.1 μm^2, and the values decrease to 3.1 μm^2 with the TMC1, and then to 2.0 μm^2 with the TMC2. This indicates that a trace amount of LaB_6 could cause a significant grain refinement on the LPBF TC4 alloy. However, the average grain size of the TMC3 was found to unexpectedly increase to 2.6 μm^2. This trend is consistent with the findings of Refs. [31,32]. These studies found that, although the addition of LaB_6 caused significant grain refinement in the TC4 alloy, the La_2O_3 particles tended to aggregate as the mass fraction exceeded a certain value, which had a negative influence on the grain refinement. This implies that there is an optimal mass fraction of LaB_6 since the reinforcing particles range from 0.2 wt.% to 1.0 wt.%. Based on the results of this study, the mass fraction of 0.5 wt.% LaB_6 is concluded as the optimal value for the LPBF of TC4/LaB_6 composites.

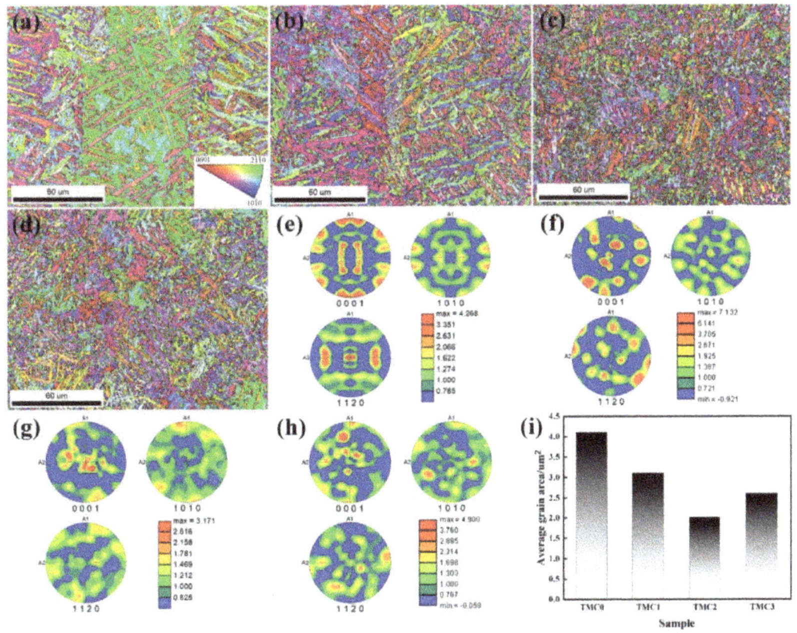

Figure 5. EBSD inverse pole figures (IPF), pole figures (PF), and the average grain size of LPBF samples: (**a**,**e**) TMC0, (**b**,**f**) TMC1, (**c**,**g**) TMC2, (**d**,**h**) TMC3, and (**i**) average grain size.

3.2. Compression Mechanical Properties

3.2.1. Quasi-Static Compressive Properties

The quasi-static stress–strain curves of LPBF TC4/LaB_6 composites with different mass fractions are shown in Figure 6. In Figure 6a, it is observed that the specimens remain consistent within the elastic deformation stage, while a significant difference in the plastic deformation stage was observed. The yield strength of the samples is significantly enhanced due to the addition of LaB_6. It was also observed that even a 0.2 wt.% addition could increase the yield strength from 857 to 954 MPa. However, when the mass fraction increases to 1.0 wt.%, the yield strength suddenly decreases. As mentioned before, LaB_6 particles cause a remarkable grain refinement on the prior β columnar grains and the acicular α'/α. According to the Hall–Petch relationship, the yield strength increases as the grain size decreases, and this is called the refinement strengthening effect [33]. Moreover, the finer grains imply that there are more grain boundaries in the samples, and more grain boundaries effectively prevent the movement of dislocations and thereby improve the yield strength. The average grain size decreases from 4.1 to 2.0 μm^2 in the TMC2 when compared to the TMC0, and the yield strength increases from 857 to 1312 MPa as a consequence. Thereafter, with the further addition of LaB_6, the agglomeration of La_2O_3 is found to make the average grain size of the TMC3 increase to 2.6 μm^2, resulting in a decrease in the yield strength. Figure 6b illustrates that the fracture strain decreases monotonously with the increasing LaB_6. This is due to an increase in the mass fraction of LaB_6 causing a greater distribution of the reinforced particles in the samples. Furthermore, the corresponding dispersion strengthening effect is more apparent, which makes the yield strength evenly increase and the grain size coarsen. Simultaneously, the compressive load tends to cause a concentration of stress on the interface between the particle and matrix. when the stress exceeds the yield limit, the generation, growth and connection of microcracks are bound to happen on the interface, which makes the sample more fragile and prone to brittle failure.

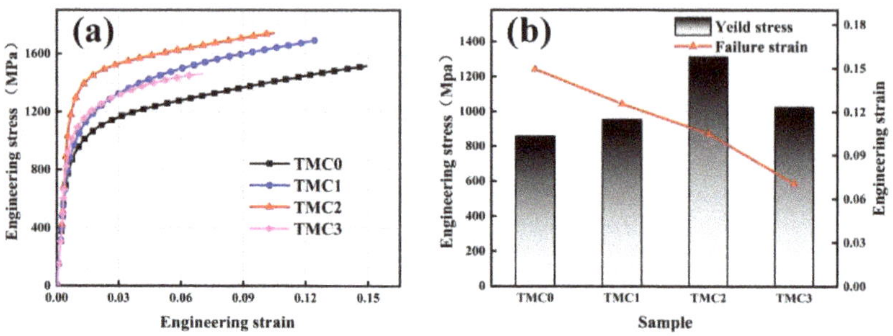

Figure 6. Quasi-static compressive properties of TC4/LaB6 with different mass fractions: (**a**) engineering stress–strain curves; (**b**) compressive 0.2% proof stress and failure strain.

3.2.2. Dynamic Compression Properties

The dynamic compressive stress–strain curves of the TC4 and TC4/LaB_6 samples at strain rates of 1800/s, 2500/s, and 3000/s are shown in Figure 7. All curves have the same shape, and the fluctuation in the curves is caused by the repeated oscillations of stress waves in sample [34]. Moreover, it was found that the work hardening rates (UCS-YS) at high strain rates are significantly higher than those under quasi-static conditions. The excellent work hardening ability is advantageous in structural applications in order to guarantee a large safety margin before the fracture [35,36]. Although the dynamic YS are lower than those under quasi-static conditions, the ultimate compressive strength (UCS) is apparently higher than that of the former (compared with Figure 6), showing a significant strain rate sensitivity. Similar to the quasi-static condition, at the three strain rates, the dynamic compressive strength is found to initially increase as the mass fraction of LaB_6 increases from 0 to 0.5 wt.%, and then drops as LaB_6 further increases to 1.0 wt.%. From Figure 7b,d,f, it is observed that the change trends in dynamic YS, UCS, and fracture strain coincided well with the trend of the grain size. Unlike the quasi-static condition, the fracture strain of the TMC3 under the dynamic condition does not continue to decrease, but instead increases.

It is also found that the fracture strain decreased at high strain rates, compared with Figure 6. This is due to the plastic localization of titanium alloy under impact loading, thus deteriorating plastic deformation [26,37]. Furthermore, when the mass fraction of LaB_6 is relatively higher, more micro pores are formed in the TC4/LaB_6. These pores and cracks play an important role in dynamic compressive properties. Under high-speed impact loading, the release of the stress is caused due to the nucleation of micro cracks and increases fracture toughness [38], whereas, since the impact resistance of alloys is less sensitive to pores and microcracks than tensile properties [39], some pores that are close to each other may even collapse, merge, and close, thereby improving the fracture strain of the alloys under impact loading conditions.

3.2.3. Influence of Grain Size on the UCS

The Hall–Petch laws define the relationship between grain size ($d^{-1/2}$) and yield strength. In Figures 6 and 7, it is observed that a positive relationship exists between the grain size and strength of LPBF of TC4 and TC4/LaB_6 samples. However, there is no definite yield point in the stress–strain curves under dynamic conditions. Hence, the UCS is selected as the basis for comparison instead, as shown in Table 2 and Figure 8. Under the influence of strain rate strengthening [36], the dynamic UCS is higher than the quasi-static state, but the change trend is similar on the whole. With the decrease in grain size, UCS increases monotonously. Then, the grain size of the TMC3 increases due to La_2O_3 agglomeration, and UCS decreases accordingly. It is also indicated that UCS is closely related to grain size. As shown in Figure 5i, the addition of LaB_6 effectively reduced the grain size. This is mainly attributed to the enrichment of La and its oxides at the grain boundaries, which hindered the growth of β-Ti, known as pinning effect. Moreover, they could serve as heterogeneous nucleation sites for the α phase and promote its nucleation and growth [31,32]. The effective slip distance of dislocation decreased, and the hindrance effect of grain boundary increased due to the grain refinement caused by the introduction of the reinforcing phase. The reinforced particles also hindered the dislocation, and thereby the energy required for the dislocation slip increased. Moreover, due to the limited slip systems of the α phase with an HCP structure, twinning is usually caused in places where stress is concentrated and when the slip is impeded under high-strain-rate deformation [40,41]. Yu et al. [42] suggested that the Hall–Petch slope k of deformation twinning is larger than that of the dislocation slip in titanium matrix composites. According to Hall's dislocation pile-up model [43], dislocations piled up in the vicinity of the grain boundary, and the stress concentration from the dislocation pile-up produced a dislocation source in an adjacent grain [44]. Grain refinement resulted in the increase in grain boundaries of the α phase, so the activation energy required for dislocation propagation in the adjacent grains enhanced. Thus, the UCS of TC4/LaB_6 effectively improved. This was consistent with the results of Wang et al. [45], who found that the strength of titanium matrix composites can be significantly improved after matrix grain refinement.

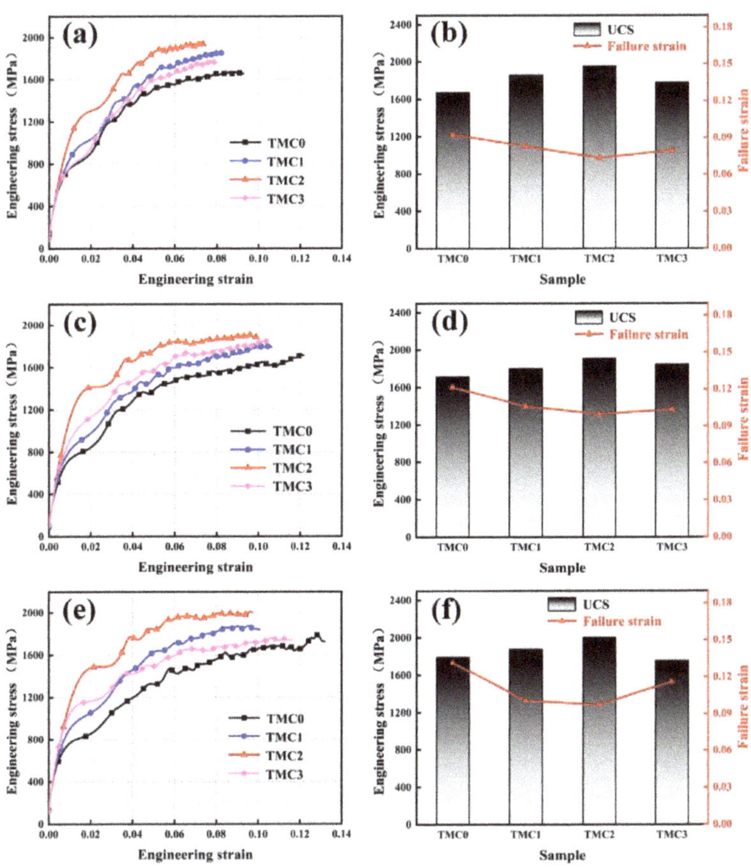

Figure 7. Dynamic compressive properties of LPBF of TC4/LaB6 composites with different mass fractions at different strain rates (1800/s, 2500/s and 3000/s): (**a**,**c**,**e**) engineering stress–strain curves; (**b**,**d**,**f**) ultimate compressive strength and failure strain.

Table 2. Ultimate compressive strength and grain size for strain rates of 10^{-3}/s and 2500/s.

Name	UCS (MPa) Quasi-Static (10^{-3}/s)	UCS (MPa) Dynamic (2500/s)	Grain Size $d^{-1/2}$ ($\mu m^{-1/2}$)
TMC0	1532.7 ± 32.7	1733.6 ± 16.4	4.1
TMC1	1630.6 ± 19.8	1800.1 ± 20.1	3.1
TMC2	1736.3 ± 13.6	1956.3 ± 49.5	2
TMC3	1525.4 ± 24.5	1801.7 ± 43.4	2.6

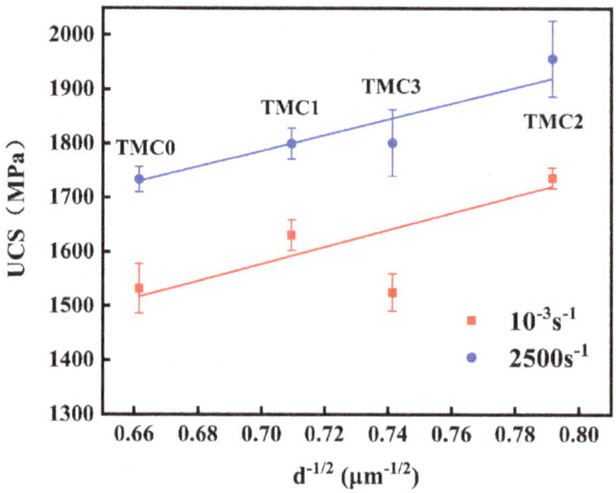

Figure 8. Ultimate compressive strength for strain rates of 10^{-3}/s and 2500/s.

3.3. Fracture Characteristics

The SEM images of the fracture surfaces of the TMC0 and TMC2 are exhibited in Figure 9. A staggered distribution of dimples and smooth shear zones are observed in the TMC0 at low magnification (Figure 9a), which is a typical mixed fracture mode of ductile and brittle fracture [46]. The same phenomenon was reported in the EBM of TC4 alloy [47,48] because the α' phase has a higher strength but lower ductility, thus showing brittleness [49]. At a higher magnification, it is found that the dimples had a elongated shapes and sizes of 2~20 μm (Figure 9b), which were caused by the shear stress in the shear direction [50]. As shown in Figure 9c, cleavage steps and planes are alternately distributed on the whole fracture surface of the TMC2, which is a typical cleavage fracture. At higher magnification (Figure 9d), small dimples with smaller sizes are observed near the step surface, most of which are found to be less than 1 μm and uniformly distributed due to the uniform distribution of LaB_6 particles in the matrix. The size of the dimple is a critical indicator for assessing material ductility, reflecting the material's energy absorption capability and toughness characteristics. Larger dimples indicate better energy absorption and enhanced ductility performance [26,36]. In this study, TC4/LaB_6 exhibited smaller dimples compared to LPBF of TC4, indicating a lower ductility, which is consistent with the failure strain results shown in Figure 6b.

The 3D morphologies of the samples under an impact load at a strain rate of 2500/s are depicted in Figure 10, where blue arrows indicate convexity and black arrows indicate concavity. After the impact compression, all the samples are shear fractured with different fracture characteristics. The TMC0 was found to be incompletely fractured (Figure 10a; external force is needed to separate the two parts). Additionally, the TMC1 and TMC2 were fully fractured along 45° (the direction of maximum shear stress), and the samples fractured into two fragments (Figure 10b,c). The two fragments of the TMC1 have rough convex and concave surfaces, while those of the TMC2 have relatively flat fracture surfaces.

Correspondingly, the TMC3 shear fractures into several fragments with different sizes, in which only the two largest pieces were selected after the fracture, as shown in Figure 10d. Some larger-size fracture plateaus and undulating surfaces on the fracture surface of the TMC3 were also found.

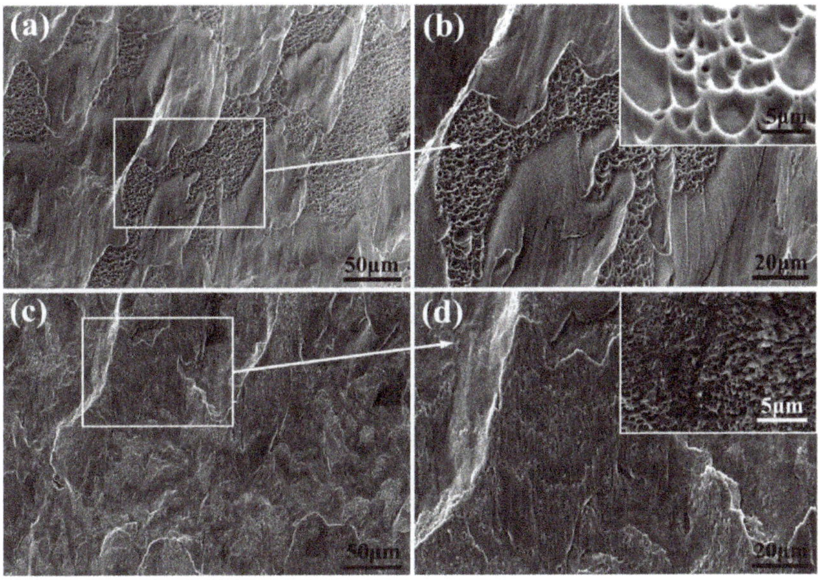

Figure 9. The fracture surfaces of compressed samples of TMC0 and TMC2: (**a**,**b**) the fracture surface of TMC0; (**c**,**d**) the fracture surface of TMC2. The inset showed detailed dimple morphologies at higher magnifications.

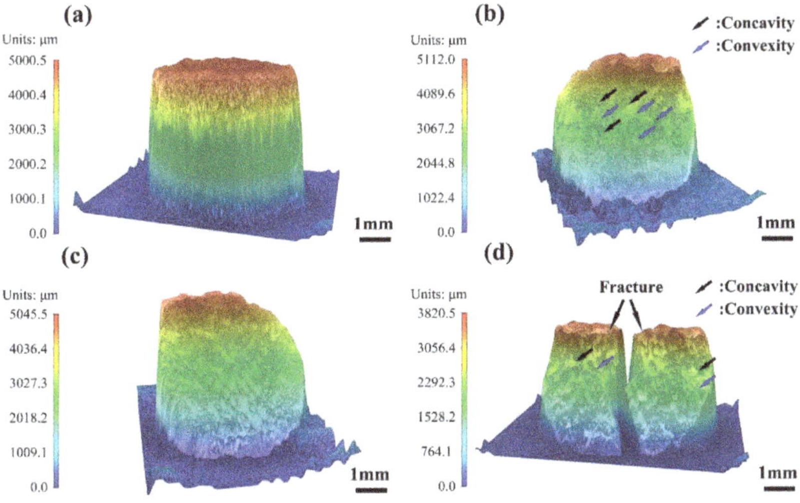

Figure 10. Thre-dimensional topographies of samples with different mass fractions after shock compression at a strain rate of 2500/s: (**a**) TMC0, (**b**) TMC1, (**c**) TMC2, and (**d**) TMC3.

In order to characterize the fracture models of LPBF of TC4 and TC4/LaB_6 under impact loading conditions, a scanning electron microscope (SEM) was used to further observe the fracture surfaces at a strain rate of 2500/s, as shown in Figure 11. It was observed that the morphology of the TMC0 is quite different from those of TC4/LaB_6 composites. As shown in Figure 11a, large dimple areas and smooth areas are alternatively distributed on the fracture surface. Dimples with sizes of 2~10 μm (less than those at 10^{-3}/s) are elongated in the shear direction, indicating that the TMC0 is mainly characterized by mixed ductile/brittle fracture. This kind of fracture model is also observed in some studies on the AM of a TC4 alloy [20].

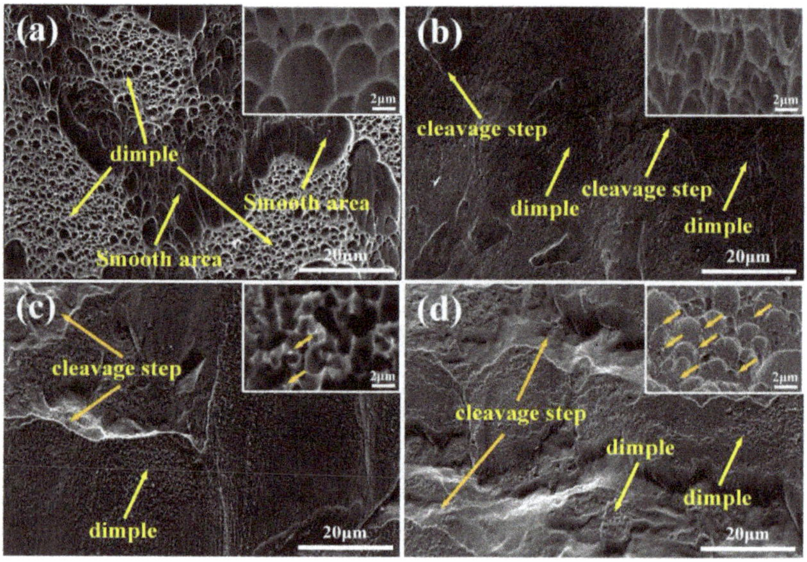

Figure 11. SEM images of samples with different mass fractions after an impact load at a strain rate of 2500/s: (**a**) TMC0, (**b**) TMC1, (**c**) TMC2, and (**d**) TMC3. The inset shows detailed dimple morphologies at higher magnifications.

Correspondingly, a large smooth dimple area is no longer found in the TC4/LaB_6 samples; instead, large-sized cleavage steps and cleavage planes are visible, and the surfaces are no longer flat and smooth. The higher-magnification SEM images show that dense featureless dimples are distributed on the cleavage planes. These dimples are so fine (less than 2 μm) that TC4/LaB_6 cannot be characterized under ductile fracture. The dimple size is associated with the grain size because the grain boundaries function as crack nucleation sites that influence the size of the dimples [27,51]. When compared with the TCM0, TMC1 starts to show cleavage steps, and the dimples on it become dense and irregular (Figure 11b), indicating that the reinforcing particles hinder crack slippage during the dynamic loading conditions [10,52]. It is also observed that the TMC2 surface starts to show cleavage steps with a larger size, whereas higher-magnification images show small and deep featureless dimples (Figure 11c) and particulate matter (yellow arrows in Figure 11c) present on cleavage planes. This indicates that the hindering effect of the reinforcing particles on the crack slip during dynamic loading is further enhanced. The TMC3 surface exhibited the most cleavage fracture characteristics. The SEM image at high magnification shows the presence of rough dimples on the cleavage planes (Figure 11d), and a large number of particles filled inside (yellow arrows in Figure 11d). Too many particles reduced the mechanical properties of the TMC3, which coincides with the reduction in its UCS.

4. Conclusions

This paper exposes the impact of LaB_6 content on the microstructure, quasi-static characteristics, and dynamic properties of titanium matrix composites, focusing primarily on the TC4 and TC4/LaB_6 composites manufactured using LPBF. Based on the results, we made following conclusions:

(1) A small amount of LaB_6 had a considerable effect on the microstructure of the titanium matrix composites, and 0.2 wt.% LaB_6 weakened and blurred the prior β-columnar grain boundaries and acicular α'/α. LaB_6 reached a critical point at 0.5 wt.%, where the boundaries of the prior β-columnar grains became indistinguishable, and the acicular α'/α decreased rapidly. When it increased to 1.0 wt.%, the influence on the change in microstructure was significantly diminished. The addition of LaB_6 had a significant grain refinement effect, but when its mass fraction exceeded 0.5 wt.%, the nano-reinforced particulate tended to agglomerate, which had a negative impact on grain refinement.

(2) Benefiting from the fine-grain strengthening and dispersion strengthening of LaB_6, the quasi-static compressive strength of the TC4/LaB_6 was found to be significantly higher than that of the pure TC4. However, it also led to a decrease in plasticity, especially when the LaB_6 fraction was relatively high (1.0 wt.%), and the specimens fractured, even shortly after the end of the elastic segment.

(3) The performance of TC4/LaB_6 was related to the mass fraction of LaB_6, but higher did not mean better. A critical mass fraction of 0.5 wt.% was found in this study. Excessive LaB_6 significantly increased the brittle fracture characteristics of the material, showing a more significant terrace-like fracture surface after extrusion and friction.

Author Contributions: Conceptualization, Y.L.; Investigation, X.L.; Data curation, X.L.; Writing—original draft, X.L.; Writing—review & editing, Y.L.; Funding acquisition, Y.L. All authors have read and agreed to the published version of the manuscript.

Funding: This study was supported by Natural Science Foundation of Ningbo City (Grant No. 2023J008) and the Self-innovation Research Funding Project of Hanjiang Laboratory (No. HJL202202A006).

Data Availability Statement: Data are contained within the article.

Conflicts of Interest: The authors declare no conflict of interest.

References

1. Guo, A.X.Y.; Cheng, L.; Zhan, S.; Zhang, S.; Xiong, W.; Wang, Z.; Wang, G.; Cao, S.C. Biomedical applications of the powder-based 3D printed titanium alloys: A review. *J. Mater. Sci. Technol.* **2022**, *125*, 252–264. [CrossRef]
2. Li, G.X.; Chandra, S.; Rashid, R.A.R.; Palanisamy, S.; Ding, S.L. Machinability of additively manufactured titanium alloys: A comprehensive review. *J. Manuf. Process.* **2022**, *75*, 72–99. [CrossRef]
3. Nguyen, H.D.; Pramanik, A.; Basak, A.K.; Dong, Y.; Prakash, C.; Debnath, S.; Shankar, S.; Jawahir, I.S.; Dixit, S.; Buddhi, D. A critical review on additive manufacturing of Ti-6Al-4V alloy: Microstructure and mechanical properties. *J. Mater. Res. Technol.* **2022**, *18*, 4641–4661. [CrossRef]
4. Liu, S.; Shin, Y.C. Additive manufacturing of Ti6Al4V alloy: A review. *Mater. Des.* **2019**, *164*, 107552. [CrossRef]
5. Kang, L.; Yang, C. A Review on High-Strength Titanium Alloys: Microstructure, Strengthening, and Properties. *Adv. Eng. Mater.* **2019**, *21*, 1801359. [CrossRef]
6. Sen, I.; Tamirisakandala, S.; Miracle, D.B.; Ramamurty, U. Microstructural effects on the mechanical behavior of B-modified Ti-6Al-4V alloys. *Acta Mater.* **2007**, *55*, 4983–4993. [CrossRef]
7. Williams, J.C.; Boyer, R.R. Opportunities and issues in the application of titanium alloys for aerospace components. *Metals* **2020**, *10*, 705. [CrossRef]
8. Cui, C.; Hu, B.; Zhao, L.; Liu, S. Titanium alloy production technology, market prospects and industry development. *Mater. Des.* **2011**, *32*, 1684–1691. [CrossRef]
9. Huang, X.; Gao, Y.; Wang, Z.; Yi, Y.; Wang, Y. Microstructure, mechanical properties and strengthening mechanisms of in-situ prepared (Ti5Si3+TiC0. 67)/TC4 composites. *J. Alloys Compd.* **2019**, *792*, 907–917. [CrossRef]
10. Hayat, M.D.; Singh, H.; He, Z.; Cao, P. Titanium metal matrix composites: An overview. *Compos. Part A* **2019**, *121*, 418–438. [CrossRef]
11. Ropars, L.; Dehmas, M.; Gourdet, S.; Delfosse, J.; Tricker, D.; Aeby-Gautier, E. Structure evolutions in a Ti-6Al-4V matrix composite reinforced with TiB, characterised using high energy X-ray diffraction. *J. Alloys Compd.* **2015**, *624*, 179–188. [CrossRef]

12. Lagos, M.A.; Agote, I.; Atxaga, G.; Adarraga, O.; Pambaguian, L. Fabrication and characterisation of Titanium Matrix Composites obtained using a combination of Self propagating High temperature Synthesis and Spark Plasma Sintering. *Mater. Sci. Eng. A* **2016**, *655*, 44–49. [CrossRef]
13. Sing, S.L.; Huang, S.; Goh, G.D.; Goh, G.L.; Tey, C.F.; Tan, J.H.K.; Yeong, W.Y. Emerging metallic systems for additive manufacturing: In-situ alloying and multi-metal processing in laser powder bed fusion. *Prog. Mater. Sci.* **2021**, *119*, 100795. [CrossRef]
14. Liu, M.; Liu, S.; Chen, W.; Chen, C.; Lv, Y.; Zhang, X.; Lei, P.; Lin, Y.; Zhou, K. Effect of trace lanthanum hexaboride on the phase, grain structure, and texture of electron beam melted Ti-6Al-4V. *Addit. Manuf.* **2019**, *30*, 100873. [CrossRef]
15. Barriobero-Vila, P.; Gussone, J.; Stark, A.; Schell, N.; Haubrich, J.; Requena, G. Peritectic titanium alloys for 3D printing. *Nat. Commun.* **2018**, *9*, 3426. [CrossRef] [PubMed]
16. Bermingham, M.J.; Kent, D.; Zhan, H.; StJohn, D.H.; Dargusch, M.S. Controlling the microstructure and properties of wire arc additive manufactured Ti-6Al-4V with trace boron additions. *Acta Mater.* **2015**, *91*, 289–303. [CrossRef]
17. Bermingham, M.J.; McDonald, S.D.; Dargusch, M.S. Effect of trace lanthanum hexaboride and boron additions on microstructure, tensile properties and anisotropy of Ti-6Al-4V produced by additive manufacturing. *Mater. Sci. Eng. A* **2018**, *719*, 1–11. [CrossRef]
18. Ye, T.; Xu, Y.; Ren, J. Effects of SiC particle size on mechanical properties of SiC particle reinforced aluminum metal matrix composite. *Mater. Sci. Eng. A* **2019**, *753*, 146–155. [CrossRef]
19. Mohammadhosseini, A.; Masood, S.H.; Fraser, D.; Jahedi, M. Dynamic compressive behaviour of Ti-6Al-4V alloy processed by electron beam melting under high strain rate loading. *Adv. Manuf.* **2015**, *3*, 232–243. [CrossRef]
20. Liu, Y.; Xu, H.; Zhu, L.; Wang, X.; Han, Q.; Li, S.; Wang, Y.; Setchi, R.; Wang, D. Investigation into the microstructure and dynamic compressive properties of selective laser melted Ti-6Al-4V alloy with different heating treatments. *Mater. Sci. Eng. A* **2021**, *805*, 140561. [CrossRef]
21. Alaghmandfard, R.; Chalasani, D.; Hadadzadeh, A.; Amirkhiz, B.S.; Odeshi, A.; Mohammadi, M. Dynamic compressive response of electron beam melted Ti–6Al–4V under elevated strain rates: Microstructure and constitutive models. *Addit. Manuf.* **2020**, *35*, 101347. [CrossRef]
22. Alaghmandfard, R.; Dharmendra, C.; Odeshi, A.G.; Mohammadi, M. Dynamic mechanical properties and failure characteristics of electron beam melted Ti-6Al-4V under high strain rate impact loadings. *Mater. Sci. Eng. A* **2020**, *793*, 139794. [CrossRef]
23. Cho, J.Y.; Xu, W.; Brandt, M.; Qian, M. Selective laser melting-fabricated Ti-6Al-4V alloy: Microstructural inhomogeneity, consequent variations in elastic modulus and implications. *Opt. Laser Technol.* **2019**, *111*, 664–670. [CrossRef]
24. Mills, K.C.; Keene, B.J. Factors affecting variable weld penetration. *Int. Mater. Rev.* **1990**, *35*, 185–216. [CrossRef]
25. Shirali, A.A.; Mills, K.C.J.W.J. The effect of welding parameters on penetration in GTA welds. *Weld. J.* **1993**, *72*, 7.
26. Liu, Y.; Pang, Z.; Li, M.; Wang, Y.; Wang, D.; Song, C.; Li, S. Investigation into the dynamic mechanical properties of selective laser melted Ti-6Al-4V alloy at high strain rate tensile loading. *Mater. Sci. Eng. A* **2019**, *745*, 440–449. [CrossRef]
27. Wang, X.; Zhang, L.; Ning, J.; Li, S.; Zhang, L.; Long, J. Hierarchical grain refinement during the laser additive manufacturing of Ti-6Al-4V alloys by the addition of micron-sized refractory particles. *Addit. Manuf.* **2021**, *45*, 102045. [CrossRef]
28. Li, W.; Yang, Y.; Liu, J.; Zhou, Y.; Li, M.; Wen, S.; Wei, Q.; Yan, C.; Shi, Y. Enhanced nanohardness and new insights into texture evolution and phase transformation of TiAl/TiB_2 in-situ metal matrix composites prepared via selective laser melting. *Acta Mater.* **2017**, *136*, 90–104. [CrossRef]
29. He, D.; Wang, H.; Huang, W.; Chen, X.; Lian, G.; Wang, Y. Microstructure and mechanical properties of LaB6/Ti-6Al-4V composites fabricated by selective laser melting. *Metals* **2023**, *13*, 264. [CrossRef]
30. Roy, S.; Suwas, S.; Tamirisakandala, S.; Miracle, D.B.; Srinivasan, R. Development of solidification microstructure in boron-modified alloy Ti-6Al-4V-0.1B. *Acta Mater.* **2011**, *59*, 5494–5510. [CrossRef]
31. Feng, Y.; Feng, K.; Yao, C.; Li, Z. Effect of LaB_6 addition on the microstructure and properties of (Ti_3Al+TiB)/Ti composites by laser cladding. *Mater. Des.* **2019**, *181*, 107959. [CrossRef]
32. Liang, J.; Yin, X.; Lin, Z.; Chen, S.; Liu, C.; Yan, S.; Dong, S. Effects of LaB6 on microstructure evolution and properties of in-situ synthetic TiC+TiBx reinforced titanium matrix composite coatings prepared by laser cladding. *Surf. Coat. Technol.* **2020**, *403*, 126409. [CrossRef]
33. Chong, Y.; Deng, G.; Gao, S.; Yi, J.; Shibata, A.; Tsuji, N. Yielding nature and Hall-Petch relationships in Ti-6Al-4V alloy with fully equiaxed and bimodal microstructures. *Scr. Mater.* **2019**, *172*, 77–82. [CrossRef]
34. Field, J.E.; Walley, S.M.; Proud, W.G.; Goldrein, H.T.; Siviour, C.R. Review of experimental techniques for high rate deformation and shock studies. *Int. J. Impact Eng.* **2004**, *30*, 725–775. [CrossRef]
35. Chen, G.; Ren, C.; Qin, X.; Li, J. Temperature dependent work hardening in Ti-6Al-4V alloy over large temperature and strain rate ranges: Experiments and constitutive modeling. *Mater. Des.* **2015**, *83*, 598–610. [CrossRef]
36. Liu, Y.; Meng, J.; Zhu, L.; Chen, H.; Li, Z.; Li, S.; Wang, D.; Wang, Y.; Kosiba, K. Dynamic compressive properties and underlying failure mechanisms of selective laser melted Ti-6Al-4V alloy under high temperature and strain rate conditions. *Addit. Manuf.* **2022**, *54*, 102772. [CrossRef]
37. Liu, R.; Hui, S.; Ye, W.; Li, C.; Fu, Y.; Yu, Y.; Song, X. Dynamic stress–strain properties of Ti-Al-V titanium alloys with various element contents. *Rare Met.* **2013**, *32*, 555–559. [CrossRef]
38. Zhang, X.M.; Zhang, Y.J.; Mao, X.N. TiC particles for Ti matrix composites by SHS method. *Tai Gongye Jinzhan* **2003**, *2*, 18–21. [CrossRef]

39. Fu, H.; Li, K.; Li, T.; Wu, T.; Zhang, G. Experiment of Cavity Collapse Process in Plastic-Bonded Explosives under Shock Loading. *Gaoya Wuli Xuebao* **2015**, *29*, 268–272. [CrossRef]
40. Kuang, L.; Chen, Z.; Jiang, Y.; Wang, Z.; Wang, R.; Liu, C. Adiabatic shear behaviors in rolled and annealed pure titanium subjected to dynamic impact loading. *Mater. Sci. Eng. A* **2017**, *685*, 95–106. [CrossRef]
41. Ren, Y.; Zhang, X.; Xia, T.; Sun, Q.; Liu, Q. Microstructural and textural evolution of high-purity titanium under dynamic loading. *Mater. Des.* **2017**, *126*, 123–134. [CrossRef]
42. Yu, Q.; Shan, Z.-W.; Li, J.; Huang, X.; Xiao, L.; Sun, J.; Ma, E. Strong crystal size effect on deformation twinning. *Nature* **2010**, *463*, 335–338. [CrossRef] [PubMed]
43. Hall, E.O. The Deformation and Ageing of Mild Steel: III Discussion of Results. *Proc. Phys. Soc. Sect. B* **1951**, *64*, 747. [CrossRef]
44. Wang, C.L.; Yu, D.P.; Niu, Z.Q.; Zhou, W.L.; Chen, G.Q.; Li, Z.Q.; Fu, X.S. The role of pyramidal <c plus a> dislocations in the grain refinement mechanism in Ti-6Al-4V alloy processed by severe plastic deformation. *Acta Mater.* **2020**, *200*, 101–115. [CrossRef]
45. Wang, L.; Lu, W.; Qin, J.; Zhang, F.; Zhang, D. Tensile properties of in situ synthesized ($TiB+La_2O_3$)/β-Ti composite. *Mater. Sci. Eng. C* **2009**, *29*, 1897–1900. [CrossRef]
46. Zhang, B.; Fu, H.; Sha, P.; Zhu, Z.; Dong, C.; Zhang, H.; Hu, Z. Anisotropic compressive deformation behaviors of tungsten fiber reinforced Zr-based metallic glass composites. *Mater. Sci. Eng. A* **2013**, *566*, 16–21. [CrossRef]
47. Zhao, X.; Li, S.; Zhang, M.; Liu, Y.; Sercombe, T.B.; Wang, S.; Hao, Y.; Yang, R.; Murr, L.E. Comparison of the microstructures and mechanical properties of Ti-6Al-4V fabricated by selective laser melting and electron beam melting. *Mater. Des.* **2016**, *95*, 21–31. [CrossRef]
48. Fadida, R.; Shirizly, A.; Rittel, D. Static and dynamic shear-compression response of additively manufactured Ti_6Al_4V specimens with embedded voids. *Mech. Mater.* **2020**, *147*, 103413. [CrossRef]
49. Zhang, T.; Huang, Z.; Yang, T.; Kong, H.; Luan, J.; Wang, A.; Wang, D.; Kuo, W.; Wang, Y.; Liu, C.T. In situ design of advanced titanium alloy with concentration modulations by additive manufacturing. *Science* **2021**, *374*, 478–482. [CrossRef] [PubMed]
50. Markovsky, P.E.; Janiszewski, J.; Dekhtyar, O.; Mecklenburg, M.; Prikhodko, S.V. Deformation Mechanism and Structural Changes in the Globular Ti-6Al-4V Alloy under Quasi-Static and Dynamic Compression: To the Question of the Controlling Phase in the Deformation of $\alpha+\beta$ Titanium Alloys. *Crystals* **2022**, *12*, 645. [CrossRef]
51. Tabachnikova, E.D.; Podolskiy, A.V.; Bengus, V.Z.; Smirnov, S.N.; Bidylo, M.I.; Csach, K.; Miskuf, J.; Saitova, L.R.; Semenova, I.P.; Valiev, R.Z. Mechanical characteristics, failure regularities, and dimple structures on failure surfaces of Ti-6Al-4V 'ELI' ultrafine-grained alloy at temperatures from 300 to 4.2K. *Mater. Sci. Eng. A* **2009**, *503*, 106–109. [CrossRef]
52. Song, W.; Ning, J.; Mao, X. Meso-Mechanical Behavior of TiC Particle-Reinforced Titanium Matrix Composites under Dynamic Loadings. *Xiyou Jinshu Cailiao Yu Gongcheng* **2011**, *40*, 1555–1560.

micromachines MDPI

Article

Microstructure, Tensile Properties, and Fracture Toughness of an In Situ Rolling Hybrid with Wire Arc Additive Manufacturing AerMet100 Steel

Lei Lei, Linda Ke *, Yibo Xiong, Siyu Liu, Lei Du, Mengfan Chen, Meili Xiao, Yanfei Fu, Fei Yao, Fan Yang, Kun Wang and Baohui Li

Shanghai Engineering Technology Research Center of Near-Net-Shape Forming for Metallic Materials, Shanghai Spaceflight Precision Machinery Institute, Shanghai 201600, China; leileix5@126.com (L.L.)
* Correspondence: kelinda_casc@163.com

Abstract: As a type of ultra-high strength steel, AerMet100 steel is used in the aerospace and military industries. Due to the fact that AerMet100 steel is difficult to machine, people have been exploring the process of additive manufacturing to fabricate AerMet100 steel. In this study, AerMet100 steel was produced using an in situ rolling hybrid with wire arc additive manufacturing. Microstructure, tensile properties, and fracture toughness of as-deposited and heat-treated AerMet100 steel were evaluated in different directions. The results reveal that the manufacturing process leads to grain fragmentation and obvious microstructural refinement of the AerMet100 steel, and weakens the anisotropy of the mechanical properties. After heat treatment, the microstructure of the AerMet100 steel is mainly composed of lath martensite and reversed austenite. Alloy carbides are precipitated within the martensitic matrix, and a high density of dislocations is the primary strengthening mechanism. The existence of film-like austenite among the martensite matrix enhances the toughness of AerMet100 steel, which coordinates stress distribution and restrains crack propagation, resulting in an excellent balance between strength and toughness. The AerMet100 steel with in situ rolling is isotropy and achieves the following values: an average ultimate strength of 1747.7 ± 16.3 MPa, yield strength of 1615 ± 40.6 MPa, elongation of 8.3 ± 0.2% in deposition direction, and corresponding values in the building direction are 1821.3 ± 22.1 MPa, 1624 ± 84.5 MPa, and 7.6 ± 1.7%, and the K_{IC} value up to 70.6 $MPa/m^{0.5}$.

Keywords: in situ rolling; wire arc additive manufacturing; AerMet100 steel; tensile property; fracture toughness

Citation: Lei, L.; Ke, L.; Xiong, Y.; Liu, S.; Du, L.; Chen, M.; Xiao, M.; Fu, Y.; Yao, F.; Yang, F.; et al. Microstructure, Tensile Properties, and Fracture Toughness of an In Situ Rolling Hybrid with Wire Arc Additive Manufacturing AerMet100 Steel. *Micromachines* **2024**, *15*, 494. https://doi.org/10.3390/mi15040494

Academic Editor: Saulius Juodkazis

Received: 29 December 2023
Revised: 19 January 2024
Accepted: 21 January 2024
Published: 3 April 2024

1. Introduction

AerMet100 is a high alloyed secondary hardening steel with ultra-high strength, featuring remarkable plasticity and fatigue strength [1,2]. The exceptional properties of this steel make it an excellent choice for the manufacture of high-performance components in the aerospace and military industries [3–5]. In recent years, extensive research has been carried out on AerMet100 steel. The microstructure of AerMet100 steel consists of lath martensite, small amounts of residual/reversed austenite, and carbides precipitated in the martensite matrix [6,7]. The crystal morphology of as-deposited laser additive manufacturing AerMet100 steel is a columnar crystal with a size of 100–600 μm, and the mechanical properties are anisotropic [8]. After heat treatment, the prior austenite grains transformed into martensite packets, and the columnar crystals transformed into equiaxed crystals to improve the material properties [9]. The average size of AerMet100 grains fabricated using wire arc additive manufacturing (WAAM) with in situ rolling can be refined from about 150–600 μm to 20 μm after heat treatment [10]. In AerMet100 ultra-high-strength steel, the lath martensite is rich in dislocations due to high lattice strain caused by martensitic transformation after high-temperature quenching, and during the tempering process, carbides are precipitated, resulting in pinning dislocations, which is one of the most important factors in the strength of AerMet100 steel [11]. As a secondary precipitation aging steel, the tempering temperature and time will affect the type and size of precipitates, thus affecting the performance of the material [12]. M_2C carbides can be formed by the enrichment of supersaturated carbon atoms and the substitution of Mo and Cr atoms for Fe atoms during the tempering process [13]. AerMet100 steel achieves the best combination of strength, plasticity, and toughness at a tempering temperature of 482 °C because the continuous film of reversed austenite starts to precipitate at the martensite boundaries, and the M_2C carbide can grow to a suitable size [6,14]. On the basis of the above-mentioned study, the temperature of 482 °C was also chosen for the annealing treatment in this experiment in order to obtain favorable mechanical properties.

Due to the fact that AerMet100 steel presents challenges in terms of processing and forming [15], people have been exploring the process of additive manufacturing to fabricate AerMet100 steel, such as laser cladding [16], laser additive manufacturing (LAM) [17], laser direct energy deposition (LDED) [18] and friction stir processing (FSP) [18]. As a kind of hybrid deposition and microrolling (HDMR) method, in situ rolling with wire arc additive manufacturing is a novel additive manufacturing technique that combines microrolling with WAAM to produce large components. The TC4 [19] alloy produced using in situ rolling has a higher dislocation density and larger α-bars than the alloy produced using WAAM, which improves the tensile strength and plasticity of TC4. Al-Mg4.5Mn alloys [20] produced using in situ interlayer rolling resulted in significant grain refinement; the mean and maximum grain dimensions were reduced from 59 to 23 μm and 172 to 98 μm, respectively, and the number and size of defects were significantly reduced than those produced using WAAM. Zhao [21] found that applying external force to the AerMet100 steel surface leads to a notable refinement of the martensitic slats and the generation of significant compressive residual stresses. Lu [10] investigated the grain refinement mechanism and the microstructure evolution of in situ rolling hybrid WAAM fabricated AerMet100 steel, which obtained a fully refined and segregation-free zone. Zhi [22] et al. introduced deep residual compressive stresses in the surface layer of AerMet100 steel, which refines the surface structure and reduces fatigue notch sensitivity, thus significantly improving fatigue life and fatigue crack initiation/expansion resistance.

Fracture toughness is a critical material property that demonstrates the structural integrity and reliability of the material, taking into account the overall effect of applied loads in the event of failure [23,24]. The standard fracture test procedure demands preparation of the specimen, pre-fatigue of the notch, measurement of the fatigue crack growth rate, and evaluation of the original data [25]. Some studies on additive manufacturing have shown that material fracture toughness is anisotropic in the horizontal and vertical directions [26,27]. Tempering treatment improved the fracture toughness but not the anisotropy [28]. Lu [10] investigated the fracture toughness of heat-treated AerMet100 steel fabricated using WAAM with in situ microrolling and demonstrated that the fracture toughness under plane strain conditions is connected to the homogenization of the composition, which enhanced the K_{IC}. Ran [8] investigated the fatigue crack growth (FCG) characteristics in the deposition and building directions of AerMet100 steel produced using laser additive manufacturing and revealed that the FCG of as-deposited AerMet100 steel is anisotropic with a lower crack resistance along the direction of deposition. The fatigue crack growth resistance of the heat-treated specimens has been improved. Meanwhile, the fatigue crack growth rate of steel can be effectively reduced by introducing thicker film-like reversed austenite. Jyoti Suryawanshi [29] et al. weakened the anisotropy of the fracture toughness of martensitic aging steels by 90° rotation between successive layers during selective laser melting (SLM) printing. Up to now, there has been a lack of comparative studies on the fracture toughness of AerMet100 steel in the deposition direction and building direction.

Therefore, the main objective of this study is to understand better the properties of AerMet100 steel processed using wire arc additive manufacturing hybrid with in situ rolling. In particular, the microstructure and tensile properties of the as-deposited and heat-treated AerMet100 steel were investigated, and the plane strain fracture toughness of the AerMet100 steel was investigated considering the different directions.

2. Materials and Methods

2.1. Material Manufacturing and Heat Treatment

The raw material used is 1.2 mm diameter AerMet100 steel wire. Before carrying out the experiments, ensure that the AerMet100 steel wire is intact and free from stains and cracks. The in situ rolling hybrid with wire arc additive manufacturing platform independently developed by Huazhong University of Science and Technology was used to complete the experiment. The welding machine is equipped with a Tetrix551 DC welding power. The self-developed control system ensures the multi-axis robot and micro-roller move synchronously with the welding gun. Figure 1 shows the schematic diagram and the definitions of deposition direction (DD), building direction (BD), and X, Y, and Z directions. The experiments were carried out in an enclosed chamber containing the welding torch, roller, and robotic arm, which were gas-protected with argon, in an oxygen content reduced to less than 0.004 vol.%. Based on previous experimental experience, a combination of process parameters was selected to provide good roll force, wire feed speed, travel speed, forming current, and inter-layer temperature. The process parameters are shown in Table 1. Thin-walled specimens with dimensions of 240 mm length × 15 mm width × 105 mm height were produced, the top surface of which is smooth and flat. A series of heat treatment processes and parameters are shown in Table 2. The material composition of the as-deposited AerMet100 steel is shown in Table 3, which conforms to the requirements of the standard.

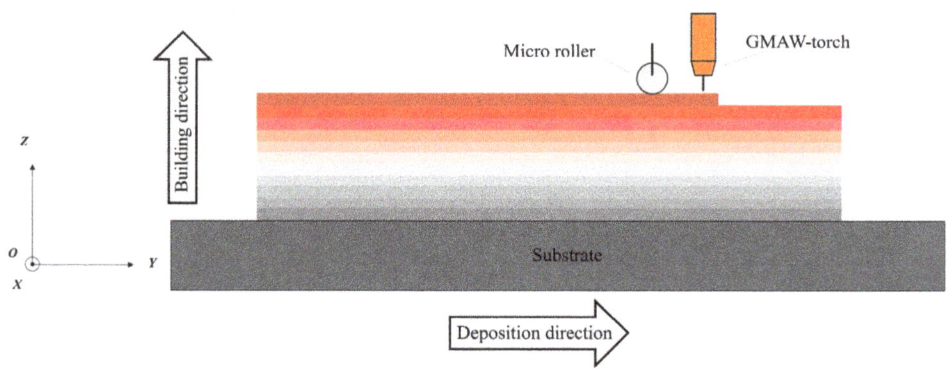

Figure 1. In situ rolling hybrid wire arc additive manufacturing AerMet100 steel process and defining direction.

Table 1. Additive manufacturing parameters for in situ rolling hybrid with WAAM.

Parameter	Value
Preheating temperature (°C)	100
Roll force (KN)	10
Wire feed speed (m/min)	1.3
Travel speed (mm/min)	120
Voltage (V)	23
Current (A)	200
Gas flow (L/min)	15

Table 2. Process parameters for heat treatment.

Process	Parameter
Normalizing	900 °C × 1 h, Air cooling
High-temperature tempering	680 °C × 16 h, Air cooling
Quenching	885 °C × 1 h, Oil cooling
Cryogenic treatment	−73 °C × 1 h, Warming in the air
Tempering	482 °C × 5 h, Air cooling

Table 3. Chemical composition of as-deposited AerMet100 steel samples.

Elements	C	S	P	Si	Mo	Cr	Ni
Content (wt.%)	0.21	0.019	0.010	0.04	1.16	2.94	10.57
Elements	Ti	Al	Mn	Co	O	N	Fe
Content (wt.%)	0.01	0.01	0.03	14.13	0.001	0.0004	Balance

2.2. Microstructural Characterization

Three planes perpendicular to the *Y*-axis, *X*-axis, and *Z*-axis of the as-deposited and heat-treated AerMet100 steel were chosen for microstructure examination. The cut surfaces were ground and, polished and then etched with a 4% nitric acid solution. The material microstructure was observed using a ZEISS SCOPE A1 optical microscope (OM), (Carl Zeiss, Oberkochen, Germany). The OXFORD Quanta 450 scanning electron microscope (SEM) (Oxford Instruments, Oxford, UK) was used to observe the fracture morphology. Electron backscattered diffraction (EBSD) was conducted using a GAIA3 GMU Model scanning electron microscope (TESCAN, Brno, Czech Republic) with a scanning step of 0.5 μm on heat-treated AerMet100 steel. Nanoscale microstructural features were investigated using a Talos F200X G2 transmission electron microscope (TEM) (Thermo Fisher Scientific, Waltham, MA, USA) on heat-treated AerMet100 steel.

2.3. Mechanical Property Tests and Fracture Analysis

Three tensile test rods were taken separately from as-deposited and heat-treated specimens in deposition direction and building direction. In accordance with GB/T 228.1-2010 [30], a bar specimen with a length of 75 mm and a diameter of 5 mm was used for the test. The test was performed using a CMT5305 universal electronic tensile testing machine at room temperature.

Plane strain fracture toughness testing was performed in accordance with the ASTM E399-20a standard [31]. After the heat treatment process, three compact tensile specimens were obtained for the plane strain fracture toughness test, separately in the deposition direction and the building direction. For each specimen, a fatigue pre-crack was made at approximately 3 mm from the pre-cracked notch prior to conducting the fracture toughness test. Crack opening displacement (COD) was then measured using a clip-on extensometer. Side grooves were also machined to ensure a flat crack front and to provide reliable crack length measurements. Plane strain fracture toughness testing was performed using the MTS Land Mark 50 KN hydraulic servo testing system. Standard procedures were followed to assess initial and final crack lengths. The average initial crack size was determined by measuring five equally spaced positions. Figure 2 displays the detailed dimensions of the tensile specimens and plane strain fracture toughness specimens. The fracture morphology was analyzed using an OXFORD Quanta 450 scanning electron microscopy.

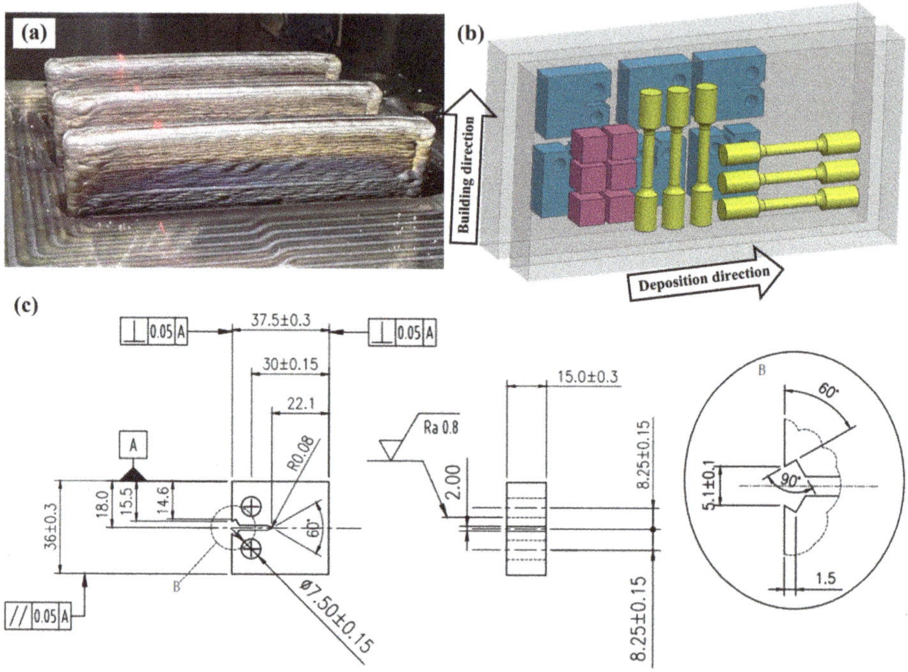

Figure 2. (**a**) Straight wall of AerMet100 steel produced using in situ rolling hybrid wire arc additive manufacturing; (**b**) sampling diagrams in different directions; (**c**) the size of the fracture toughness sample.

3. Results

3.1. Microstructure Analysis

Figure 3 shows the micromorphology in planes perpendicular to the *Y*-axis, *X*-axis, and *Z*-axis of the as-deposited AerMet100. The grain morphology appears to be highly irregular, and there is no obvious grain boundary in Figure 3a,c,e. Previous research has shown that AerMet100 steel produced using conventional additive manufacturing typically has a clear preferred orientation, and the material properties also show significant anisotropy [9]. There is no obvious difference in morphology and no obvious preferred orientation in the three directions. In this study, the in situ rolling hybrid with the WAAM process applies compressive stress to the molten pool that is not yet solidified, resulting in grain refinement and breakage, achieving the desired effect of grain refinement [21]. The microstructure of the AerMet100 steel is mainly composed of lath martensite with an irregular orientation. It can be seen in Figure 3b,d,f that a group of oriented lath martensite forms a martensite packet. There is the Kurdjumov–Sachs orientation relationship between prior austenite and martensite [14]. The most basic structural unit is lath martensite with a small angular deviation (2–5°). Lath martensite is mainly found in low-carbon steel, whereas the carbon content of AerMet100 steel is very low, only 0.21%. Figure 3b,d,f show preliminary judgments based on the micromorphology of as-deposited AerMet100 steel, with possible austenite grains circled, which probably have a grain size of around 40–60 μm.

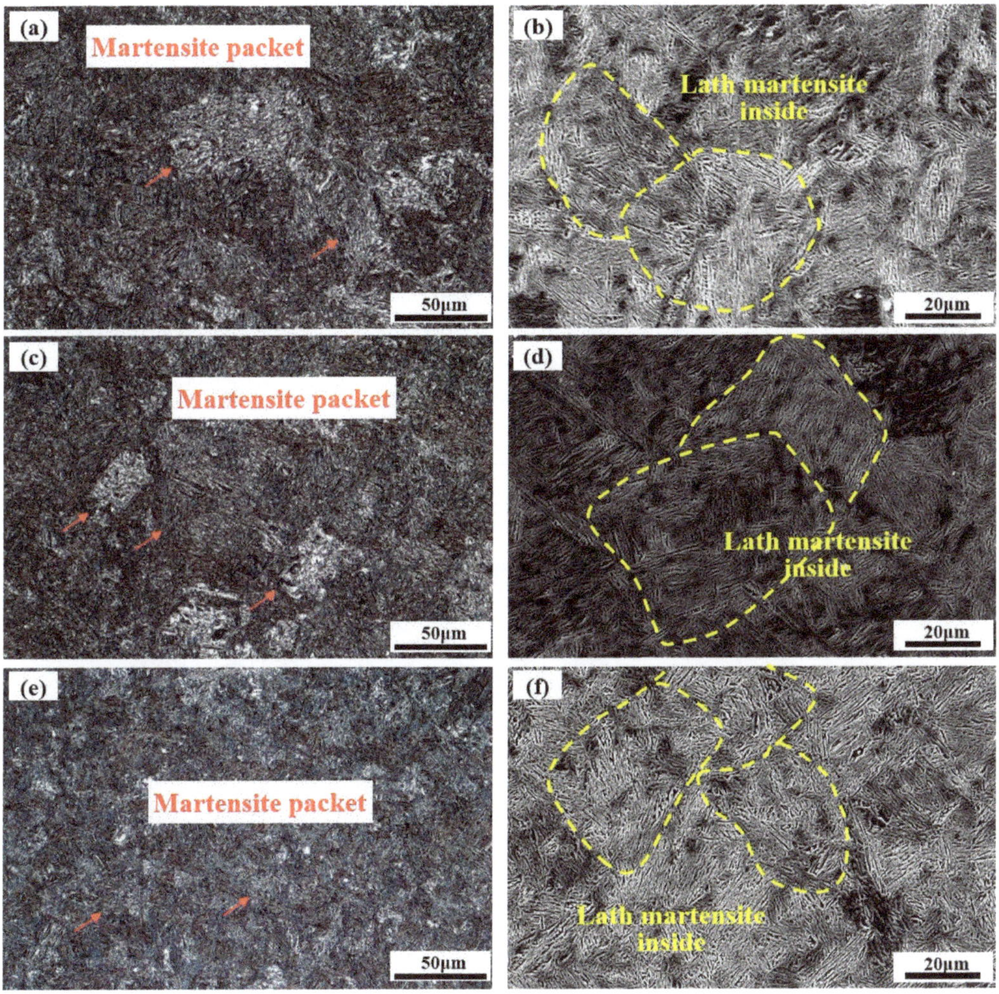

Figure 3. Microstructure of as-deposited AerMet100 steel fabricated using WAAM hybrid with in situ rolling: microstructure morphology in planes perpendicular to (**a**,**b**) *Y*-axis; (**c**,**d**) *X*-axis and (**e**,**f**) *Z*-axis.

The microstructure morphology of AerMet100 steel produced using an in situ rolling hybrid WAAM process after heat treatment is shown in Figure 4. As the original grains are crushed and refined using in situ rolling during the deposition process, the grain size of the AerMet100 steel remains small after heat treatment [32], and there is no difference in the microstructure in the three directions. In Figure 4a,c,e, it can be seen that the microstructure is still mainly composed of martensite packets, and according to previous research there may be a small amount of reversed austenite [33]. Figure 4b,d,f show that the fine particles are precipitated in the heat-treated AerMet100 steel during the heat treatment process [13], which can trap dislocations and increase the strength of the material [34].

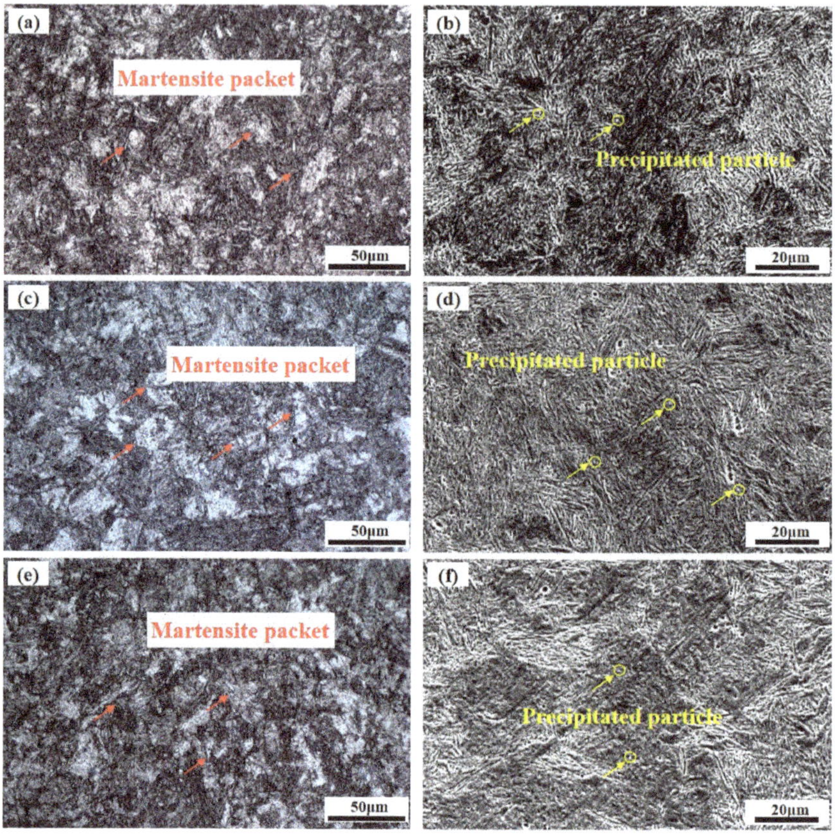

Figure 4. Microstructure morphology of AerMet100 after heat treatment produced using an in situ rolling hybrid WAAM process: microstructure morphology in planes perpendicular to the (**a**,**b**) *Y*-axis; (**c**,**d**) *X*-axis and (**e**,**f**) *Z*-axis.

3.2. Tensile Properties Analysis

Figure 5 shows the tensile results of the as-deposited AerMet100 specimens in deposition and building directions. In Figure 5a, the tensile mechanical behavior of the as-deposited specimens in two directions showed the same trend. The curves in both directions are consistent and show no obvious difference. Figure 5b shows the ultimate tensile strength, yield strength, and elongation after fracture in the deposition direction, which are 1280 MPa, 912 MPa, and 14.0%, respectively. Meanwhile, these data in building direction are 1277 MPa, 908 MPa, and 15.3%. The tensile properties of the AerMet100 in building direction and deposition direction show little difference, which differs from traditional additive manufacturing [9]. The introduction of a microrolling process into each melting deposition layer can significantly prevent the development of columnar crystals [35], compact the pore defects, weaken the texture [10,36], promote the formation of equiaxed crystals, and increase the tensile strength of materials. This process can eliminate the differences in tensile properties between the deposition direction and the building direction. In conventional additive manufacturing, the grain aligns with the greatest temperature gradient, resulting in the development of columnar crystals following the direction of heat flow [37]. This phenomenon results in the tensile properties of additively manufactured materials showing significant variations in the deposition and building directions.

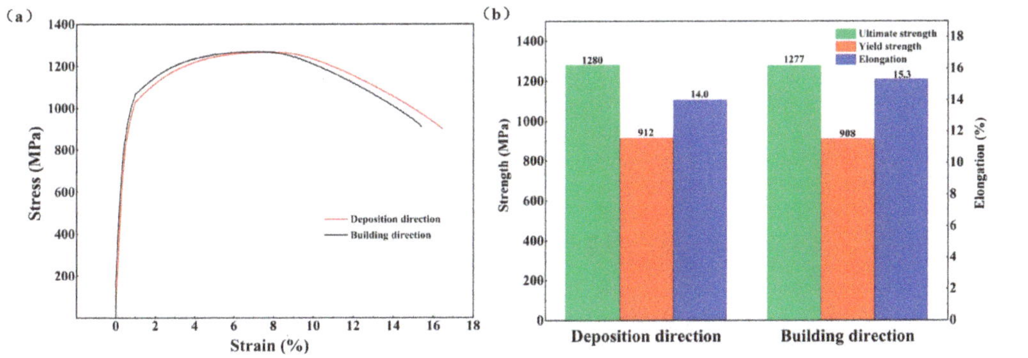

Figure 5. Tensile results of the as-deposited AerMet100 specimens in different directions: (**a**) stress–strain curves and (**b**) the corresponding ultimate tensile strength, yield strength, and elongation.

Figure 6 shows the tensile results of the heat-treated AerMet100 specimens in deposition and building directions. In Figure 6a, the tensile mechanical behavior of the heat-treated AerMet100 steel shows the same trend in both directions: the short uniform plastic deformation stage and the long local necking stage (after reaching the ultimate tensile strength) [34]. The stress–strain curves in the two directions are consistent and show no obvious difference. Figure 6b shows the ultimate tensile strength, yield strength, and elongation after fracture in deposition direction are 1747.7 ± 16.3 MPa, 1615 ± 40.6 MPa, and 8.3 ± 0.2%, respectively. Meanwhile, these data in building direction are 1821.3 ± 22.1 MPa, 1624 ± 84.5 MPa, 7.6 ± 1.7%. The tensile properties of the AerMet100 in building direction and deposition direction show little difference, and the ultimate tensile strength of the AerMet100 steel significantly improves after heat treatment compared with the as-deposited specimens. On the contrary, The AerMet100 steel, after heat treatment, displays a lower elongation compared with the as-deposited.

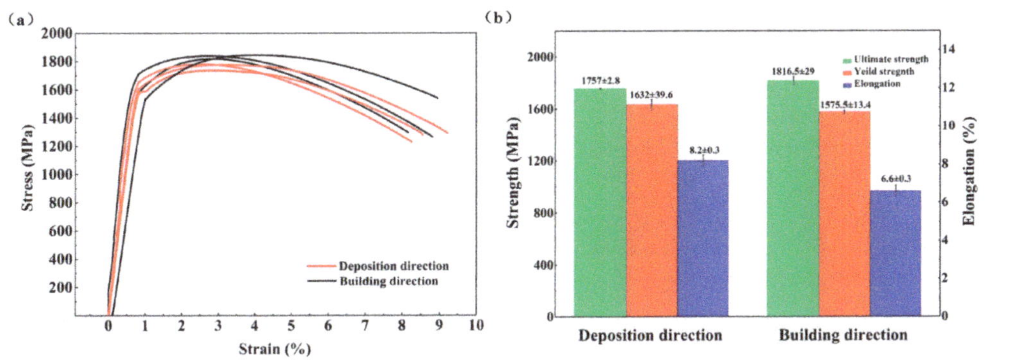

Figure 6. Tensile results of the heat-treated AerMet100 specimens in different directions: (**a**) stress–strain curves and (**b**) the corresponding ultimate tensile strength, yield strength, and elongation.

It is widely accepted that during the quenching process, lath martensite develops with high-density dislocations is formed. During the tempering process, many tiny and dispersed M_2C carbides would be precipitated, which is the main mechanism for the secondary hardening effect of AerMet100 steel [13]. The high density of dislocations and dispersed M_2C carbides in the matrix of the lath martensite leads to the high-level strength of this steel [38]. Furthermore, the crystal structure of the lath martensite enables several slip systems to be activated during deformation, which accounts for its plasticity behavior.

The fracture morphology of the as-deposited and heat-treated AerMet100 tensile specimen in the deposition direction and building direction is shown in Figure 7. From the fracture morphology, it can be seen that both the as-deposited and heat-treated AerMet100 steel have a large number of dimples, indicating that the material is a ductile fracture. In addition, the fracture morphology of the as-deposited specimen is more uneven with tearing edges in Figure 7a,b, and the sizes of the dimples in the as-deposited specimen are smaller, indicating that the plasticity of the as-deposited specimen is better, which is consistent with the higher elongation compared with the heat-treated specimen. In Figure 7c,d, the second phase particles or carbides appear at the bottom of the dimples. In precipitation-hardened alloys, the presence of secondary nanoparticles could promote microvoid nucleation during tensile fracture. During the tensile loading process, the voids would aggregate and connect, ultimately leading to material fracture. As the aging process progresses, more secondary nanoparticles may be precipitated, resulting in the formation of more crack sources. Therefore, the elongation of AerMet100 steel in the heat-treated state decreases compared with that in the deposited state.

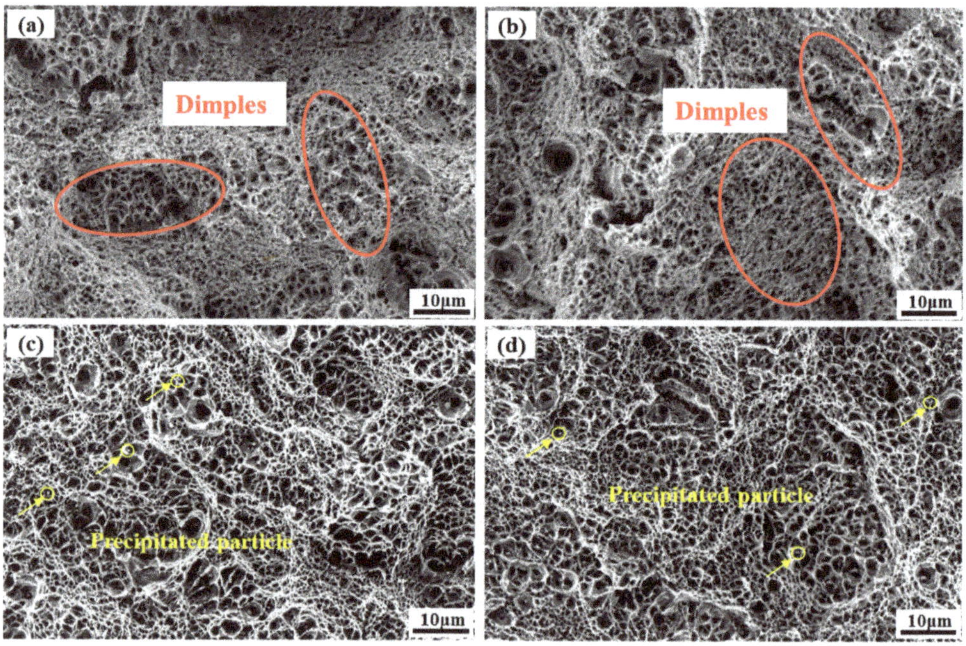

Figure 7. Tensile fracture morphology of AerMet100 steel in different directions: (**a**) as-deposited specimen in DD; (**b**) as-deposited specimen in BD; (**c**) heat-treated specimen in DD; (**d**) heat-treated specimen in BD.

3.3. Plain Strain Fracture Toughness Test and Fracture

Results from the plane strain fracture toughness test of heat-treated AerMet100 are presented in Table 4, and all the tested specimens attained valid K_{IC} test results. The mean K_{IC} value for the deposition direction was 68.5 MPa/m$^{0.5}$, while the value for the building direction was 59.25 MPa/m$^{0.5}$. The K_{IC} values differ slightly in both directions, but not significantly.

Table 4. Plain–strain fracture toughness properties of AerMet100 steel after heat treatment fabricated using an in situ rolling hybrid WAAM process.

Sample Direction	K_{IC} (MPa/m$^{0.5}$)
Deposition direction	70.6 66.4
Building direction	58.5 60

Figure 8 shows the macroscopic and microscopic fracture morphology of the fracture toughness specimens in the deposition direction and building direction. As shown in Figure 8a,b, the macroscopic fracture surface morphology of the K_{IC} specimen consists of pre-fatigue fracture, stable crack growth regions, and sheer lip. The obviously shear lip can be observed, which illustrates the high toughness [39]. Scanning electron microscopy in Figure 8c,d can clearly distinguish the toughness dimples. The surfaces of the fracture are not smooth, with many tearing edges. Energy absorption during fracture is one of the measures used to evaluate fracture toughness. The microscopic fracture surface morphology of materials would, therefore, be used to judge fracture toughness [40].

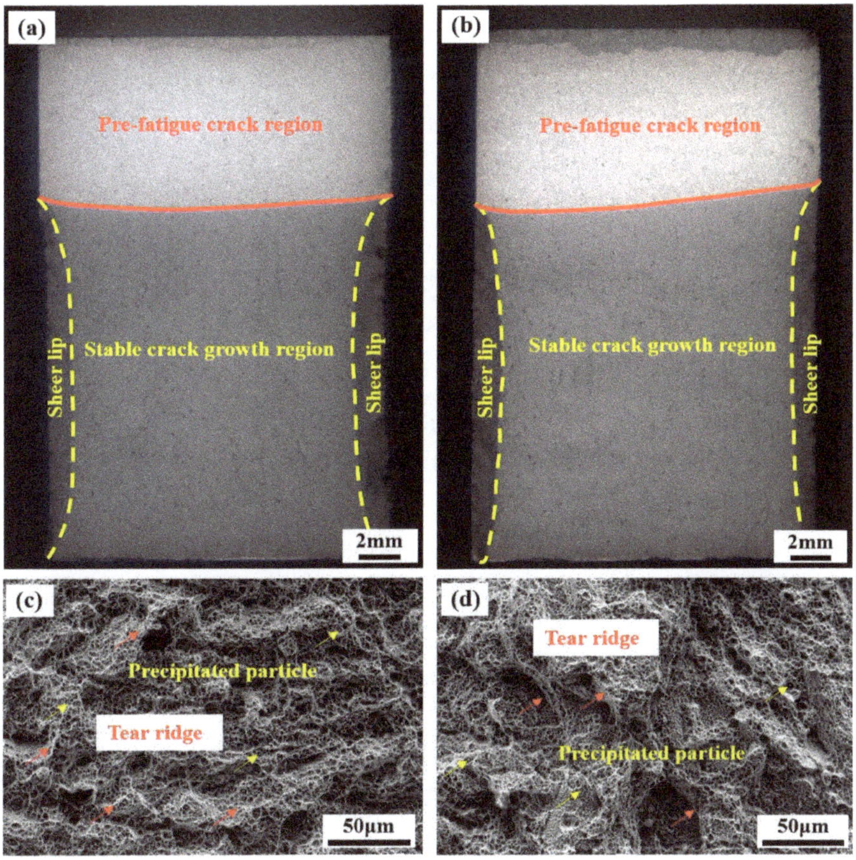

Figure 8. Fracture surface morphology of the plain–strain fracture toughness specimen: (**a**) macrostructure in DD; (**b**) macrostructure in BD; (**c**) microstructure in DD; (**d**) microstructure in BD.

4. Discussion

4.1. Effect of Hybrid Manufacturing Process on Strengthening of Mechanical Properties

Figure 9 shows the summary chart of AerMet100 steel mechanical properties. The mechanical performance data associated with the additive manufacturing of AerMet100 steel are shown in Figure 9. Among them, the light blue background represents the performance of the deposited state and the heat-treated state. The light red background represents the mechanical properties in different directions under the same process conditions. It can be seen that the ultimate tensile strength of heat-treated AerMet100 steel is significantly improved compared with the as-deposited material. Our work has a high elongation in as-deposited, which is significantly higher than other manufacturing processes, which may be due to the grain refinement caused by the introduction of microrolling. Compared with the mechanical properties of the LAM process, which exhibit anisotropy in both the deposited and heat-treated states, it can be seen that our material is isotropic in the deposited and heat-treated states. Compared with Lu's work [10], it can be seen that our work is isotropy in both deposited and heat-treated states, indicating that our process does not eliminate anisotropy due to heat treatment but rather introduces in situ microrolling manufacturing in the deposited state and maintaining isotropy after heat treatment.

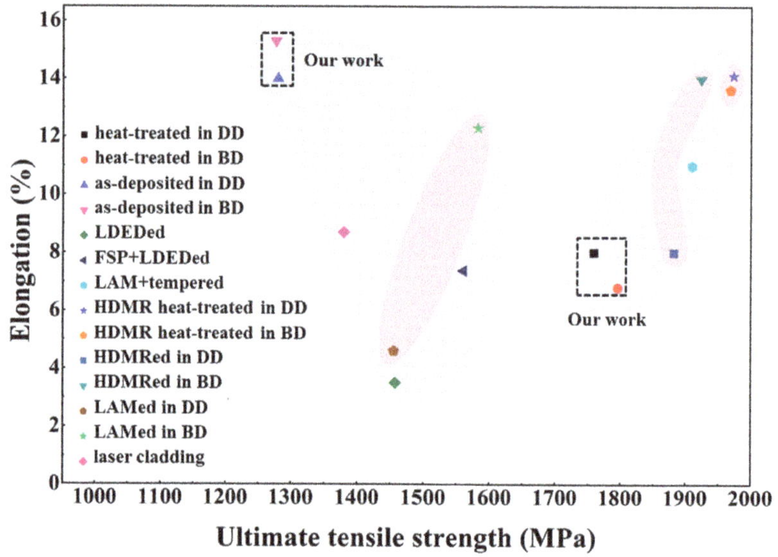

Figure 9. A summary chart of AerMet100 mechanical properties.

In the tempering process, the microstructure evolution behavior of AerMet100 steel is dependent on austenite decomposition, martensite formation, M_2C carbide formation and coarsening, and film-like reversed austenite formation and thickening [12]. The strengthening mechanism of AerMet100 steel can be divided into precipitation strengthening, grain boundary strengthening, and dislocation strengthening.

As can be seen in Figure 10a, a large number of lath martensites have different orientations within prior austenite grains. Due to grain fragmentation and random orientation, visualization of the austenite grains is challenging in the WAAM hybrid with an in situ rolling process [41]. AZtecCrystal software 2.1 was used to reconstruct the austenite grain based on the K-S orientation relationship for the martensite microstructure in the heat-treated AerMet100 steel. As can be seen in Figure 10b, the prior austenite grains are approximately 40–60 μm in size, and the grain shape is irregular. By means of the inverse pole figure map in Figure 10c, we can learn that although some grains have (101) directions parallel to the Y1 direction of the sample coordinate system, the selective orientation is not obvious due to Mmax = 1.82. There is no apparent texture visible in the sample shown in Figure 10. The process of utilizing in situ rolling hybrid with wire arc additive manufacturing proves to be effective in breaking up prior austenite grains and eliminating texture, leading to a random orientation distribution and no obvious anisotropy of the materials. According to the Hall–Petch equation [42,43]:

$$\sigma_H = \sigma_0 + k/D^{1/2} \quad (1)$$

where σ_H is the material strength, σ_0 is the frictional stress, k is a constant, and D is the grain diameter. Based on this equation, σ_H increases as the grain size decreases. Grain refinement can enhance mechanical properties attributable to the inhibition of dislocation movement by grain boundary interactions. Due to the different orientations of the grains on either side of the boundary, a stress field is generated when dislocations approach the boundary, hindering the movement of other dislocations. The smaller the grain size, the higher the stress required for dislocation movement in neighboring grains with different orientations. As the grain size decreases, the percentage of grain boundaries rises, potentially creating additional impediments for dislocation across these boundaries and enhancing the strength of the material [20].

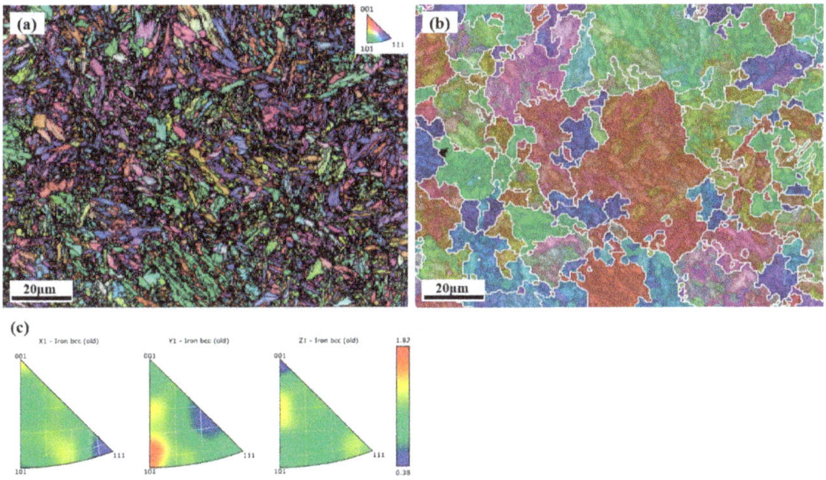

Figure 10. (**a**) Martensite lath in heat-treated AerMet100 steel; (**b**) austenite reconstruction of the microstructure in (**a**); (**c**) inverse pole figure of (**a**).

As shown in Figure 11a, the lath martensite structure of AerMet100 steel displays high-density dislocation. The strengthening process is also influenced significantly by carbide precipitation and dislocation [44]. The high-density dislocation entanglement in the AerMet100 steel hinders the movement of slip systems, making plastic deformation difficult and thus improving the strength [45,46]. Figure 11b shows that the precipitated particles are spheroidal distributed in the lath martensite. There are two kinds of precipitated particles: big spheroidal particles with a size of 74.58 nm and small-sized precipitated phases with a size of 11.25 nm. Nanoscale fine and dispersed carbide precipitates in lath martensite strengthen dislocation pinning and hinder the movement of the slip system, effectively forming dislocation delivery or twinning, thereby enhancing strength through precipitation strengthening [47–50]. The mechanism of precipitation enhancement depends on the size, density, and type of the nanoscale precipitate, in addition to the interaction between the dislocation and the precipitate [51]. For precipitation-hardened steel, there is an Orowan bypass hardening mechanism [52–54]. During the aging process, the martensite matrix is strengthened by the formation of uniform and dense alloy carbide particles, which act as precipitates. The Orowan strengthening mechanism indicates that dislocation lines cannot directly cut second-phase particles during plastic deformation, but external forces can bend around them. The bending of dislocation lines amplifies the lattice distortion energy in the dislocation region, leading to an increase in dislocation line resistance and slip resistance.

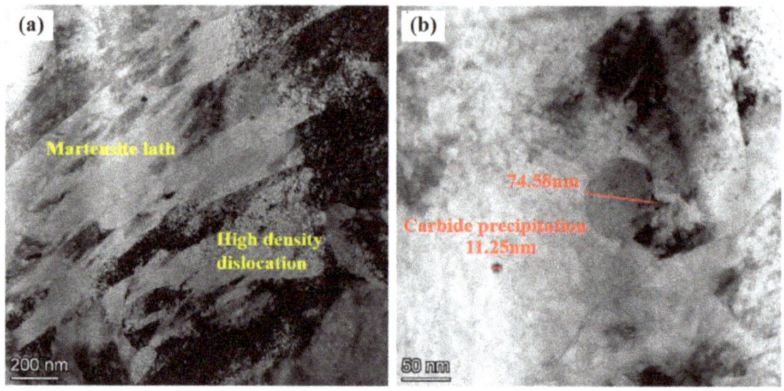

Figure 11. TEM photographs of heat-treated AerMet100 steel: (**a**) the lath martensite with high-density dislocation in it, (**b**) the carbide precipitation.

4.2. Fracture Behavior and Crack Propagation Mechanisms

As a result, similar fracture toughness values were observed in both the building direction and deposition direction. In addition, compressive residual stresses are beneficial in inhibiting crack initiation and early expansion, thus improving crack resistance [22].

Ductile fracture occured due to the growth and aggregation of micropore, resulting in slow and steady crack extension. This macroscopic fracture mode had a continuous process of ductile tearing, which absorbed more energy. Theoretically, the better the fracture toughness, the more tortuous the macroscopic fracture path of a specimen [39]. To better understand the crack extension mechanism, Figure 8 examines the macroscopic crack extension paths of fracture toughness specimens in both the deposition and building directions.

The crack extension can be divided into two phases: the pre-fatigue crack and the crack instability extension. In the fatigue crack region, as shown in Figure 12a,b, the crack extension path is relatively flat. However, in the unstable stage, the angle of the crack extension path increases significantly, which means that the sample with high fracture toughness will experience greater plastic deformation during the fracture process.

The morphology of the fracture surface is composed of small and large dimples, as shown in Figure 7. The dominant crack would join the microvoids near the crack tip in a stress-dominated mode, thereby propagating the crack rapidly. Meanwhile, the microvoids in front of the crack tip would grow and connect, causing the dominant to propagate in a strain-dominated manner, leading to ductile fracture [39]. Figure 13a demonstrates the phase composition of AerMet100 steel after heat treatment by EBSD analysis. A minimal proportion of reversed austenite is noticeable in the specimen, which correlates with the TEM discoveries in Figure 13b. Figure 13b presents a TEM bright field micrograph of heat-treated AerMet100 steel, showing film-like reversed austenite in the martensite interlayers of the steels. During the tempering at 482 °C, the continuous film of reversed austenite started to precipitate at lath martensite boundaries. It has been reported that fine film-like reversed austenite prevents crack initiation and propagation by deformation and transformation [55], slowing the rate of crack extension in steel, releasing stress conditions, and blunting the crack tip, resulting in higher resistance to crack extension [56]. On the one hand, the existence of an austenite film between the lath martensite can decrease the possibility of local stress concentration and the potential for micro-crack formation during the deformation of AerMet100 steel, thus significantly improving the plasticity [2]. On the other hand, due to the excellent flexibility and stability of the reversed austenite, fracture toughness is improved by promoting the "zigzag" path of crack deflection [15].

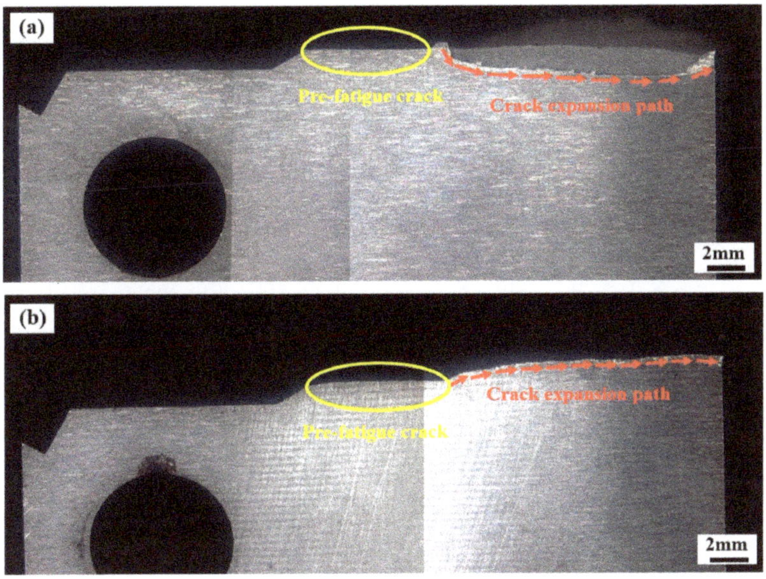

Figure 12. Crack expansion path: (**a**) in the deposition direction; (**b**) in the building direction.

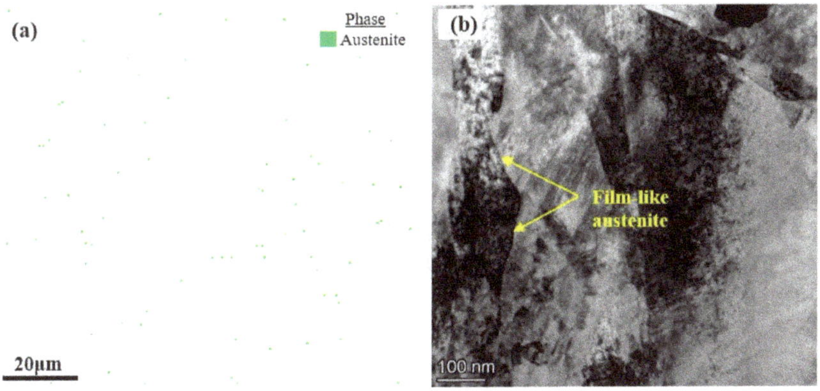

Figure 13. Film-like austenite distribution at the boundary of lath martensite in heat-treated AerMet100 steel: (**a**) EBSD phase composition characterization; (**b**) TEM morphology.

5. Conclusions

In this study, AerMet100 steel was produced using an in situ rolling hybrid with a wire arc additive manufacturing process. The microstructure, tensile properties, and fracture toughness of the as-deposited and heat-treated AerMet100 steel were evaluated in different directions. The results show that our method can not only produce AerMet100 steel with isotropic mechanical properties but also achieve near-net forming and reduced machining, which can be used for the production of parts with complex shapes and relatively high mechanical property requirements, such as landing gears.

(1) The primary microstructure of the AerMet100 steel produced using in situ rolling hybrid with wire arc additive manufacturing is a martensite packet with lath marten-

site inside, and after heat treatment, the martensite, carbides precipitation, and the film-like austenite. The microstructure and mechanical properties of the AerMet100 steel show isotropy in as-deposited state and remain isotropy after heat treatment.

(2) The AerMet100 steel exhibited ductile fracture, with micro-hole nucleation and coalescence being the predominant mechanisms. In addition, alloy carbide was precipitated in the matrix, and the strength of the material was enhanced by the precipitation-strengthening mechanism. This material exhibits isotropic tensile properties, where the ultimate tensile strength, yield strength, and elongation after fracture in the deposition direction are 1747.7 ± 16.3 MPa, 1615 ± 40.6 MPa, and $8.3 \pm 0.2\%$, respectively, while the corresponding values in the building direction are 1821.3 ± 22.1 MPa, 1624 ± 84.5 MPa, and $7.6 \pm 1.7\%$.

(3) The austenite film, the film-like structure formed by tempering at 482 °C, is located among the martensite matrix, which can turn the crack growth path and absorb the energy consumed in the crack growth process, thus playing a role in increasing the toughness and improving the fracture toughness to 70.6 $MPa/m^{0.5}$.

Author Contributions: Conceptualization, L.L. and S.L.; writing—original draft preparation, L.L.; writing—review and editing, L.K.; formal analysis, Y.X.; visualization, L.D. and M.C.; project administration, M.X., F.Y. (Fei Yao), K.W. and B.L.; data curation, Y.F. and F.Y. (Fan Yang); funding acquisition, L.K. All authors have read and agreed to the published version of the manuscript.

Funding: This research received no external funding.

Data Availability Statement: Data are contained within the article.

Conflicts of Interest: The authors declare no conflict of interest.

References

1. Wang, W.Z.; Feng, S.S.; Li, Z.M. Microstructure and properties of micro-arc oxidation ceramic films on AerMet100 steel. *J. Mater. Res. Technol.* **2020**, *9*, 6014–6027. [CrossRef]
2. Zhang, B.; He, B.; Wang, H.M. Microstructures and tensile properties of laser cladded AerMet100 steel coating on 300 M steel. *Surf. Coat. Technol.* **2022**, *440*, 128498. [CrossRef]
3. Wu, H.J.; Wang, K.H.; Yang, H. Effects of gradient nanostructures on the tribological properties and projectile abrasion during high-speed penetration in AerMet100 steel. *J. Mater. Res. Technol.* **2023**, *25*, 5871–5887. [CrossRef]
4. Chen, B.F.; Yan, F.Y.; Yan, M.F. Nitriding behavior and mechanical properties of AerMet100 steel and first-principles calculations of phase interfaces. *J. Mater. Res. Technol.* **2022**, *19*, 46–60. [CrossRef]
5. Zhong, J.Y.; Chen, Z.; Yang, S.L. Effect of Solution and Aging Temperatures on Microstructure and Mechanical Properties of 10Cr13Co13Mo5Ni3W1VE(S280) Steel. *Micromachines* **2021**, *12*, 566. [CrossRef] [PubMed]
6. Liu, J.; Li, J.; Cheng, X. Effect of dilution and macrosegregation on corrosion resistance of laser clad AerMet100 steel coating on 300M steel substrate. *Surf. Coat. Technol.* **2017**, *325*, 352–359. [CrossRef]
7. Ji, G.L.; Li, F.G.; Li, Q.H. Research on the dynamic recrystallization kinetics of AerMet100 steel. *Mater. Sci. Eng. A* **2010**, *527*, 2350–2355. [CrossRef]
8. Ran, X.Z.; Liu, D.; Li, J. Effects of microstructures on the fatigue crack growth behavior of laser additive manufactured ultrahigh-strength AerMet100 steel. *Mater. Sci. Eng. A* **2018**, *721*, 251–262. [CrossRef]
9. Ran, X.Z.; Liu, D.; Li, A. Microstructure characterization and mechanical behavior of laser additive manufactured ultrahigh-strength AerMet100 steel. *Mater. Sci. Eng. A* **2016**, *663*, 69–77. [CrossRef]
10. Lu, Y.F.; Wang, G.L.; Zhang, M.B. Microstructures, heat treatments and mechanical properties of AerMet100 steel fabricated by hybrid directed energy deposition. *Addit. Manuf.* **2022**, *56*, 1028855. [CrossRef]
11. Wang, Y.; Deng, T.Q.; Zheng, J. Unusual precipitation and its effect on mechanical properties for AerMet100 steel during electropulsing ageing. *Mater. Sci. Eng. A* **2023**, *871*, 144884. [CrossRef]
12. Shi, L.Q.; Ran, X.Z.; Zhai, Y.M. Influence of isothermal tempering on microstructures and hydrogen-environmentally embrittlement susceptibility of laser additively manufactured ultra-high strength AerMet100 steel. *Mater. Sci. Eng. A* **2023**, *876*, 145167. [CrossRef]
13. Wang, H.L.; Zhang, J.; Zhu, J.C. Structures of M2C carbides and its influence on strengthening in AerMet100 steel at the typical tempering temperature 482 °C. *Vacuum* **2023**, *214*, 112209. [CrossRef]
14. Shi, X.H.; Zeng, W.D.; Zhao, Q.Y. Study on the microstructure and mechanical properties of Aermet 100 steel at the tempering temperature around 482 °C. *J. Alloys Compd.* **2016**, *679*, 184–190. [CrossRef]
15. Zeng, H.H.; Hu, X.T.; Yang, D. Analytical modeling of residual stresses in laser-assisted milling AerMet100 steel. *Opt. Laser Technol.* **2023**, *158*, 108931. [CrossRef]

16. Yao, C.W.; Pang, X.T.; Gong, Q.F. Effect of laser remelting on the microstructure and mechanical properties of AerMet100 steel fabricated by laser cladding. *Mater. Sci. Eng. A* **2022**, *840*, 142951. [CrossRef]
17. Ran, X.Z.; Zhang, S.Q.; Liu, D. Role of Microstructural Characteristics in Combination of Strength and Fracture Toughness of Laser Additively Manufactured Ultrahigh-Strength AerMet100 Steel. *Metall. Mater. Trans. A* **2021**, *52*, 1248. [CrossRef]
18. Lu, J.Y.; Li, W.Y. Effects of friction stir processing on strength and ductility of laser directed energy deposited AerMet100 steel. *Mater. Sci. Eng. A* **2023**, *886*, 145740. [CrossRef]
19. Hu, Y.N.; Ao, N.; Wu, S.C. Influence of in situ micro-rolling on the improved strength and ductility of hybrid additively manufactured metals. *Eng. Fract. Mech.* **2021**, *253*, 107868. [CrossRef]
20. Xie, C.; Wu, S.C.; Yu, Y.K. Defect-correlated fatigue resistance of additively manufactured Al-Mg4.5Mn alloy with in situ micro-rolling. *J. Mater. Process. Technol.* **2021**, *291*, 117039. [CrossRef]
21. Zhao, W.D.; Liu, D.D.; Yang, J. Improving plain and fretting fatigue resistance of A100 steel using ultrasonic nanocrystal surface modification. *Int. J. Fatigue* **2021**, *148*, 106204. [CrossRef]
22. Zhi, Y.L.; Zhang, X.H.; Liu, X. Improvement of traction-traction fatigue properties of A100 steel plate-hole-structure by double shot peening. *Int. J. Fatigue* **2022**, *162*, 106925. [CrossRef]
23. Becker, T.H.; Kumar, P.; Ramamurty, U. Fracture and fatigue in additively manufactured metals. *Acta Mater.* **2021**, *219*, 117240. [CrossRef]
24. Zhu, X.K.; James, A.J. Review of fracture toughness (G, K, J, CTOD, CTOA) testing and standardization. *Eng. Fract. Mech.* **2012**, *85*, 1–46. [CrossRef]
25. Neelakantha, V.L.; Jayaraju, T.; Naik, P. Determination of fracture toughness and fatigue crack growth rate using circumferentially cracked round bar specimens of Al2014T651. *Aerosp. Sci. Technol.* **2015**, *47*, 92–97. [CrossRef]
26. Ramadas, H.; Nath, A.K.; Sarkar, S. Fatigue crack growth rate and fracture toughness evaluation of 15-5 precipitation hardening stainless steel fabricated by laser powder bed fusion process. *Mater. Sci. Eng. A* **2022**, *861*, 144356. [CrossRef]
27. Abdelwahed, M.; Bengtsson, S.; Boniardi, M. An investigation on the plane-strain fracture toughness of a water atomized 4130 low-alloy steel processed by laser powder bed fusion. *Mater. Sci. Eng. A* **2022**, *855*, 143941. [CrossRef]
28. Jiang, C.M.; Ho, J.R.; Tung, P.C. Anisotropic fracture toughness of a selective laser melted martensitic stainless steel. *Eng. Fract. Mech.* **2023**, *287*, 109348. [CrossRef]
29. Suryawanshi, J.; Prashanth, K.G.; Ramamurty, U. Tensile, fracture, and fatigue crack growth properties of a 3D printed maraging steel through selective laser melting. *J. Alloys Compd.* **2017**, *725*, 355–364. [CrossRef]
30. *GB/T 228.1-2010;* Metallic Materials—Tensile Testing—Part 1: Method of Test at Room Temperature. Standards Press of China: Beijing, China, 2011.
31. *ASTM E399-20a;* Standard Test Method for Linear-Elastic Plane-Strain Fracture Toughness of Metallic Materials. ASTM International: West Conshohocken, PA, USA, 1996.
32. Zhang, C.Y.; Wang, Q.F.; Ren, J.X. Effect of martensitic morphology on mechanical properties of an as-quenched and tempered 25CrMo48V steel. *Mater. Sci. Eng. A* **2012**, *534*, 339–346. [CrossRef]
33. Zhang, Y.P.; Zhan, D.P.; Qi, X.W. Austenite and precipitation in secondary-hardening ultra-high-strength stainless steel. *Mater. Charact.* **2018**, *144*, 393–399. [CrossRef]
34. Ran, X.Z.; Liu, D.; Li, J. Effects of post homogeneity heat treatment processes on microstructure evolution behavior and tensile mechanical properties of laser additive manufactured ultrahigh-strength AerMet100 steel. *Mater. Sci. Eng. A* **2018**, *723*, 8–21. [CrossRef]
35. Tian, X.N.; Zhu, Y.M.; Lim, C.V.S. Isotropic and improved tensile properties of Ti-6Al-4V achieved by in-situ rolling in direct energy deposition. *Addit. Manuf.* **2021**, *46*, 102151. [CrossRef]
36. Yuan, D.; Shao, S.; Guo, C.; Jiang, F.; Wang, J. Grain refining of Ti-6Al-4V alloy fabricated by laser and wire additive manufacturing assisted with ultrasonic vibration. *Ultrason. Sonochem.* **2021**, *73*, 105472. [CrossRef]
37. Le, V.T.; Mai, D.S.; Doan, T.K. Wire and arc additive manufacturing of 308L stainless steel components: Optimization of processing parameters and material properties. *Eng. Sci. Technol. Int. J.* **2021**, *24*, 1015–1026. [CrossRef]
38. Morito, S.; Huang, X.; Furuhara, T. The morphology and crystallography of lath martensite in alloy steels. *Acta Mater.* **2006**, *54*, 5323–5331. [CrossRef]
39. Liu, Z.D.; Du, Z.X.; Jiang, H.Y. Controlling the microstructure and fracture toughness of the Ti–5Al–5Mo–5V–1Cr–1Fe alloy by multiple heat treatments. *J. Mater. Res. Technol.* **2022**, *17*, 2528–2539. [CrossRef]
40. Li, H.F.; Wang, S.G.; Zhang, P. Crack propagation mechanisms of AISI 4340 steels with different strength and toughness. *Mater. Sci. Eng. A* **2018**, *729*, 130–140. [CrossRef]
41. Vaughan, M.W.; Samimi, P.; Gibbons, S.L. Exploring performance limits of a new martensitic high strength steel by ausforming via equal channel angular pressing. *Scr. Mater.* **2020**, *184*, 63–69. [CrossRef]
42. Li, X.; Li, X.T. A modified Hall-Petch relation for predicting size-induced weakening effect on yield strength of coarse-grained thin films and wires. *Mater. Lett.* **2023**, *344*, 134461. [CrossRef]
43. Xu, G.; Song, C.C.; Zhang, H.M. Spatially heterogeneous microstructure in in-situ TiO-reinforced Ti6Al4V/316L. *Addit. Manuf.* **2022**, *59*, 103178.
44. Zhu, J.; Lin, G.T.; Zhang, Z.H. The martensitic crystallography and strengthening mechanisms of ultra-high strength rare earth H13 steel. *Mater. Sci. Eng. A* **2020**, *797*, 140139. [CrossRef]

45. Wei, H.; Chen, Y.L.; Li, Z.L. Microstructure evolution and dislocation strengthening mechanism of Cu–Ni–Co–Si alloy. *Mater. Sci. Eng. A* **2021**, *826*, 142023. [CrossRef]
46. Nong, X.D.; Zhou, X.L.; Li, J.H. Selective laser melting and heat treatment of precipitation hardening stainless steel with a refined microstructure and excellent mechanical properties. *Scr. Mater.* **2020**, *178*, 7–12. [CrossRef]
47. Dong, D.D.; Wang, J.; Chen, Y. Influence of Aging Treatment Regimes on Microstructure and Mechanical Properties of Selective Laser Melted 17-4 PH Steel. *Micromachines* **2023**, *14*, 871. [CrossRef] [PubMed]
48. Fan, H.D.; Zhu, Y.X.; Awady, J.A. Precipitation hardening effects on extension twinning in magnesium alloys. *Int. J. Plast.* **2018**, *106*, 186–202. [CrossRef]
49. Cui, Y.N.; Po, G.; Ghoniem, N.M. A coupled dislocation dynamics-continuum barrier field model with application to irradiated materials. *Int. J. Plast.* **2018**, *104*, 54–67. [CrossRef]
50. Jung, J.G.; Farkoosh, A.R.; Seidman, D.N. Microstructure and mechanical properties of a precipitation-strengthened Al-Zr-Sc-Er-Si alloy with a very small Sc content. *Acta Mater.* **2023**, *257*, 119167. [CrossRef]
51. Li, Y.; Li, W.; Liu, W.Q. The austenite reversion and co-precipitation behavior of an ultra-low carbon medium manganese quenching-partitioning-tempering steel. *Acta Mater.* **2018**, *146*, 126–141. [CrossRef]
52. Tan, C.L.; Zhou, K.S.; Ma, W.Y. Microstructural evolution, nanoprecipitation behavior and mechanical properties of selective laser melted high-performance grade 300 maraging steel. *Mater. Des.* **2017**, *134*, 23–34. [CrossRef]
53. Queyreau, S.; Monnet, G.; Devincre, B. Orowan strengthening and forest hardening superposition examined by dislocation dynamics simulations. *Acta Mater.* **2010**, *58*, 5586–5595. [CrossRef]
54. Yang, G.; Deng, F.B.; Zhou, S.Y. Microstructure and mechanical properties of a novel Cu-reinforced maraging steel for wire arc additive manufacturing. *Mater. Sci. Eng. A* **2021**, *825*, 141894. [CrossRef]
55. Liu, B.G.; Li, W.; Lu, X.W. The effect of retained austenite stability on impact-abrasion wear resistance in carbide-free bainitic steels. *Wear* **2019**, *428–429*, 127–136. [CrossRef]
56. Wang, C.C.; Zhang, C.; Yang, Z.G. Multi-scale simulation of hydrogen influenced critical stress intensity in high Co-Ni secondary hardening steel. *Mater. Des.* **2015**, *87*, 501–506. [CrossRef]

micromachines

MDPI

Article

Process Optimization and Tailored Mechanical Properties of a Nuclear Zr-4 Alloy Fabricated via Laser Powder Bed Fusion

Changhui Song [1], Zhuang Zou [1], Zhongwei Yan [1], Feng Liu [2], Yongqiang Yang [1], Ming Yan [3] and Changjun Han [1,*]

[1] School of Mechanical and Automotive Engineering, South China University of Technology, Guangzhou 510641, China
[2] Institute of Reactor Waste and Radiochemistry Research, China Nuclear Power Technology Research Institute Co., Ltd., Shenzhen 518028, China
[3] Department of Materials Science and Engineering, Southern University of Science and Technology, Shenzhen 518055, China
* Correspondence: cjhan@scut.edu.cn

Abstract: A nuclear Zr-4 alloy with a near full density was fabricated via laser powder bed fusion (LPBF). The influences of process parameters on the printability, surface roughness, and mechanical properties of the LPBF-printed Zr-4 alloy were investigated. The results showed that the relative density of the Zr-4 alloy samples was greater than 99.3% with the laser power range of 120–160 W and the scanning speed range of 600–1000 mm/s. Under a moderate laser power in the range of 120–140 W, the printed Zr-4 alloy possessed excellent surface molding quality with a surface roughness less than 10 μm. The microstructure of the printed Zr-4 alloy was an acicular α phase with an average grain size of about 1 μm. The Zr-4 alloy printed with a laser power of 130 W and a scanning speed of 400 mm/s exhibited the highest compression strength of 1980 MPa and the highest compression strain of 28%. The findings demonstrate the potential in the fabrication of complex Zr-4 alloy parts by LPBF for industrial applications.

Keywords: additive manufacturing; laser powder bed fusion; zirconium alloys; Zr-4; process optimization

Citation: Song, C.; Zou, Z.; Yan, Z.; Liu, F.; Yang, Y.; Yan, M.; Han, C. Process Optimization and Tailored Mechanical Properties of a Nuclear Zr-4 Alloy Fabricated via Laser Powder Bed Fusion. *Micromachines* **2023**, *14*, 556. https://doi.org/10.3390/mi14030556

Academic Editors: Jie Yin, Yang Liu, Linda Ke and Kai Guan

Received: 22 January 2023
Revised: 22 February 2023
Accepted: 24 February 2023
Published: 27 February 2023

1. Introduction

Zirconium and its alloys have been indispensable materials for the development of the atomic energy industry due to their small thermal neutron-capturing cross-sections and outstanding nuclear properties [1]. In particular, Zr-4 alloys, the key materials of nuclear reactors, possess excellent nuclear properties, mechanical properties, corrosion resistance, and thermal stability in high-temperature environments, which can be used in fuel envelope positioning lattices of pressurized water reactors, component boxes, and heat exchangers [2,3]. Zr alloys often consist of a single α phase at room temperature. The α phase can be obtained from the transformation of a β phase, when the phase transition temperature is lower than 865 °C. In addition, non-equilibrium phase transitions (such as martensitic phase transition) may occur in Zr alloys to form α′/α″ phases [4].

Traditional manufacturing efforts have been made on Zr alloys fabricated by casting. Fuloria et al. [5] studied the mechanical properties and microstructure changes of a multi-axial forged Zr-4 alloy under different cumulative strains at low temperatures. The results showed that the ultimate tensile strength and hardness of 5.91 cumulative strain increased from 474 MPa to 717 MPa and from 190 HV to 238 HV, respectively. Such increases in strength and hardness are influenced by the grain size effect and high dislocation density. Li [6] investigated the microstructure evolution of a rolled Zr-4 alloy followed by annealing. After annealing at 480−580 °C for 2 h, the grain boundary morphology of the Zr-4 alloy evolved from relatively blurred to equiaxial. Qiu [7] identified the second phase in the

Zr-4 alloy as a $Zr_x(Fe_2Cr)$ compound, which was distributed between the dendrites of the α-Zr matrix.

Laser additive manufacturing has been used for fabricating Zr alloys, which overcomes challenges faced by traditional processes to manufacture fully dense metal parts with complex geometries and good mechanical properties [8]. Su [9] studied the formation of a Zr-based bulk metallic glass by laser-directed energy deposition additive manufacturing. A crack-free Zr-based bulk metallic glass was successfully fabricated with a sufficient high ultimate tensile strength of 880 MPa, as well as a high yield strength of 835 MPa, which was similar to that of the as-cast alloy.

Laser powder bed fusion (LPBF) is another laser additive manufacturing technology suitable for fabricating bulk parts from metal powders. LPBF uses a high-energy laser beam to melt metal powders [8]. First, a thin layer of metal powder is laid on a substrate. Then, the laser scans and melts the powder layer through a predefined path. The powder layers are continuously melted layer by layer, resulting in accumulation to obtain the geometrically complex parts with a high accuracy [10,11]. Sun [12] focused on the effect of annealing on the mechanical properties of a Zr-1Mo alloy fabricated by LPBF. The ultimate tensile strength and elongation of the alloy were 964 MPa and 11.0%, respectively, after annealing at 803 K for 2 h. A good balance of the strength and toughness was achieved, when α phase experienced stress relief. Currently, there has been limited research works on Zr-4 alloys fabricated by LPBF.

During LPBF, the volumetric energy density is determined by the laser power, scanning speed, hatch spacing, and layer thickness, which has a great influence on the relative density of printed parts. Particularly, the laser power and scanning speed are the main factors affecting the energy density. The hatch spacing may affect the overlapping ratio of melt tracks. A large hatch could cause microholes and unmelted powder particles [13]. The layer thickness determines the interlayer combination, and an excessive layer thickness leads to the lack of fusion between layers and severe spattering [14]. Low energy densities are prone to resulting in unmelted powder particles due to insufficient energy input and balling phenomena [15]. Comparatively, high energy densities may lead to excessive recoil pressure and form deep melt pools. The gas from the powder or atmospheric gas inside the chamber is thus difficult to escape and trapped inside the melt pools to form gas pores [16]. Additionally, high energy densities may lead to stress cracking that results from large residual stress caused by large temperature gradients [17]. During the solidification process, hot tearing may occur due to the shrinkage of melt pools [18].

The surface roughness is greatly influenced by process parameters. The laser power, scanning speed, and their combined interaction (energy density) are often presented as the primary parameters affecting the surface roughness of LPBF-printed parts [19]. A high laser power tends to reduce the surface roughness, as the recoil pressure flattens out the melt pools and reduces the balling phenomenon by increasing wettability of melt pools. Low scanning speeds reduce top surface roughness but increase the side surface roughness [20]. The low hatch spacing ensures overlapping between melt tracks and leads to low surface roughness [21]. A small layer thickness ensures finer particle sizes and facilitates a more complete melting process due to the higher surface-to-volume ratio compared to a large layer thickness, which reduces the surface roughness [22]. Additionally, the energy density has a great influence on the surface roughness. A low energy density leads to discontinuous melt tracks caused by lacking wettability and formation of balling on melt tracks, resulting in poor surface roughness. A moderate energy density leads to low surface roughness because of the formation of continuous and smooth melt tracks with good wettability. However, an excessive energy density leads to the evaporation of powder and severe spattering, which seriously hampers the surface quality [23].

In conclusion, this work primarily investigated the influence of process parameters on the relative density, surface roughness, and mechanical properties of a Zr-4 alloy. Optimum parameters were determined to produce a near-fully dense Zr-4 alloy with low surface roughness. The microstructure, phase, and change of mechanical properties and surface roughness of Zr-4 alloy affected by process parameters are described, which provides guidance for the future research of Zr-4 alloy formation by LPBF.

2. Materials and Methods

2.1. Materials

Spherical Zr-4 alloy powder (Hebei Baoju New Material Technology Co., LTD, Hebei province, China) prepared by the plasma method was used as the raw material, and the chemical composition of the powder is shown in Table 1. Figure 1a shows the morphology of the powder with a particle size ranging from 15 to 53 μm. The powder flowability was measured as 17.5 s/50 g using a Hall flow meter.

Table 1. Chemical composition of the Zr-4 powder.

Element	Cr	Sn	Fe	Zr
Wt. %	0.12	0.89	0.19	Bal.

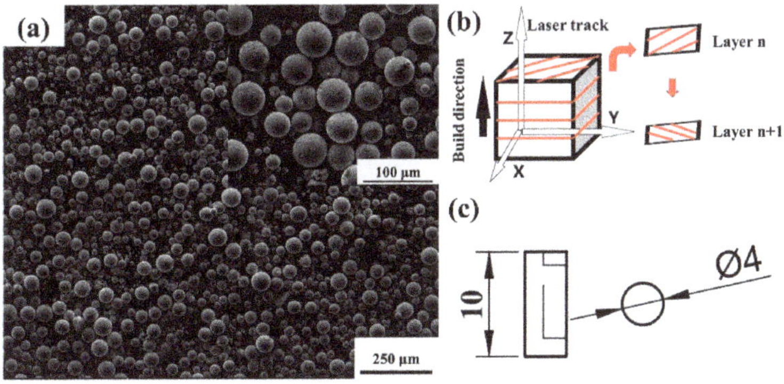

Figure 1. (**a**) Powder morphology of the Zr-4 alloy; (**b**) the schematic of the orthogonal scanning strategy; (**c**) geometries of the compressive sample model.

2.2. LPBF Process

A Dimetal-100 LPBF equipment (Laseradd, Guangzhou, China) was employed to print Zr-4 alloy samples. The LPBF chamber was filled with argon atmosphere to ensure the oxygen content less than 0.01 vol%. The relative density of the Zr-4 alloy by LPBF is affected by the laser power, scanning speed, hatch spacing, and layer thickness [24,25]. Based on our previous studies on the parameter exploration, the laser power and scanning speed were in the range of 100–200 W and 400–1400 mm/s, respectively. The hatch spacing was fixed at 0.07 mm to ensure the adjacent melt tracks could overlap, and the layer thickness was fixed at 0.03 mm to enable every layer of the powder to be completely melted [19]. Compared with the fill line scan and the chess board scan, LPBF-printed samples can obtain lower porosity by the orthogonal scanning strategy [26], as shown in Figure 1b. Such a scanning strategy was thus adopted for printing. The specific process parameters are shown in Table 2.

Table 2. Parameters to print the Zr-4 alloy via LPBF.

Laser Power (W)	**100, 110, 120, 130, 140, 150, 160, 170, 180, 190, 200**
Scanning speed (mm/s)	400, 600, 700, 800, 900, 1000, 1200, and 1400
Hatch spacing (mm)	0.07
Layer thickness (mm)	0.03
Spot diameter (mm)	0.08
Scanning strategy	Orthogonal scanning

2.3. Characterizations

Metallographic samples with dimensions of 10 × 10 × 10 mm^3 were ground, polished and eventually etched by a corrosive solution (10 vol% HF, 45 vol% HNO_3, and 45 vol% H_2O) for 15 s [27,28]. A DMI-3000M optical microscope (OM, Leica Microsystems, Weztlar, Germany) and a FEI Quanta 250 scanning electron microscope (SEM, FEI, Portland, OR, USA) were used to observe the microstructure of the samples. The WDs of the SEM images ranged from 9 mm to 10 mm, and the voltage was set as 15 Kv. The phase composition was identified through X-ray diffraction (XRD, The X'Pert3 Powder diffractometer; Malvern Panalytical, Almelo, Holland). The printed samples were polished for the XRD characterization. The degree of 2 theta was from 30° to 100°, and the scan rate was 0.013°/min. The Archimedean drainage method was used to test the relative density, i.e., the ratio of the test density to the theoretical density (6.55 g/cm3). The Zr-4 material was insoluble in water, and thus the relative density of the Zr-4 alloy can be measured by the method. Three effective data were measured from the samples to determine the average relative density value. The top surface roughness of the Zr-4 alloy was measured by an ultra-depth of field microscope (VHX-5000, KEYENCE, Osaka, Japan). The compression test was carried out by an AG-IC 50kN electronic universal testing machine (SHIMADZU, Tokyo, Japan) with a compression rate of 0.2 mm/min. Three compressive samples for each set were used to estimate the compression properties according to the GB/T 7314-2017 standard. The models of Zr-4 samples for testing are shown in Figure 1c. The samples were printed without any support structures, which were removed from the baseplate by wire cutting. The fractures of the compressed samples were observed using SEM. The SPSS software (IBM, Almonk, NY, USA) was used to conduct the correlation analysis.

3. Results and Discussion

3.1. Influence of Laser Parameters on the Relative Density

Figure 2a shows the variation of relative density of the LPBF-printed Zr-4 alloy with the laser power and scanning speed. When the scanning speed ranged from 600 mm/s to 1000 mm/s, the relative density of the alloy increased but then decreased with the increase in the laser power. When the scanning speed of 600 mm/s was applied, the relative density reached a peak of 99.40% ± 0.01% at the laser power of 130 W. At the scanning speed of 1000 mm/s, the highest relative density of 99.64% ± 0.02% could be achieved at the laser power of 160 W. The peak value of the relative density gradually increased and shifted to a higher scanning speed with the increase in the laser power. With the scanning speed of 1100 mm/s, the relative density of the alloy increased from 97.81% ± 0.02% to 99.18% ± 0.02%, as the laser power increased from 100 W to 180 W. Figure 2b exhibits the iso-density diagram of the LPBF-printed Zr-4 alloy with the laser power and scanning speed. When the laser power and scanning speed were in the ranges of 120–160 W and 600–1000 mm/s, respectively, the relative density of the alloy could remain larger than 99.3%, indicating an excellent densification. Noted that when the scanning speed exceeded 900 mm/s, the laser power should be increased to 170 W to ensure a high relative density.

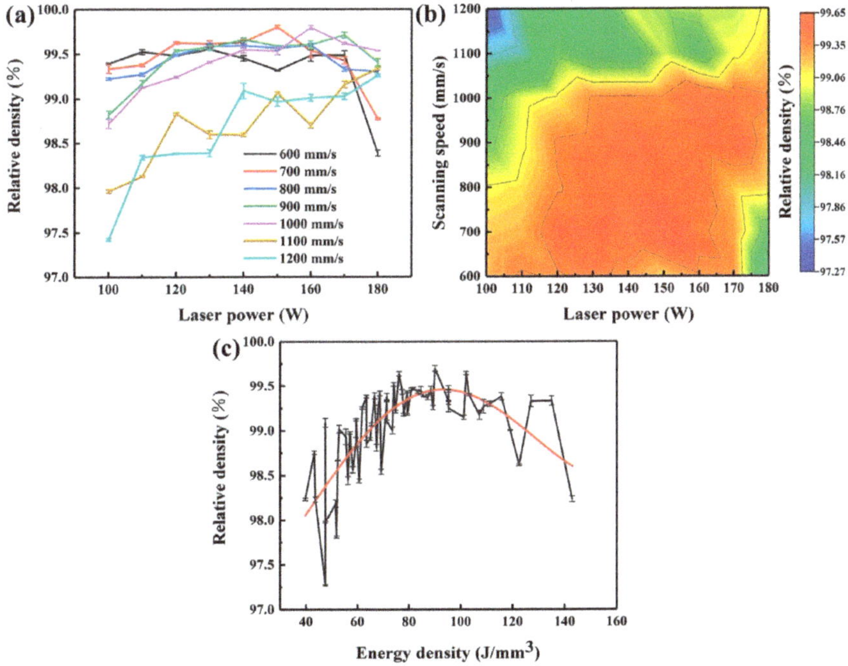

Figure 2. Effects of process parameters on the relative density of the Zr-4 alloy: (**a**) variation of relative density against the laser power and scanning speed; (**b**) iso-density diagram; (**c**) variation of the relative density against the energy density.

The Archimedean drainage method has been widely used in the density measurement of LPBF-printed parts [29,30]. The LPBF-printed Zr-4 samples were fully dense with a relative density exceeding 99%. However, the limitation of the Archimedean drainage method is the accuracy to estimate true weight, as surface tension of liquids used in the method tends to increase the weight to affect the density [31].

The energy density is dominant to influence the relative density of the printed alloy, which can be described by:

$$E = \frac{P}{VSH} \tag{1}$$

where P is the laser power (W), V is the scanning speed (mm/s), S is the hatch spacing (mm), and H is the layer thickness (mm). Figure 2c indicates the relationship between the relative density and the energy density. It can be concluded that the relative density tended to increase with the increasing energy density and deceased when the energy density exceeded 120 J/mm^3. A high laser power and a low laser scanning speed can result in the accumulation of the laser energy in melt pools and the surge of the melt pool energy, causing excessive heat accumulation. Moreover, metallurgical defects such as cracks and pores could be induced. With the increase of the scanning speed, the energy of the melt pools gradually decreases. The melt pools become stable under a moderate laser power. A further increase in the scanning speed under a low laser power leads to insufficient energy within the melt pools, resulting in the increase of unmelted powder particles and pores, thus exhibiting a decreasing trend of the relative density [32–34]. As shown in Table 3, the Pearson correlation analysis was used to analyze the correlation between the energy density, the laser power, and the laser scanning speed via SPSS. The results showed the Pearson index between the laser power and the relative density was 0.244 while it was

−0.517 between the laser scanning speed and the relative density, indicating the relative density of the LPBF-printed Zr-4 alloy was more sensitive to the laser scanning speed.

Table 3. Pearson correlation analysis between the energy density, the laser power, and the laser scanning speed.

		P	Relative Density			*V*	Relative Density
P	Pearson correlation	1	0.244	*V*	Pearson correlation	1	−0.517
	Significant		0.054		Significant		0
	number	63	63		number	63	63
Relative density	Pearson correlation	0.244	1	Relative density	Pearson correlation	−0.517	1
	Significant	0.054			Significant	0	
	number	63	63		number	63	63

3.2. Influence of Laser Parameters on the Surface Roughness

The macroscopic morphologies of Zr-4 alloys by LPBF under different parameters are shown in Figure 3. A low laser power and a high scanning speed led to a smooth surface of the printed alloy. An overburn phenomenon occurred, when the laser power was greater than 150 W with a scanning speed of 600 mm/s. When the laser power was increased to 170 W, the printed alloy gradually obtained overburn surfaces at a scanning speed of 1000 mm/s. The generation of overburn surfaces can be attributed to the high laser energy density produced at high laser power and low scanning speed values [35,36].

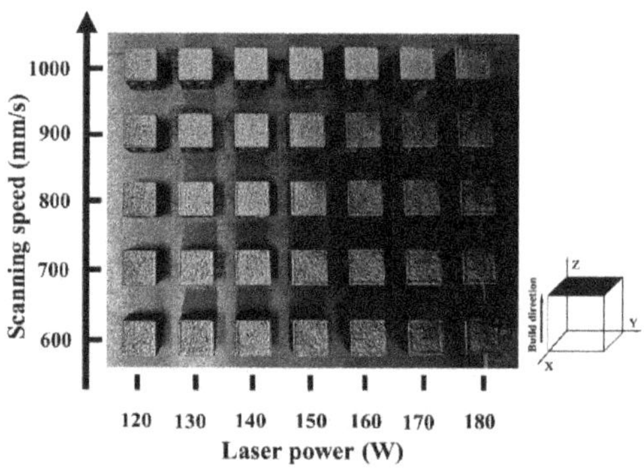

Figure 3. Macroscopic morphologies of the LPBF-printed Zr-4 samples with different process parameters.

Table 4 shows the surface roughness on the side for the samples printed with a laser power of 130 W and a scanning speed of 600–1200 mm/s. The results showed the influence of the scanning speed on the roughness of the side surface was not obvious. Generally, the side surface roughness is mainly affected by the layer thickness [19], which was fixed as 0.03 mm in this work. Additionally, the procedure of the contour scan with the same parameters used at each layer further reduced the distinction of the side surface roughness between different process parameters.

Table 4. The side surface roughness of the Zr-4 alloy (the laser power was 130 W).

Scanning Speed (mm/s)	600	800	900	1000	1100	1200
Roughness (μm)	8.63 ± 0.55	9.25 ± 0.6	9.14 ± 1.57	9.89 ± 0.23	10.12 ± 0.59	9.49 ± 0.76

As shown in Figure 4a, with the increase of the laser power at a certain scanning speed, the top surface roughness of the alloy tended to reduce. When the scanning speed of

1000 mm/s and the laser power of 100 W were applied, the maximum surface roughness of 14.58 ± 0.86 μm was obtained. The surface roughness decreased to the minimum value of 7.09 ± 0.38 μm, when the laser power increased to 130 W. As the laser power increased, the roughness began to rise to an Ra value of 12.47 ± 0.69 μm at 150 W. By further increasing the laser power to 180 W, the roughness value decreased to 8.24 ± 0.80 μm. Figure 4b shows the iso-roughness diagram of the top surface of the printed Zr-4 alloy. The low roughness of the alloy was achieved with a laser power of 120–140 W and a scanning speed of 1000–1100 mm/s. However, the laser power greater than 160 W and the scanning speed greater than 1100 mm/s also led to a low top surface roughness. In Figure 4c, the trend of the surface roughness against the energy density was similar with that of the laser power. A low surface roughness was obtained with the energy density in the range of 50–80 J/mm^3 and exceeding 130 J/mm^3.

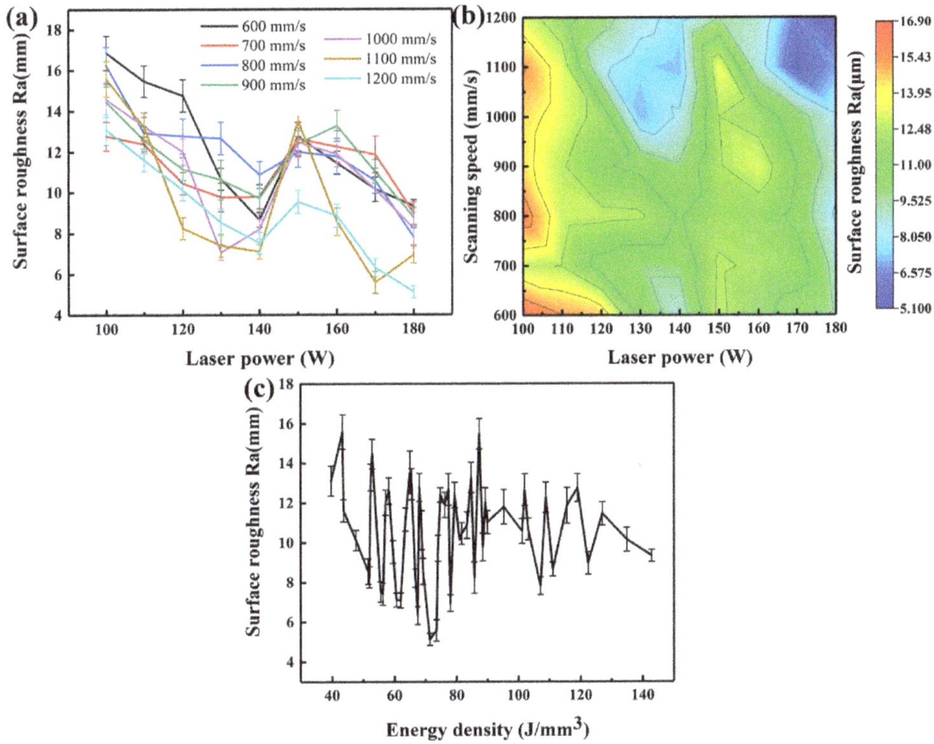

Figure 4. Effects of process parameters on the top surface roughness of the LPBF-printed Zr-4 alloy: (**a**) variation of the top surface roughness against the laser power and scanning speed; (**b**) iso-roughness diagram; (**c**) variation of the top surface roughness against the energy density.

Figure 5 shows the profiles of the top surface of the printed Zr-4 alloy with different laser power values. A lower laser power led to an insufficient laser input energy, and the Zr-4 powders could not be fully melted, resulting in the existence of bulges. Meanwhile, the viscosity of the low-energy melt pools increased; thus, the molten tracks became rugged and discontinuous [37]. Such factors led to the increase of the surface roughness (Figure 5a). With a further increase in the laser power to 140 W, the powder was completely melted, and the liquid molten pools were spread smoothly under the moderate power. Therefore, a low surface roughness of the alloy was obtained (Figure 5b). The uneven morphology was affected by the surface tension of liquid melt pools [38]. When the laser power increased to 160 W, it was easy to produce the periodizing effect and increase the surface protrusion due to the generation of the large temperature gradient between the center and the edge of the melt pools [36], resulting in the high surface roughness (Figure 5c). When the laser power was greater than 160 W, an obvious overburn phenomenon appeared on the surface [35], indicating that the laser energy was excessive. In this case, adjacent melt pools fused together before their solidification, and the lap between the melt pools was reduced. As a result, a giant bulge was created at the edge of the top surface (Figure 5d), which caused a low top surface roughness but warping deformation on all sides with a laser power of 180 W. Therefore, it can be concluded that the ideal surface roughness was obtained with the laser power in the range of 120–140 W.

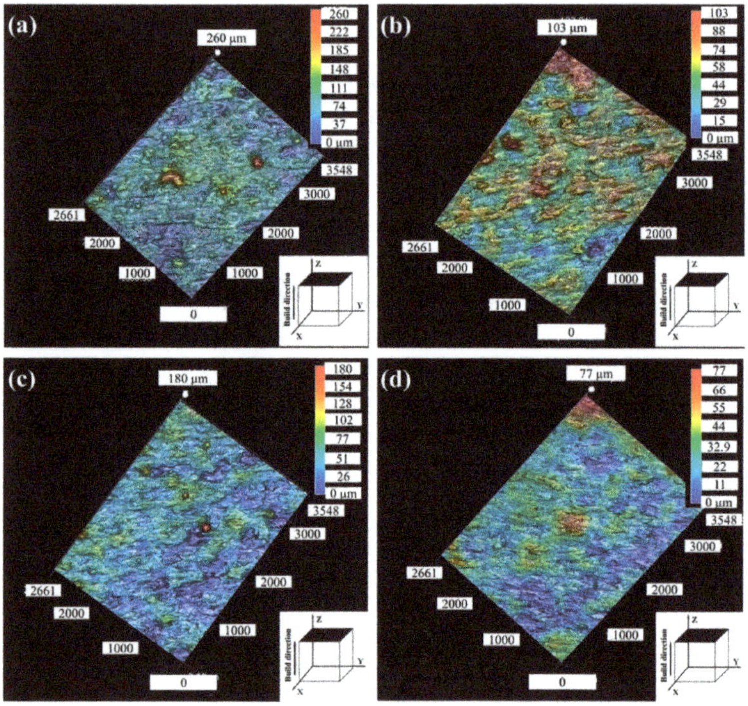

Figure 5. Top surface profiles under different laser powers (the scanning speed was fixed as 1100 mm/s): (**a**) P = 100 W; (**b**) P = 130 W; (**c**) P = 150 W; (**d**) P = 180 W.

3.3. Microstructure

Figure 6 shows the OM morphology of the LPBF-printed Zr-4 alloy with a laser power of 130 W and a scanning speed of 1000 mm/s. As observed in Figure 6a, the morphology of the alloy along the transverse plane exhibited acicular grains with a size of ~1 μm. Figure 6b exhibits overlapped molten pools without obvious metallurgical defects along the longitudinal plane.

Figure 7 shows the XRD diffraction pattern of the printed Zr-4 alloy with a laser power of 130 W and a scanning speed of 1000 mm/s. According to the standard diffraction pattern, an α-Zr phase could be identified in the printed Zr-4 alloy. During LPBF, the powder melted rapidly to form a liquid melt pool and solidified rapidly to form acicular martensite. In the rapid cooling stage, β-Zr, a stable phase at high temperatures in Zr alloys, undergoes the phase transition from β to α. Additionally, non-equilibrium phase transformation often occurs during the cooling process, producing metastable structures such as α'/α" phases. Because their crystal structures are similar to the α phase obtained by the equilibrium phase transformation, they cannot be separated in XRD patterns [39–41].

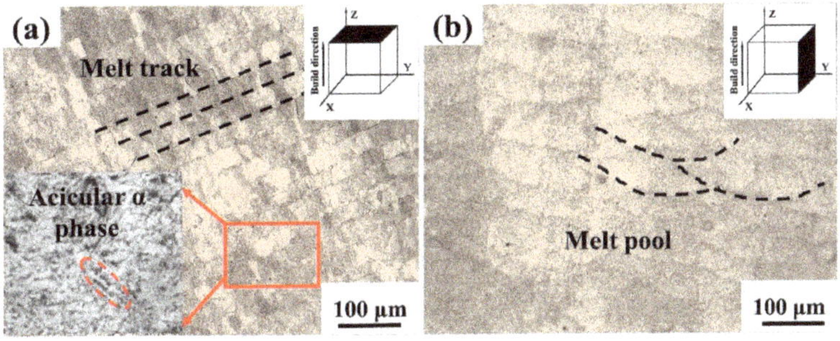

Figure 6. Metallographic images of the Zr-4 alloy printed using a laser power of 130 W and a scanning speed of 1000 mm/s along different planes: (**a**) the transverse plane; (**b**) the longitudinal plane.

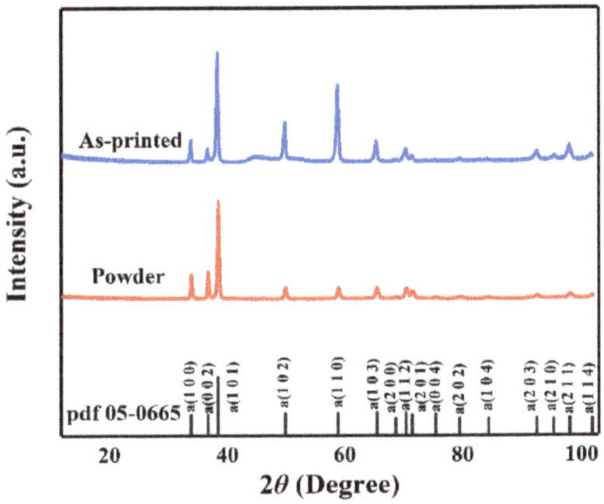

Figure 7. XRD pattern of the LPBF-printed Zr-4 alloy obtained using a laser power of 130 W and a scanning speed of 1000 mm/s.

3.4. Compression Properties

Our previous studies on the tensile properties showed the printed Zr-4 tensile coupons exhibited a quite low ductility. Even with the optimized parameters, the tensile coupons were still fractured at the very early stage of the testing. Therefore, the compression test was selected. Figure 8 shows the compressive properties of the LPBF-printed Zr-4 alloys with different parameters. When a laser power of 130 W was applied, both the compression strength and compression strain of the printed Zr-4 alloy decreased with the increase in the scanning speed (Figure 8a–b). The compressive strength gradually decreased from 1950 MPa with a scanning speed of 400 mm/s to 1540 MPa with a scanning speed of 1400 mm/s, while the compressive strain decreased from 28.3% to 15.5%. When the scanning speed was kept as 1000 mm/s, the compression strength of the printed alloy remained about 1650 MPa with an increase in the laser power from 100 W to 200 W (Figure 8c,d). When the laser power increased from 100 to 180W, the compression strain increased from 15% to 22.5% and then decreased to 20.5% as the laser power further increased to 200 W.

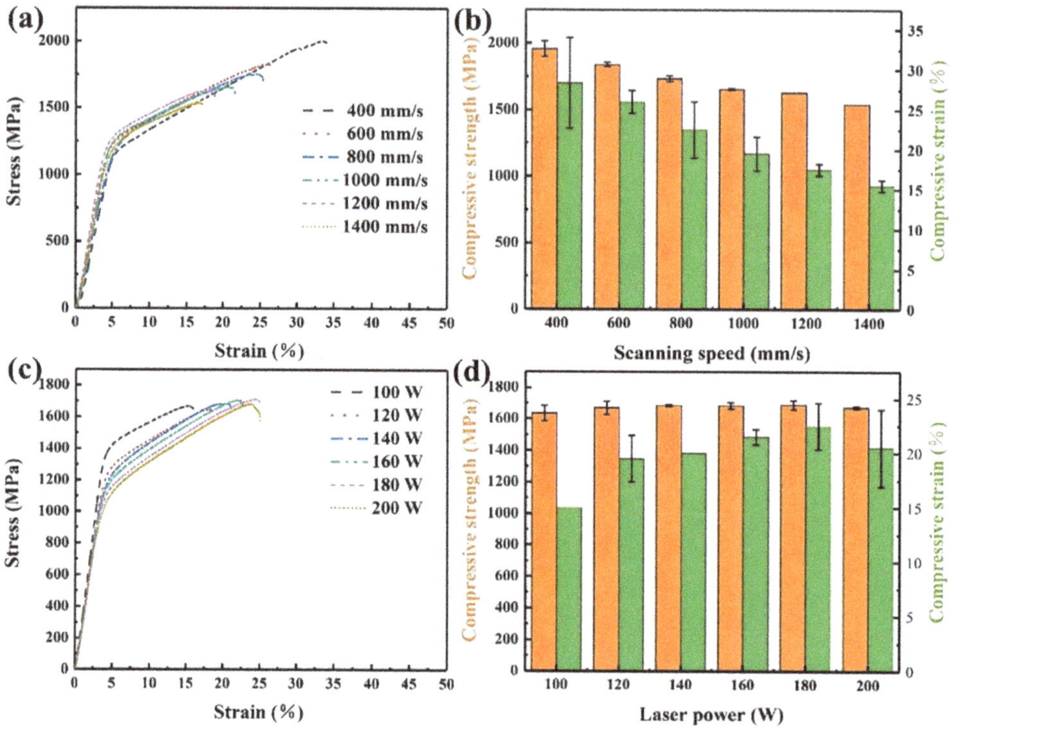

Figure 8. Compressive properties of the Zr-4 alloy at different laser powers and scanning speeds: (**a**) compressive curves at different scanning speeds; (**b**) variation of the compressive strength against the scanning speed (P = 130 W); (**c**) compressive curves at different laser powers; (**d**) variation of the compressive strength against the laser power (V = 1000 mm/s).

Figure 9 shows the compression fracture morphologies of the LPBF-printed Zr-4 alloys using different process parameters. The fracture surface of the printed alloy was mainly composed of the quasi-cleavage planes with a laser power of 130 W and a scanning speed of 400 mm/s (Figure 9a,d). When the scanning speed increased to 1000 mm/s, the fracture surface was composed of cleavage and quasi-cleavage planes (Figure 9b,e). When the scanning speed reached 1400 mm/s, obvious tearing edges and quasi-cleavage regions between them were found on the fracture surface (Figure 9c,f). With the increase in the scanning speed, the fracture surface gradually evolved from cleavage to semi-cleavage and semi-quasi-cleavage plane and finally became a dominated cleavage fracture. Such an increase in the cleavage region and a decrease in the quasi-cleavage and dimple regions are the main reasons for the deterioration of ductility [42].

Comparatively, as the scanning speed was fixed as 1000 mm/s, the fracture surface was relatively flat at the laser power of 100 W (Figure 9g). Cleavage planes could be observed with a few quasi-cleavage areas (Figure 9i). When the laser power increased to 130 W, the quasi-cleavage regions increased. When the laser power reached 200 W, the fracture surface presented an angle of 45° between the fracture surface and the horizontal plane (Figure 9h). Additionally, all the surfaces were presented as quasi-cleavage planes, where some possessed lamellar shedding boundaries (Figure 9j). The increase of the quasi-cleavage regions led to an increase in the compression strain with the increase of the laser power.

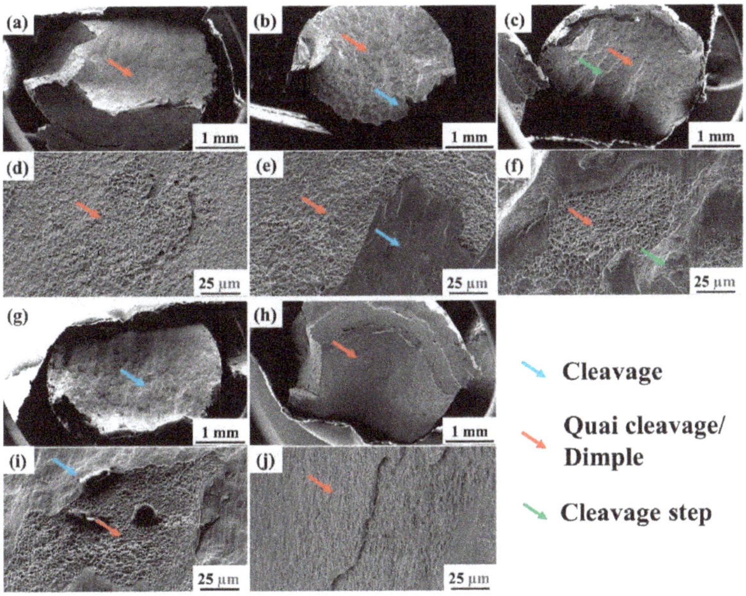

Figure 9. Compression fractures of the Zr-4 alloy under different laser parameters: (**a**,**d**) P = 130 W and V = 400 mm/s; (**b**,**e**) P = 130 W and V = 1000 mm/s; (**c**,**f**) P = 130 W and V = 1400 mm/s; (**g**,**i**) P = 100 W and V = 1000 mm/s; (**h**,**j**) P = 200 W and V = 1000 mm/s.

The changes in laser power and scanning speed led to the variation of the energy density. When a low laser power or high scanning speed was used, the resulted low laser energy density caused the formation of defects such as cracks and pores. For example, samples with a laser power of 130 W and a scanning speed of 1400 mm/s (energy density of 44 J/mm^3) exhibited a relatively lower compressive strength (1540 MPa) and compressive strain (15%), corresponding to a low relative density of ~98.5% (Figure 2c). Therefore, with an increase in the laser power and a decrease in the scanning speed, the compressive strength of the printed alloy was improved. Additionally, the scanning speed played an important role in the interaction between the laser and the powder. A long interaction period tended to cause oxidation and nitriding, which limited the mobility of dislocation and hampered the ductility [43,44]. Consequently, a high energy density (80–120 J/mm^3) and a low scanning speed (below 800 mm/s) were recommended during LPBF.

4. Conclusions

A Zr-4 alloy with a high relative density and excellent mechanical properties was manufactured by LPBF. Process optimization, surface roughness measurement, microstructure observation, and mechanical properties analysis were conducted. The main findings are presented as follows:

1. The LPBF process parameters for printing the Zr-4 alloy were optimized as the laser power of 120–160 W and the scanning speed of 600–1000 mm/s. A highest relative density greater than 99.3% could be obtained with the energy density of 70–110 J/mm^3.
2. The top surface roughness of the alloy tended to reduce with the increase of the laser power as well as the energy density at a certain scanning speed. The low roughness of the alloy was achieved with laser powers of 120–140 W and scanning speeds of 1000–1100 mm/s and energy densities of 60–80 J/mm^3.

3. The microstructure of the printed Zr-4 alloy consisted of an α-Zr phase with an average grain size of about 1 μm. The α-Zr existed in the form of crisscross acicular grains, determined by the temperature gradient of molten pools.
4. The laser power and scanning speed showed remarkable influences on the compressive strength and compressive strain of the Zr-4 alloys. Under the laser power of 130 W, the compressive strength gradually decreased from 1950 MPa to 1540 MPa and the compressive strain decreased from 28.3% to 15.5% with the increase in the scanning speed from 400 mm/s to 1400 mm/s. When the scanning speed was 1000 mm/s and the laser power increased from 100 W to 200 W, the compression strength remained about 1650 MPa while the compressive strain tended to increase from 15% to 22.5%. A high energy density (80–120 J/mm^3) and a low scanning speed (below 800 mm/s) were recommended during LPBF.

Author Contributions: Conceptualization, C.S., M.Y., Y.Y., and C.H.; methodology, Z.Y., Z.Z., and C.H.; formal analysis, Z.Y. and Z.Z.; writing—original draft preparation, Z.Z.; writing—review and editing, Z.Y., Z.Z., C.S., and C.H.; investigation, C.S., C.H., and Z.Z.; visualization, Z.Z. and C.H.; supervision, Y.Y., C.S., C.H., M.Y., and F.L.; project administration, C.S.; funding acquisition, C.S. All authors have read and agreed to the published version of the manuscript.

Funding: This research was funded by Guangdong Basic and Applied Basic Research Foundation (No. 2020B1515120013), Guangdong Province Science and Technology Project (No. 2020B090924002), National Natural Science Foundation of China (No. U2001218), Special Support Foundation of Guangdong Province (No. 2019TQ05Z110), Capital Health Development Scientific Research Project (No. A5210020), and Shenzhen Science and Technology Plan Project (No. JSGG20210802153809029).

Data Availability Statement: Not applicable.

Conflicts of Interest: The authors declare that they have no known competing financial interests or personal relationships that could have appeared to influence the work reported in this paper.

References

1. Abe, K. Ninth International Symposium on Zirconium in the Nuclear Industry. *J. Nucl. Sci. Technol.* **1991**, *28*, 369–374. [CrossRef]
2. Linga Murty, K.; Charit, I. Texture development and anisotropic deformation of zircaloys. *Prog. Nucl. Energy* **2006**, *48*, 325–359. [CrossRef]
3. Zinkle, S.J.; Was, G.S. Materials challenges in nuclear energy. *Acta Mater.* **2013**, *61*, 735–758. [CrossRef]
4. Singh, R.N.; Bind, A.K.; Singh, J.B.; Chakravartty, J.K.; Paul, V.T.; Madhusoodnan, K.; Suwas, S.; Saroja, S.; Suri, A.K.; Banerjee, S. Development and Characterization of Microstructure and Mechanical Properties of Heat-Treated Zr–2.5Nb Alloy for AHWR Pressure Tubes. *Mater. Perform. Charact.* **2013**, *2*, 120–133. [CrossRef]
5. Fuloria, D.; Goel, S.; Jayaganthan, R.; Srivastava, D.; Dey, G.K.; Saibaba, N. Mechanical properties and microstructural evolution of ultrafine grained zircaloy-4 processed through multiaxial forging at cryogenic temperature. *Trans. Nonferrous Met. Soc. China* **2015**, *25*, 2221–2229. [CrossRef]
6. Li, J.; Jiao, Y.; Yang, F. Effect of annealing temperature on properties of Zr-4 alloy. *Shandong Chem. Ind.* **2018**, *47*, 30–31. [CrossRef]
7. Qiu, J.; Yuan, J.; Tian, Z. Study on the second phase of zirconium 4 alloy ingot casting. *Rare Met. Mater. Eng.* **1986**, *02*, 17–18+12–89.
8. Ngo, T.D.; Kashani, A.; Imbalzano, G.; Nguyen, K.T.Q.; Hui, D. Additive manufacturing (3D printing): A review of materials, methods, applications and challenges. *Compos. Part B Eng.* **2018**, *143*, 172–196. [CrossRef]
9. Su, S.; Lu, Y. Laser directed energy deposition of Zr-based bulk metallic glass composite with tensile strength. *Mater. Lett.* **2019**, *247*, 79–81. [CrossRef]
10. Kranz, J.; Herzog, D.; Emmelmann, C. Design guidelines for laser additive manufacturing of lightweight structures in TiAl6V4. *J. Laser Appl.* **2015**, *27*, S14001. [CrossRef]
11. Chen, J.; Yang, Y.; Song, C.; Zhang, M.; Wu, S.; Wang, D. Interfacial microstructure and mechanical properties of 316L /CuSn10 multi-material bimetallic structure fabricated by selective laser melting. *Mater. Sci. Eng. A* **2019**, *752*, 75–85. [CrossRef]
12. Sun, X.; Liu, D.; Chen, M.; Zhou, W.; Nomura, N.; Hanawa, T. Influence of annealing treatment on the microstructure, mechanical performance and magnetic susceptibility of low magnetic Zr–1Mo parts manufactured via laser additive manufacturing. *Mater. Sci. Eng. A* **2021**, *804*, 140740. [CrossRef]
13. Cui, L.; Peng, Z.; Chang, Y.; He, D.; Cao, Q.; Guo, X.; Zeng, Y. Porosity, microstructure and mechanical property of welded joints produced by different laser welding processes in selective laser melting AlSi10Mg alloys. *Opt. Laser Technol.* **2022**, *150*, 107952. [CrossRef]

14. Zhang, D.; Cai, Q.; Liu, J.; Li, R. Research on Process and Microstructure Formation of W-Ni-Fe Alloy Fabricated by Selective Laser Melting. *J. Mater. Eng. Perform.* **2010**, *20*, 1049–1054. [CrossRef]
15. Chen, J.; Liao, X.; Shu, J.; Zhou, L.; Li, C.; Ren, Y.; Niu, Y. Microstructure tailoring of Ti–15Mo alloy fabricated by selective laser melting with high strength and ductility. *Mater. Sci. Eng. A* **2021**, *826*, 141962. [CrossRef]
16. Guraya, T.; Singamneni, S.; Chen, Z.W. Microstructure formed during selective laser melting of IN738LC in keyhole mode. *J. Alloy. Compd.* **2019**, *792*, 151–160. [CrossRef]
17. Agius, D.; Kourousis, K.; Wallbrink, C. A Review of the As-Built SLM Ti-6Al-4V Mechanical Properties towards Achieving Fatigue Resistant Designs. *Metals* **2018**, *8*, 75. [CrossRef]
18. Lu, Y.; Bartlett, L.N.; O'Malley, R.J. A Review on Hot Tearing of Steels. *Int. J. Met.* **2021**, *16*, 45–61. [CrossRef]
19. Tian, Y.; Tomus, D.; Rometsch, P.; Wu, X. Influences of processing parameters on surface roughness of Hastelloy X produced by selective laser melting. *Addit. Manuf.* **2017**, *13*, 103–112. [CrossRef]
20. Mumtaz, K.; Hopkinson, N. Top surface and side roughness of Inconel 625 parts processed using selective laser melting. *Rapid Prototyp. J.* **2009**, *15*, 96–103. [CrossRef]
21. Pupo, Y.; Monroy, K.P.; Ciurana, J. Influence of process parameters on surface quality of CoCrMo produced by selective laser melting. *Int. J. Adv. Manuf. Technol.* **2015**, *80*, 985–995. [CrossRef]
22. Kruth, J.P.; Levy, G.; Klocke, F.; Childs, T.H.C. Consolidation phenomena in laser and powder-bed based layered manufacturing. *CIRP Ann.* **2007**, *56*, 730–759. [CrossRef]
23. Wang, D.; Liu, Y.; Yang, Y.; Xiao, D. Theoretical and experimental study on surface roughness of 316L stainless steel metal parts obtained through selective laser melting. *Rapid Prototyp. J.* **2016**, *22*, 706–716. [CrossRef]
24. Sun, X.; Zhou, W.; Kikuchi, K.; Nomura, N.; Kawasaki, A.; Doi, H.; Tsutsumi, Y.; Hanawa, T. Fabrication and Characterization of a Low Magnetic Zr-1Mo Alloy by Powder Bed Fusion Using a Fiber Laser. *Metals* **2017**, *7*, 501. [CrossRef]
25. Scipioni Bertoli, U.; Wolfer, A.J.; Matthews, M.J.; Delplanque, J.-P.R.; Schoenung, J.M. On the limitations of Volumetric Energy Density as a design parameter for Selective Laser Melting. *Mater. Des.* **2017**, *113*, 331–340. [CrossRef]
26. Pauly, S.; Schricker, C.; Scudino, S.; Deng, L.; Kühn, U. Processing a glass-forming Zr-based alloy by selective laser melting. *Mater. Des.* **2017**, *135*, 133–141. [CrossRef]
27. Zhang, F.; Ma, Y.; Zhou, D. Preparation Method of Zirconium and Zirconium Alloy Metallographic Samples. CN104535394A, 19 January 2015.
28. Han, D.; Zhang, J. Metallographic Sample Preparation and Display Technology. *Phys. Chem. Exam.* **2006**, *042*, 87.
29. Peters, R.; Beiss, P.; Lindlohr, S. Density Determination of Sintered Steel—Results of a Round Robin Test. *Prakt. Metallogr. Pract. Metallogr.* **2019**, *56*, 731–746. [CrossRef]
30. Weidmann, J.; Großmann, A.; Mittelstedt, C. Laser powder bed fusion manufacturing of aluminum honeycomb structures: Theory and testing. *Int. J. Mech. Sci.* **2020**, *180*, 105639. [CrossRef]
31. Wang, Y.; Morita, K. Measurement of CaO-SiO2-SaCl2 slag density by an improved Archimedean method. *J. Min. Metall. Sect. B Metall.* **2015**, *51*, 113–116. [CrossRef]
32. Carter, L.N.; Wang, X.; Read, N.; Khan, R.; Aristizabal, M.; Essa, K.; Attallah, M.M. Process optimisation of selective laser melting using energy density model for nickel based superalloys. *Mater. Sci. Technol.* **2016**, *32*, 657–661. [CrossRef]
33. Hu, L.; Qiu, C.; Chen, Y.; Li, H.; Liu, H. Influence of Laser Energy Density on Interfacial Diffusion Bonding and Surface Density of Chromium Coating by Multi-Arc Ion Plating on Zirconium Alloy. *Coatings* **2020**, *10*, 565. [CrossRef]
34. Skamat, J.; Cernasejus, O.; Zhetessova, G.; Nikonova, T.; Zharkevich, O.; Visniakov, N. Effect of Laser Processing Parameters on Microstructure, Hardness and Tribology of NiCrCoFeCBSi/WC Coatings. *Materials* **2021**, *14*, 6034. [CrossRef] [PubMed]
35. Strano, G.; Hao, L.; Everson, R.M.; Evans, K.E. Surface roughness analysis, modelling and prediction in selective laser melting. *J. Mater. Process. Technol.* **2013**, *213*, 589–597. [CrossRef]
36. Chen, Z.; Wu, X.; Tomus, D.; Davies, C.H.J. Surface roughness of Selective Laser Melted Ti-6Al-4V alloy components. *Addit. Manuf.* **2018**, *21*, 91–103. [CrossRef]
37. Brinksmeier, E.; Levy, G.; Meyer, D.; Spierings, A.B. Surface integrity of selective-laser-melted components. *CIRP Ann.* **2010**, *59*, 601–606. [CrossRef]
38. Gheribi, A.E.; Chartrand, P. Temperature and oxygen adsorption coupling effects upon the surface tension of liquid metals. *Sci. Rep.* **2019**, *9*, 7113. [CrossRef]
39. Rogachev, S.O.; Nikulin, S.A.; Khatkevich, V.M.; Gorshenkov, M.V.; Sundeev, R.V.; Veligzhanin, A.A. Effect of annealing on structural and phase transformations and mechanical properties of ultrafine-grained E125 zirconium alloy obtained by high-pressure torsion. *Mater. Lett.* **2017**, *206*, 26–29. [CrossRef]
40. Zhang, M.; Li, Y.N.; Zhang, F.C.; Wang, X.B.; Chen, L.Y.; Yang, Z.N. Effect of annealing treatment on the microstructure and mechanical properties of a duplex Zr-2.5Nb alloy. *Mater. Sci. Eng. A* **2017**, *706*, 236–241. [CrossRef]
41. Devi, Y.P.; Donthula, H.; Keskar, N.; Sarkar, A.; Vaibhaw, K.; Krishna, K.V.M. Microstructural evolution in (α + βZr) region of Zr-2.5 wt% Nb annealed at different temperatures: Effect on mechanical properties. *J. Nucl. Mater.* **2020**, *530*, 151978. [CrossRef]
42. Ma, W.; Wang, F.; Chen, B.; Li, B.; Zhang, X.; Ma, M.; Liu, R. Thermal compression behavior and microstructural evolution of Ti-30-5-3 alloys in lower $\alpha + \beta$ region. *Mater. Lett.* **2021**, *297*, 129876. [CrossRef]

43. Na, T.-W.; Kim, W.R.; Yang, S.-M.; Kwon, O.; Park, J.M.; Kim, G.-H.; Jung, K.-H.; Lee, C.-W.; Park, H.-K.; Kim, H.G. Effect of laser power on oxygen and nitrogen concentration of commercially pure titanium manufactured by selective laser melting. *Mater. Charact.* **2018**, *143*, 110–117. [CrossRef]
44. Sun, X.; Liu, D.; Zhou, W.; Nomura, N.; Tsutsumi, Y.; Hanawa, T. Effects of process parameters on the mechanical properties of additively manufactured Zr-1Mo alloy builds. *J. Mech. Behav. Biomed. Mater.* **2020**, *104*, 103655. [CrossRef] [PubMed]

micromachines

Article

Laser Powder Bed Fusion of 316L Stainless Steel: Effect of Laser Polishing on the Surface Morphology and Corrosion Behavior

Jun Liu [1], Haojun Ma [2], Lingjian Meng [2], Huan Yang [2,*], Can Yang [2], Shuangchen Ruan [2], Deqin Ouyang [2], Shuwen Mei [3], Leimin Deng [4], Jie Chen [5] and Yu Cao [1,*]

1 Zhejiang Provincial Key Laboratory of Laser Processing Robotics, College of Mechanical & Electrical Engineering, Wenzhou University, Wenzhou 325035, China; 20461439087@stu.wzu.edu.cn
2 Sino-German College of Intelligent Manufacturing, Shenzhen Technology University, Shenzhen 518118, China; 2110412013@stumail.sztu.edu.cn (H.M.); 2110412020@stumail.sztu.edu.cn (L.M.); yangcan@sztu.edu.cn (C.Y.); ouyangdeqin@sztu.edu.cn (D.O.)
3 Nantong Jinyuan Intelligent Technology Co., Nantong 226007, China; meishuwen@jyznjs.com
4 Wuhan National Research Center for Optoelectronics, Huazhong University of Science and Technology, Wuhan 430074, China; dlm@hust.edu.cn
5 Wenzhou University Rui'an Graduate College, Wenzhou University, Ruian 325207, China; chenjie@wzu.edu.cn
* Correspondence: yanghuan@sztu.edu.cn (H.Y.); yucao@wzu.edu.cn (Y.C.); Tel.: +86-177-226-20530 (H.Y.); +86-136-465-59970 (Y.C.)

Abstract: Recently, laser polishing, as an effective post-treatment technology for metal parts fabricated by laser powder bed fusion (LPBF), has received much attention. In this paper, LPBF-ed 316L stainless steel samples were polished by three different types of lasers. The effect of laser pulse width on surface morphology and corrosion resistance was investigated. The experimental results show that, compared to the nanosecond (NS) and femtosecond (FS) lasers, the surface material's sufficient remelting realized by the continuous wave (CW) laser results in a significant improvement in roughness. The surface hardness is increased and the corrosion resistance is the best. The microcracks on the NS laser-polished surface lead to a decrease in the microhardness and corrosion resistance. The FS laser does not significantly improve surface roughness. The ultrafast laser-induced micro-nanostructures increase the contact area of the electrochemical reaction, resulting in a decrease in corrosion resistance.

Keywords: selective laser melting; laser polishing; 316L stainless steel; surface morphology; electrochemical corrosion

Citation: Liu, J.; Ma, H.; Meng, L.; Yang, H.; Yang, C.; Ruan, S.; Ouyang, D.; Mei, S.; Deng, L.; Chen, J.; Cao, Y. Laser Powder Bed Fusion of 316L Stainless Steel: Effect of Laser Polishing on the Surface Morphology and Corrosion Behavior. *Micromachines* **2023**, *14*, 850. https://doi.org/10.3390/mi14040850

Academic Editor: Kunal Mitra

Received: 9 March 2023
Revised: 30 March 2023
Accepted: 12 April 2023
Published: 14 April 2023

1. Introduction

Additive manufacturing (AM), as an advanced technology, has received increasing attention for its significant advantages in manufacturing high-strength and complex parts. Compared with machining and other traditional manufacturing technologies, this technology shows significant advantages, such as high flexibility, no molds, material saving, and reduced design cycle time [1–3]. It can manufacture digital models into solid parts directly by accumulating materials layer-by-layer. As a representative AM technique, laser powder bed fusion (LPBF) melts the metal powder in a predetermined scanning path by a high-power laser beam and then shapes it after cooling and solidification [4,5].

316L stainless steel is austenitic stainless steel. With its high strength and excellent corrosion resistance, 316L stainless steel has been widely used in aerospace, medical, food, and chemical industries. For the manufacture of 316L stainless steel, LPBF shows many advantages over traditional techniques, such as higher utilization of material, integrated forming, and the ability to create complex structures [6–8]. However, the parts manufactured by LPBF are prone to defects, such as unmelted powder, porosity, and cracks [9–12]. Moreover, due to the powder source material and layered manufacturing process, the surface roughness (SR) of the LPBF-ed 316L stainless steel is very rough, which cannot meet the requirements of aerospace and biomedical use [13–15]. Therefore, post-treatment is required to improve the surface quality of LPBF-ed parts. Currently, the technology for polishing metal parts includes grinding, sand-blasting, and electropolishing. However, these processes are still insufficient in polishing quality and efficiency for complex and thin-wall structures.

Recently, laser polishing, as an effective post-treatment technology, has received much attention for the advantages of non-contact, non-pollution, and high efficiency. With a continuous wave (CW) or long-pulse laser, this technology relies on laser-induced remelting and the subsequent rapid solidification of surface materials [16–19]. Previous studies show that for the additive manufacturing parts with initial roughness below 8 μm, pulsed lasers can reduce surface roughness by 71.3%, whereas CW lasers can reduce surface roughness by 84.4% [20,21]. Chen et al. used a CW laser to polish 316L stainless steel, reducing the surface roughness by 92%. Meanwhile, the microhardness and corrosion resistance could also be improved [22]. Pakin et al. studied the effect of processing parameters such as repetition frequency, laser power, and scanning speed on the polishing quality of aluminum alloys [23]. When the laser used for polishing is ultrafast, improved surface roughness is achieved by removing convex materials. Hafiz et al. adopted a picosecond laser to polish nickel-based alloy, and surface roughness could be reduced from 0.435 μm to 0.127 μm [24].

Although laser polishing with different pulse widths can reduce the roughness of metal surfaces, there are differences in the morphologies, metallographic structures, and mechanical properties of material surfaces. The CW laser can quickly obtain a flat surface, but the ultrafast laser can avoid the remelting and oxidation of surface materials. Currently, there is still no clear conclusion on the need to use lasers with different pulse widths for polishing metal components with different degrees of roughness.

In this paper, LPBF-ed 316L stainless steel with different initial surface roughness was polished using different pulse width lasers. The polishing efficiency, surface morphology, and phase composition achieved by the lasers with different pulse widths were analyzed. The cross-sectional microstructures and microhardness were investigated. In addition, the corrosion resistance characteristics of the laser-polished stainless steel surfaces were also evaluated. This work aims to achieve flexible and fast polishing of 3D-printed metal components with high roughness so that the prepared parts can meet the application requirements of most fields.

2. Experimental Procedures

2.1. Materials Preparation

The experimental raw material was commercially available 316L powder (Table 1). The powder morphology and size distribution are shown in Figure 1. The powder is spherical, which is favorable for LPBF forming (Figure 1a). The powder size range was 15–65 μm with 39.0 μm average diameter and 8.1 μm standard deviation (Figure 1b).

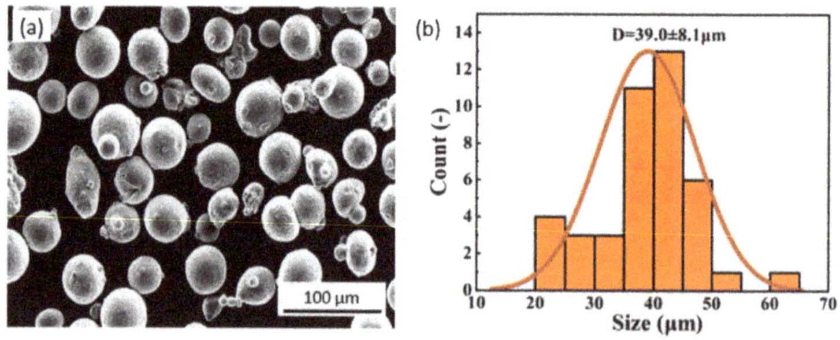

Figure 1. (**a**) SEM image and (**b**) size distribution of 316L stainless steel powders.

Table 1. Chemical composition of the 316L stainless steel powder.

Element	Ni	Cr	Mo	C	Mn	Si	Fe
Percent	10.72	16.96	2.44	0.01	0.73	0.51	balance

A batch of 15 × 15 × 15 mm^3 cubic 316L stainless steel was made by LPBF equipment (SLM-100, Han's Laser Co. Ltd., Shenzhen, China) under a nitrogen atmosphere. The LPBF process used a laser power of 180 W, a layer thickness of about 0.03 mm, and a scanning speed of 300 mm/s. The adjacent molding layers are scanned 67° apart in the direction of the LPBF process. Samples were cut from the substrate with an electric spark cutter and cleaned sequentially by sonication in ethanol and deionized water for 10 min to remove residual powder particles.

2.2. Laser Polishing Process

As shown in Figure 2b, the initial roughness of the LPBF-cube's side surface (SS) is 16.28 μm, and the top surface (TS) roughness is 8.12 μm. To study the influence on low-roughness surfaces by laser polishing, the SS was pretreated with 80-grit sandpaper, and the SR was reduced to 0.97 μm. Such a low roughness made the surface look shiny. The pretreated SS was labeled P-SS. The adopted processing parameters are shown in Table 2, and the laser polishing with an area of 5 × 5 mm^2 was carried out in an air environment (Figure 2). This study used the average surface roughness, Sa, to evaluate the laser polishing performance.

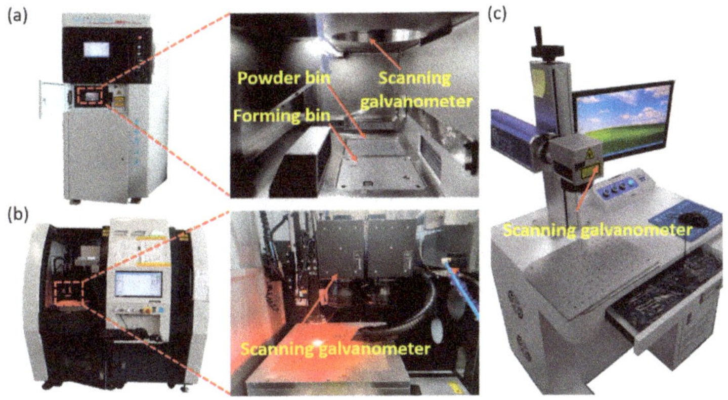

Figure 2. Actual images of the (**a**) CW, (**b**) FS, and (**c**) NS laser processing systems.

Table 2. Parameters adopted for laser polishing.

	CW Laser	NS Laser	FS Laser
Energy density (J/mm^2)	4.17	0.69	0.076
Scanning speed (mm/s)	300	100	500
Pulse frequency (kHz)	-	40	250
Number of Passes (-)	3	2	3

2.3. Microstructure and Mechanical Testing Method

The surface roughness of the samples was measured by a 3D laser confocal microscope (Olympus, OLS 5000), and the surface morphology of the samples was observed by scanning electron microscopy (SEM) (Carl Zeiss, GeminiSEM300, Oberkochen, Germany).

The crystal phase of the samples was identified using X-ray diffraction (XRD) (SmartLab, Tokyo, Japan).

The microhardness of polished LPBF-ed 316L stainless steel samples was achieved using a nanoindentation method. The nanoindentation was carried out at a distance of approximately 20 μm from the sample cross-section, with 20 μm intervals each time, and the experimental results were the average of 10 measurements. The test was performed with a 500 mN load.

The electrochemical corrosion behaviors of samples were measured in 3.5% NaCl solution through an electrochemical workstation using a three-electrode cell. The measured 316L stainless steel sample, a saturated calomel electrode (SCE), and a platinum foil were used as working, reference, and counter electrodes. Open circuit potential (OCP) measurements were performed for 1800 s to ensure that the working surface reached a relatively stable state, and potentiodynamic polarization curves were obtained on OCP in the range of -1.5 V to +1.5 V with a scan rate of 0.5 mV/s. The impedance spectrum ranges from 10^{-2}~10^{5} Hz with 10 mV AC signal amplitude. The following analytical tests were carried out on TS samples.

3. Results and Discussions

3.1. *Effect of Surface Roughness and Pulse Width on Laser Polishing Performance*

Figure 3a shows the polishing performances of the lasers with different pulse widths. For the P-SS with an initial Sa of 0.97 μm, the pulse width's effect is not apparent. Moreover, after laser polishing, the changes in roughness are also small. The FS laser polishing only reduces the Sa by 0.12 μm (Table 3), and similar results are also achieved using the CW and NS lasers. However, the CW laser's advantages come to the fore as the initial Sa increases. For the SS with the biggest Sa of 16.28 μm, the CW laser can quickly decrease the Sa to 1.15 μm, corresponding to a reduction of 92.9%. However, the Sa achieved by the NS laser is 3.11 μm, and the FS laser polished SS still possesses a Sa of 11.2 μm. Previous studies mainly focused on metal components with an initial roughness below 8 μm [15,20]. The maximum roughness of 16.28 μm is much higher than the previous research. With a high scanning speed of 300 mm/s, the CW laser can quickly reduce the roughness from 16.28 μm to about 1 μm. This reveals that the research result is very significant in polishing efficiency.

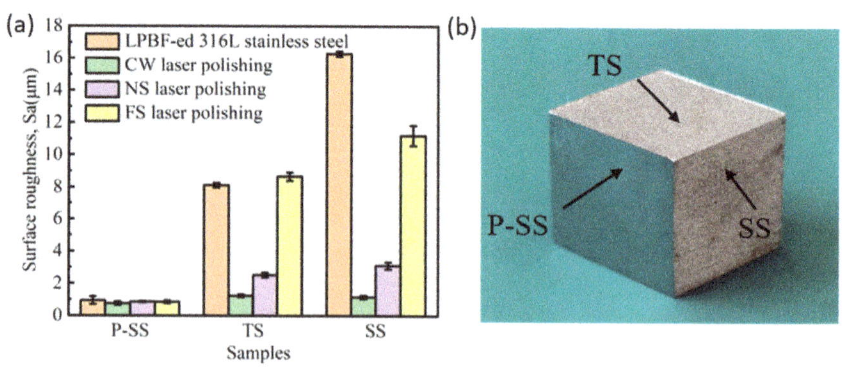

Figure 3. (**a**) Surface roughness of the polished samples with different laser pulse widths. (**b**) Object picture of the LPBF-ed 316L stainless steel.

Table 3. Surface roughness of the polished samples with different laser pulse widths.

Samples	Sa (μm)		
	P-SS	**TS**	**SS**
LPBF-ed 316L stainless steel	0.97	8.12	16.28
CW laser polishing	0.76	1.24	1.15
NS laser polishing	0.86	2.53	3.11
FS laser polishing	0.85	8.66	11.20

3.2. Effect of Laser Polishing on Microstructure

Figure 4 shows the surface morphologies of the samples before and after laser polishing. On the TS of the original LPBF-ed 316L stainless steel, some powder particles that are not entirely melted can still be observed (Figure 4a). Moreover, the laser-parallel-scanning-induced metal remelting is accompanied by sputtering. The above causes result in a ripple-like rough surface. The powder particles and ripples disappeared from the sample's surface after CW laser polishing (Figure 4b). This is attributed to the remelting of materials (Figure 5). The laser-melted material flows by gravity and then re-solidifies, flatting the rough surface. However, the thermal stresses inside the material during the cooling process cause cracks on the polished surface (Figure 4b).

Figure 4. Surface morphologies of the (**a**) original, (**b**) CW, (**c**) NS, and (**d**) FS laser-polished 316L stainless steel surfaces.

Compared to the CW laser-polished surface, the cracks become more obvious on the NS laser-polished surface (Figure 4b,c). This is because the NS laser has much less time to act on the material than the CW laser, resulting in a smaller heat conduction depth (Figure 4). The increased temperature difference along the normal direction of the material surface leads to an increase in thermal stress, making the crack more obvious. The FS laser polishing is based on removing convex materials (Figure 5). After three-laser scanning, the amount of material removal is very small (Figure 4d). Meanwhile, a new micro-nano structure is induced on the material surface, resulting in an insignificant reduction in roughness (Figure 3a). However, the cold processing effect caused by the ultrashort laser avoids the recast layer caused by the material melting, and the thermal-stress-induced microcracks are also inhibited (Figure 4d).

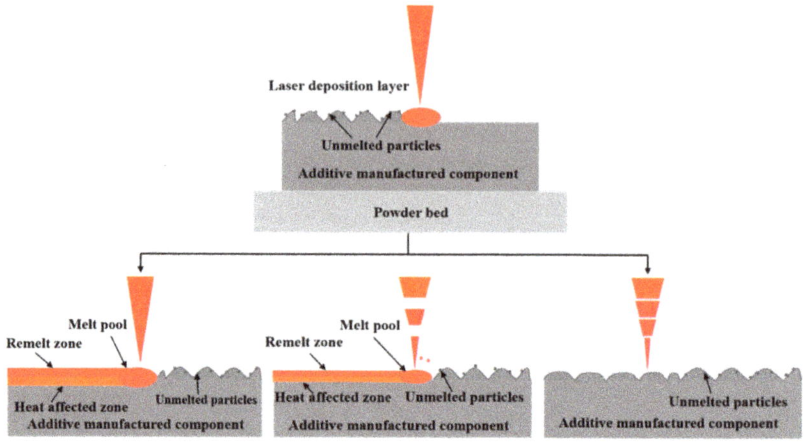

Figure 5. Schematic illustrating the laser polishing with different pulse widths.

3.3. *Surface XRD Analysis*

Figure 6 shows the surface XRD patterns of the LPBF-ed 316L stainless steel samples before and after laser polishing. It can be seen that on the original sample, diffraction peaks with different intensities appear at 43.8°, 51.1°, and 74.9°, corresponding to face-centered cubic austenite (111), (200), and (220) crystal faces, respectively. After surface polishing with the three kinds of laser, no new diffraction peaks can be observed. This indicates that no new phase is generated on the polished surface, regardless of the pulse width of the laser used. This may be attributed to the high nickel content in 316L stainless steel, and the nickel can inhibit the formation of other phases [22].

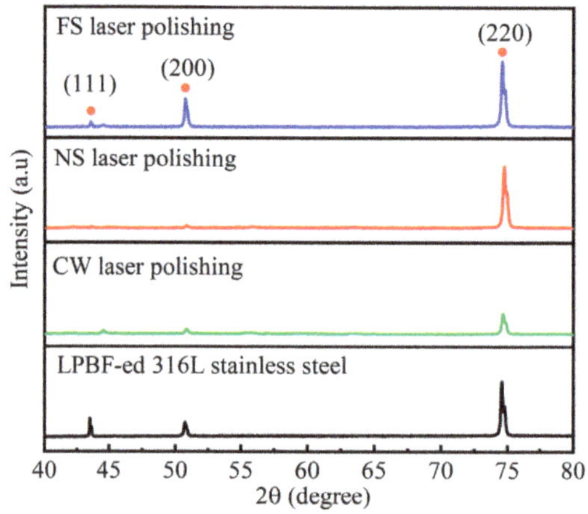

Figure 6. XRD patterns of the LPBF-ed 316L stainless steel before and after laser polishing.

3.4. *Cross-Sectional Microstructure*

Figure 7 exhibits the cross-sectional micrographs of original and laser-polished samples. The original LPBF-ed 316L stainless steel shows an uneven surface (Figure 7a). The semicircular metallographic organization caused by the unidirectional cooling of the melt pool is also clearly visible. After the CW laser polishing, the rough surface becomes flat, and a 49.08 μm remelted layer is formed (Figure 7b). Since the NS laser possesses a lower thermal effect, the remelted layer's thickness is decreased to 36.53 μm (Figure 7c). The FS laser polishing mainly depends on removing convex material (Figure 7d), so the improvement in surface roughness is limited. However, the cold processing effect caused by the ultrashort laser prevents the material from remelting.

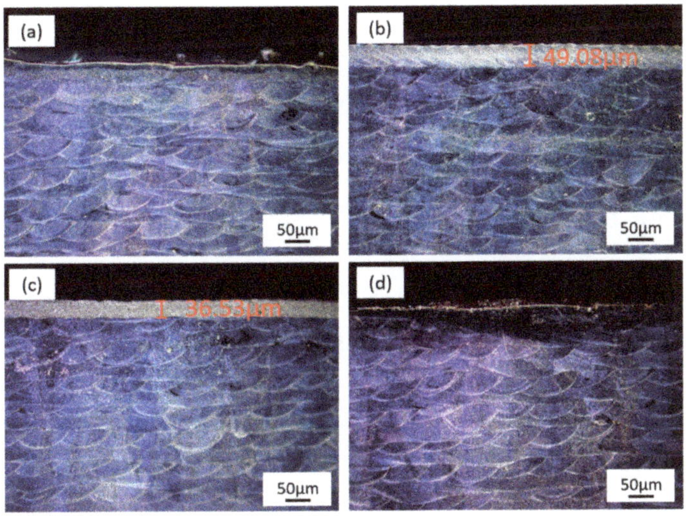

Figure 7. Cross-sectional micrographs of the (**a**) original, (**b**) CW, (**c**) NS, and (**d**) FS laser-polished 316L stainless steel surfaces.

3.5. Cross-Sectional Microhardness Measurement

To evaluate the impact of different pulse width lasers on the specimens along the depth direction, a microhardness measurement was performed on the sample's cross-sections. As shown in Figure 8, the average microhardness of the original LPBF-ed 316L stainless steel is about 3.11 Gpa, and the microhardness remained almost constant with increasing depth. The average hardness for the CW, NS, and FS laser-polished samples is 3.16 Gpa, 2.93 Gpa, and 3.04 Gpa, respectively. Figure 8a shows that the hardened layer of about 60 μm thickness is formed on the CW laser-polished sample. Related studies have shown that laser-polished additive manufactured parts produce gradient-hardened layers [25–27]. The cracks and thinned remelting layer for the NS laser-polished surface lead to decreased microhardness. Since there is no remelting layer on the FS laser-polished sample, the microhardness is almost constant along the depth direction.

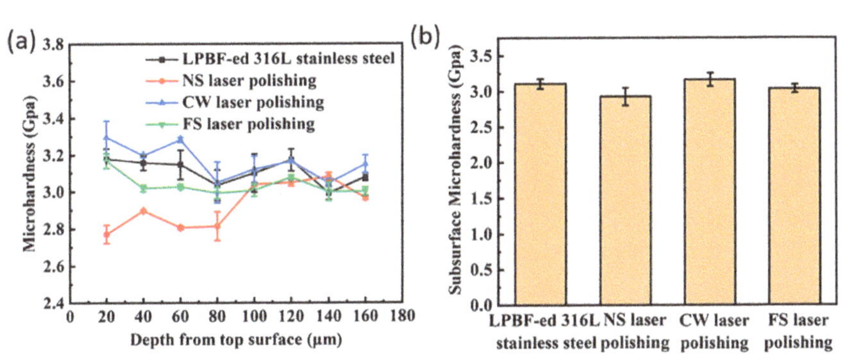

Figure 8. (**a**) Microhardness distributions and (**b**) average hardness of original and laser-polished samples.

3.6. Electrochemical Analysis

3.6.1. Potentiodynamic Polarization Studies

The polarization curves of original and laser-polished samples are shown in Figure 9, and Table 4 shows the corresponding corrosion currents and potentials calculated by the Tafel extrapolation method. The higher the corrosion potential of the sample, the better the corrosion resistance. Moreover, the corrosion current determines the corrosion rate [28]. Surface roughness is essential to the material's corrosion resistance [29]. The roughness reduction caused by CW laser polishing increases the corrosion potential and decreases the corrosion current. This indicates that the corrosion resistance is improved, and the corrosion rate is reduced [30]. The corrosion resistance of the NS laser-polished sample is not improved due to the surface cracks. During the electrochemical test, the electrolyte may penetrate the material along the cracks, resulting in corrosion inside the sample. The FS laser polishing does not obtain an improvement in surface roughness. Moreover, the ultrashort laser-induced micro-nanostructures increase the electrochemical reaction area, decreasing corrosion resistance. The polarization characteristics of the laser-polished P-SS and SS are consistent with the results shown in Figure 9 (Appendix A Figure A1). This indicates that the initial roughness does not affect the corrosion resistance of laser-polished samples with different pulse widths.

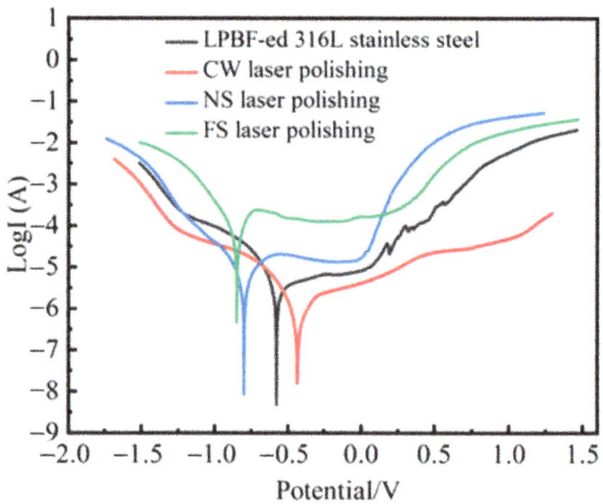

Figure 9. Potentiodynamic polarization curves of original and laser-polished samples.

Table 4. Quantitative information of four samples from the potentiodynamic polarization curves.

Samples	Corrosion Potential (V)	Corrosion Current (A)
LPBF-ed 316L stainless steel	−0.577	−7.181
CW laser polishing	−0.432	−7.722
NS laser polishing	−0.797	−6.909
FS laser polishing	−0.847	−6.038

3.6.2. Electrochemical Impedance Spectroscopic (EIS) Studies

The EIS spectra of the original and laser-polished samples are shown in Figure 10. Capacitive arcs can be observed on the Nyquist plots for all the samples (Figure 10a), which indicates that the corrosion reactions occurred at the stainless steel/electrolyte interface. The capacitive arc's radius is essential for assessing corrosion resistance [31,32]. A large radius usually indicates excellent corrosion resistance. As shown in Figure 10a, the CW laser-polished sample's radius is more significant than that of the other three samples, indicating the most vigorous corrosion resistance.

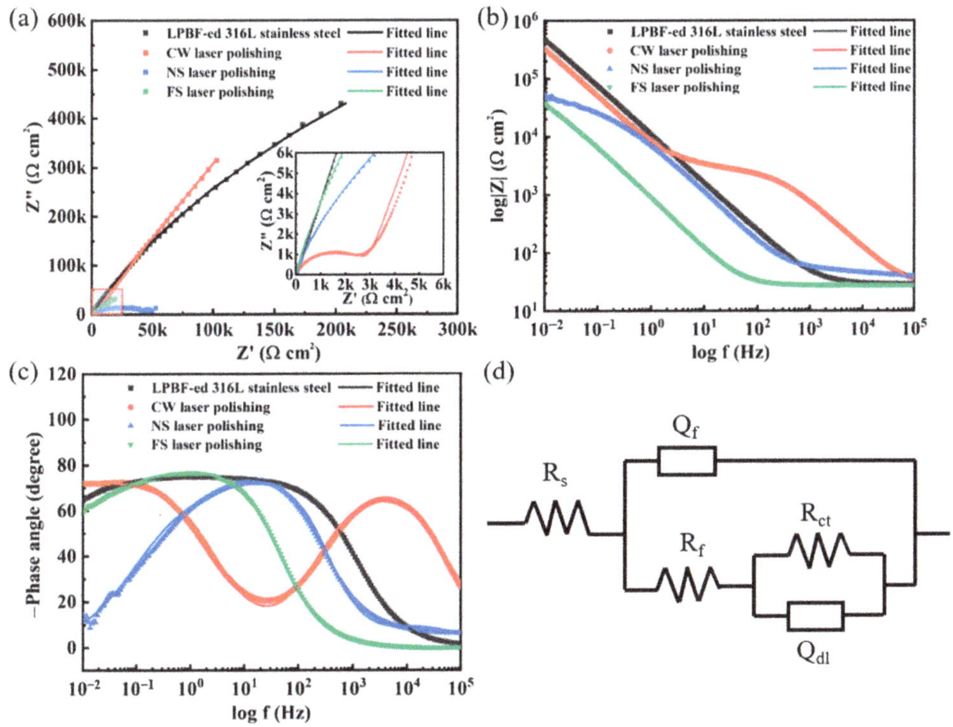

Figure 10. Measured and simulated (**a**) Nyquist curves, (**b**) Bode impedance, and (**c**) Bode phase angle of original and laser-polished samples. (**d**) Electrochemical equivalent circuit model.

To further analyze the corrosion mechanism, an equivalent circuit model, shown in Figure 10d, was used to fit the impedance data. R_s, R_f, and R_{ct} represent the resistances of electrolyte solution, passivation film, and charge transfer, respectively. Q_f and Q_{dl} represent the passivation film capacitance and the double-layer capacitance. The fitted EIS parameter is shown in Table 5. The chi-squared values (χ^2) ranged from 2.5×10^{-5} to 1.9×10^{-3}, indicating a good agreement between the EIS data and the fitting results.

Table 5. EIS fitting results of an equivalent circuit model for original and laser-polished samples.

Sample	R_S (Ω cm^2)	Q_f	R_f (Ω cm^2)	Q_{dL}	R_{ct} (Ω cm^2)	χ^2
LPBF-ed 316L stainless steel	29.14	8.67×10^{-6}	74.05	7.40×10^{-6}	4.13×10^{6}	2.5×10^{-5}
CW laser polishing	27	9.86×10^{-7}	2900	3.14×10^{-6}	4.5×10^{6}	9×10^{-4}
NS laser polishing	34.37	7.27×10^{-5}	25.15	3.20×10^{-5}	49,955	1.9×10^{-3}
FS laser polishing	24.53	3.91×10^{-5}	3.776	1.98×10^{-4}	554,950	1.0×10^{-4}

The corrosion rate r is inversely proportional to R_{ct} [33]:

$$R_{ct} \times r = K \quad (1)$$

Since K is a constant, R_{ct} depends on the charge transfer rate caused by the Faraday process of redox reactions occurring on the electrode surface. As such, R_{ct} can be used to evaluate the corrosion resistance. As can be seen in Table 5, the CW laser-polished surface with the highest R_{ct} of 4.5×10^6 Ω cm^2 shows a lower corrosion rate than the other three samples. In addition, the CW laser-polished sample also has the lowest Q_f and the highest R_f, which indicates a stable passivation film and high corrosion resistance [34,35].

3.6.3. Electrochemical Corrosion Morphology

Figure 11 shows the corrosion micromorphologies of four samples after the EIS tests. Some corrosion pits can be observed on the original surface of the LPBF-ed 316L stainless steel (Figure 11a). Compared to the original sample, the corrosion pits on the CW laser-polished specimen are significantly reduced. Only a few micro-sized corrosion pores can be observed (Figure 11b). The corrosion pits on the NS laser-polished sample are mainly due to the surface cracks (Figure 11c). A large number of bumps on the FS laser-polished sample leads to the incompleteness of the passivation film during the corrosion process. The NaCl solution tends to corrode the material interior along these surface defects, resulting in a severely corroded surface (Figure 11d).

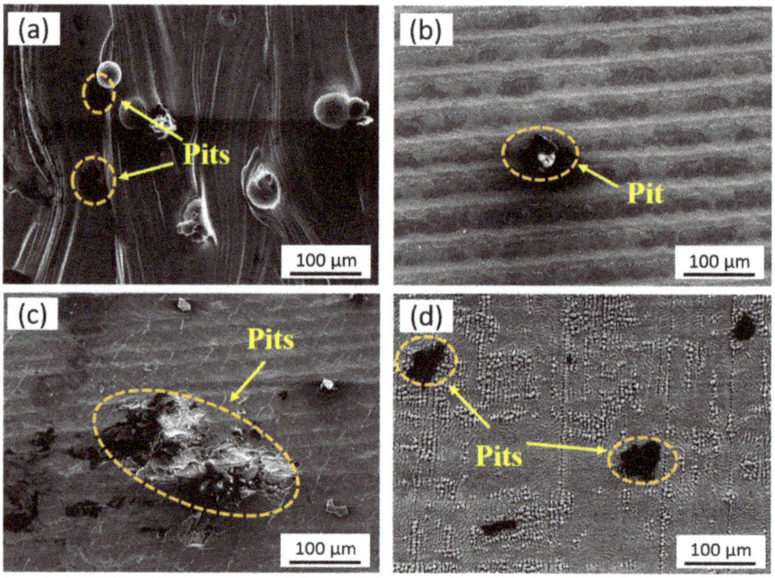

Figure 11. Surface corrosion morphologies of the (**a**) original, (**b**) CW, (**c**) NS, and (**d**) FS laser-polished 316L stainless steel samples.

3.6.4. Corrosion Mechanism

The corrosion mechanism of the original and laser-polished 316L stainless steel samples is shown in Figure 12. The oxidation reaction occurs at the anode:

$$Fe - 2e^- \rightarrow Fe^{2+} \tag{2}$$

$$Cr - 2e^- \rightarrow Cr^{2+} \tag{3}$$

The reduction reaction occurs at the cathode:

$$O_2 + 2H_2O + 4e^- \rightarrow 4OH^- \tag{4}$$

The metal cations produced by the anodic reaction readily react with Cl^- in the NaCl solution to form metal chlorides, resulting in the continued dissolution of the metal in the solution:

$$Cr^{2+} + 2Cl^- \rightarrow CrCl_2 \tag{5}$$

$$Fe^{2+} + 2Cl^- \rightarrow FeCl_2 \quad (6)$$

Next, the unstable $CrCl_2$ and $FeCl_2$ are easily transformed into Fe_2O_3 and Cr_2O_3 oxides due to the oxygen dissolved in the solution. The passivation film formed by the metal oxides can slow down the corrosion process of the material surface. However, the presence of surface microcracks leads to the erosion of Cl^- inside the material, promoting the formation of corrosion pits [36].

In summary, the rough structure on the original and FS laser-polished surface destroyed the integrity of the passivation film, leading to the formation of corrosion pits. The microcracks on the NS laser-polished sample also damaged the passivation film, resulting in a severely corroded surface. In contrast, the CW laser polishing could remove most of the defects on the LPBF-ed 316L stainless steel and obtain a flat surface, improving the corrosion resistance [37,38].

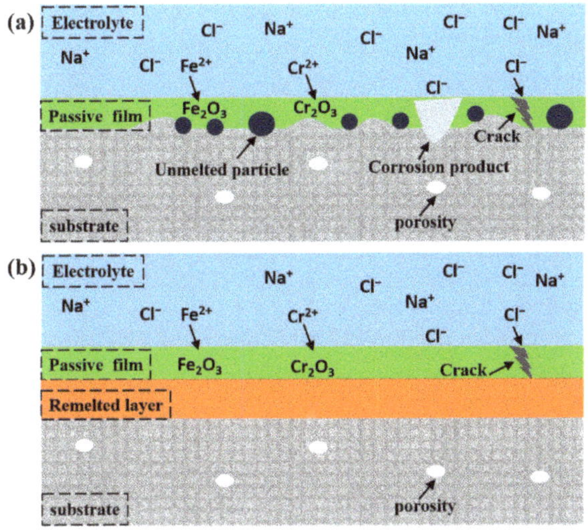

Figure 12. Corrosion mechanism diagram of the (**a**) original and (**b**) laser-polished 316L stainless steel.

4. Conclusions

In this study, the surface morphologies and corrosion behaviors of LPBF-ed 316L stainless steel polished with different laser pulse widths were investigated.

(1) Although the NS laser can also significantly reduce the surface roughness, the generated microcracks decrease the microhardness and corrosion resistance.

(2) The FS can avoid the formation of remelting layers, but its improvement in the roughness is very limited. Moreover, the ultrashort pulse laser-induced micro-nanostructures increase the contact area of the electrochemical reaction, resulting in a decrease in corrosion resistance.

(3) The demonstrated advantage of the CW laser over the FS and NS laser is decided by the initial roughness. The surface material's sufficient remelting realized by the CW laser can significantly improve the roughness. When the initial roughness of LPBF-ed stainless steel is higher than 10 μm, the CW laser can quickly reduce the surface roughness of LPBF-ed stainless steel to about 1 μm. Meanwhile, the surface hardness and corrosion can also be improved. The research results suggest that CW laser polishing presents promising applications in the pretreatment of LPBF-ed parts for ocean engineering, mechanical transmission, and punching die. However, compared to electrochemical polishing, the roughness of the laser-polished surface is still high. For applications requiring a surface

roughness below 500 nm, the laser-polished surface must be treated using chemical and mechanical methods to further reduce the roughness.

Author Contributions: Conceptualization, H.Y. and Y.C.; methodology, J.L. and J.C.; validation, S.M., J.L. and D.O.; formal analysis, J.L. and H.M.; investigation, S.R. and L.M.; writing—original draft preparation, J.L.; writing—review and editing, H.Y. and Y.C.; supervision, C.Y. and L.D.; funding acquisition, H.Y. All authors have read and agreed to the published version of the manuscript.

Funding: This research was funded by Zhejiang Provincial Natural Science Funds of China (LZ20E050003); the Natural Science Foundation of Top Talent of SZTU (2020103); Natural Science Foundation of Top Talent of Shenzhen Technology University, SZTU (2019010801005); Shenzhen key Project for Technology Development (JSGG20191129105838333); The Open Project Program of Wuhan National Laboratory for Optoelectronics (2022WNLOKF015).

Data Availability Statement: Not applicable.

Acknowledgments: The authors would like to thank the Laboratory of Advanced Additive Manufacturing, Sino-German College of Intelligent Manufacturing in SZTU, and the Analytical and Testing Centre of JNU for SEM and XRD.

Conflicts of Interest: The authors declare no conflict of interest.

Appendix A

Appendix A Figure A1 shows the polarization curves of initial and laser-polished P-SS and SS samples. The corresponding corrosion current and potential are listed in Tables A1 and A2, respectively. The experimental results show that the influence of laser pulse width on the polarization characteristics of P-SS and SS samples is consistent with the results shown in Figure 8. The CW laser-polished surfaces possess the best corrosion resistance. This indicates that the initial roughness does not affect the corrosion resistance of laser-polished samples with different pulse widths.

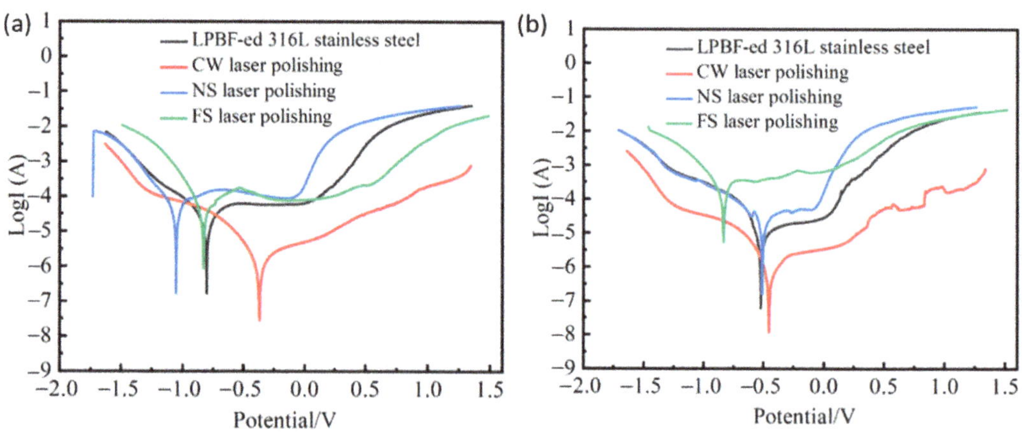

Figure A1. Comparison of polarization curves of initial and laser-polished (**a**) P-SS and (**b**) SS.

Table A1. Quantitative information of P-SS samples before and after laser polishing.

	Corrosion Potential (V)	Corrosion Current (A)
LPBF-ed 316L stainless steel	−0.805	−6.168
CW laser polishing	−0.371	−7.273
NS laser polishing	−1.054	−5.752
FS laser polishing	−0.836	−5.398

Table A2. Quantitative information of SS samples before and after laser polishing.

	Corrosion Potential (V)	Corrosion Current (A)
LPBF-ed 316L stainless steel	−0.522	−6.793
CW laser polishing	−0.453	−7.721
NS laser polishing	−0.688	−6.522
FS laser polishing	−0.930	−5.009

Figure A2 shows the influence of laser energy density, scanning speed, and processing times on the roughness of laser-polished surfaces. As shown in Appendix A Figure A2a, when the scanning speed of the CW laser was fixed at 300 mm/s, the surface roughness could decrease to 3.2 μm with the laser energy density ranging from 2.53 J/cm^2 to 5.82 J/cm^2. An energy density outside this range resulted in an increase in the surface roughness. With the optimized laser energy density of 4.17 J/cm^2, the optimal laser scanning of 300 mm/s could also be obtained (Figure A2a). The repetition frequencies of the FS laser and NS laser were fixed at 250 kHz and 40 kHz, respectively. Figure A2b shows that the optimized energy density and scanning speed for the NS laser polishing are 0.69 J/cm^2 and 100 mm/s. The optimal parameter combination for the FS laser polishing was an energy density of 0.076 J/cm^2 and a scanning speed of 500 mm/s (Figure A2c). Figure A2d shows that the optimal processing times for the CW, NS, and FS laser polishing are 3, 2, and 3, respectively.

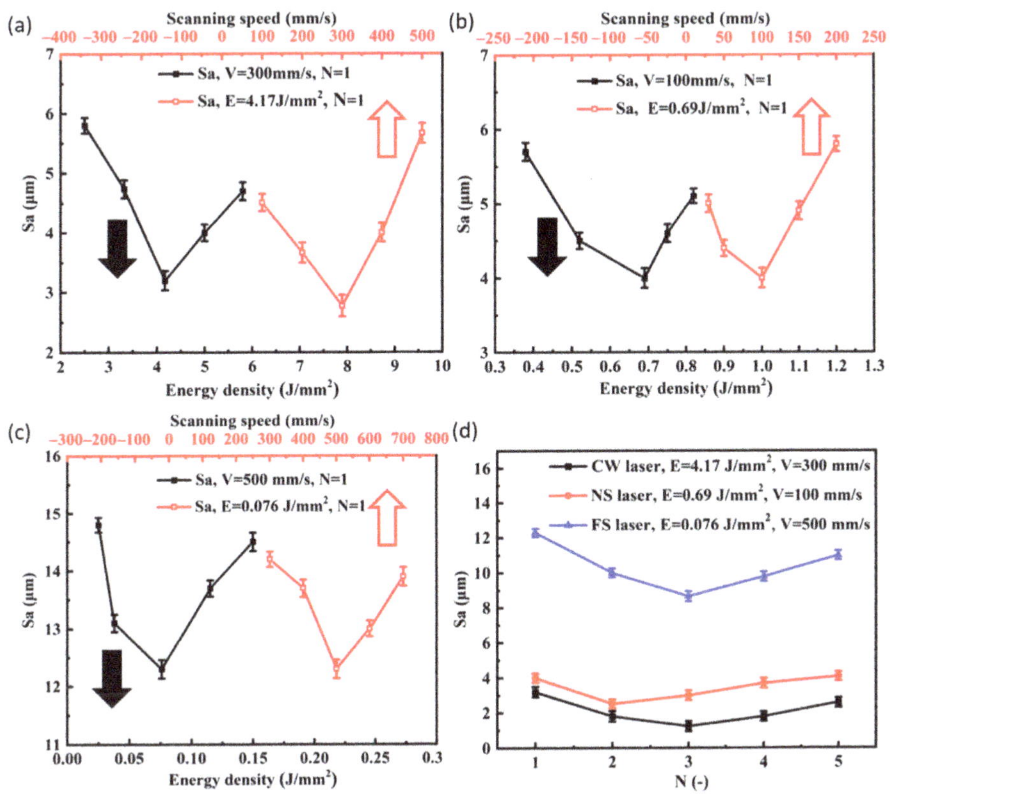

Figure A2. The influence of laser energy density and scanning speed on the roughness of (**a**) CW, (**b**) NS, and (**c**) FS laser-polished LPBF-ed stainless steel surfaces. (**d**) The influence of processing times on the roughness of laser-polished surfaces.

References

1. Sames, W.J.; List, F.A.; Pannala, S.; Dehoff, R.R.; Babu, S.S. The metallurgy and processing science of metal additive manufacturing. *Int. Mater. Rev.* **2016**, *61*, 315–360. [CrossRef]
2. Herzog, D.; Seyda, V.; Wycisk, E.; Emmelmann, C. Additive manufacturing of metals. *Acta Mater.* **2016**, *117*, 371–392. [CrossRef]
3. Khorasani, M.; Gibson, I.; Ghasemi, A.H.; Hadavi, E.; Rolfe, B. Laser subtractive and laser powder bed fusion of metals: Review of process and production features. *Rapid Prototyp. J.* 2023, *ahead-of-print*. [CrossRef]
4. Han, C.; Li, Y.; Wang, Q.; Wen, S.; Wei, Q.; Yan, C.; Hao, L.; Liu, J.; Shi, Y. Continuous functionally graded porous titanium scaffolds manufactured by selective laser melting for bone implants. *J. Mech. Behav. Biomed. Mater.* **2018**, *80*, 119–127. [CrossRef]
5. Benedetti, M.; Fontanari, V.; Bandini, M.; Zanini, F.; Carmignato, S. Low- and high-cycle fatigue resistance of Ti-6Al-4V ELI additively manufactured via selective laser melting: Mean stress and defect sensitivity. *Int. J. Fatigue* **2018**, *107*, 96–109. [CrossRef]
6. Kong, D.; Ni, X.; Dong, C.; Lei, X.; Zhang, L.; Man, C.; Yao, J.; Cheng, X.; Li, X. Bio-functional and anti-corrosive 3D printing 316L stainless steel fabricated by selective laser melting. *Mater. Des.* **2018**, *152*, 88–101. [CrossRef]
7. Mohd Yusuf, S.; Nie, M.; Chen, Y.; Yang, S.; Gao, N. Microstructure and Corrosion Performance of 316L stainless steel fabricated by Selective Laser Melting and processed through high-pressure torsion. *J. Alloys Compd.* **2018**, *763*, 360–375. [CrossRef]
8. Chen, Q.; Jing, Y.; Yin, J.; Li, Z.; Xiong, W.; Gong, P.; Zhang, L.; Li, S.; Pan, R.; Zhao, X.; et al. High Reflectivity and Thermal Conductivity Ag-Cu Multi-Material Structures Fabricated via Laser Powder Bed Fusion: Formation Mechanisms, Interfacial Characteristics, and Molten Pool Behavior. *Micromachines* **2023**, *14*, 362. [CrossRef]
9. Afazov, S.; Serjouei, A.; Hickman, G.J.; Mahal, R.; Goy, D.; Mitchell, I. Defect-based fatigue model for additive manufacturing. *Prog. Addit. Manuf.* **2022**. [CrossRef]
10. Mostafaei, A.; Zhao, C.; He, Y.; Reza Ghiaasiaan, S.; Shi, B.; Shao, S.; Shamsaei, N.; Wu, Z.; Kouraytem, N.; Sun, T.; et al. Defects and anomalies in powder bed fusion metal additive manufacturing. *Curr. Opin. Solid State Mater. Sci.* **2022**, *26*, 100974. [CrossRef]
11. Li, Z.; Li, H.; Yin, J.; Li, Y.; Nie, Z.; Li, X.; You, D.; Guan, K.; Duan, W.; Cao, L.; et al. A Review of Spatter in Laser Powder Bed Fusion Additive Manufacturing: In Situ Detection, Generation, Effects, and Countermeasures. *Micromachines* **2022**, *13*, 1366. [CrossRef]
12. Yin, J.; Zhang, W.; Ke, L.; Wei, H.; Wang, D.; Yang, L.; Zhu, H.; Dong, P.; Wang, G.; Zeng, X. Vaporization of alloying elements and explosion behavior during laser powder bed fusion of Cu–10Zn alloy. *Int. J. Mach. Tools Manuf.* **2021**, *161*, 103686. [CrossRef]
13. Kempen, K.; Yasa, E.; Thijs, L.; Kruth, J.P.; Van Humbeeck, J. Microstructure and mechanical properties of Selective Laser Melted 18Ni-300 steel. *Phys. Procedia* **2011**, *12*, 255–263. [CrossRef]
14. Lee, H.; Lim, C.H.J.; Low, M.J.; Tham, N.; Murukeshan, V.M.; Kim, Y.-J. Lasers in additive manufacturing: A review. *Int. J. Precis. Eng. Manuf.-Green Technol.* **2017**, *4*, 307–322. [CrossRef]
15. Schmidt, M.; Merklein, M.; Bourell, D.; Dimitrov, D.; Hausotte, T.; Wegener, K.; Overmeyer, L.; Vollertsen, F.; Levy, G.N. Laser based additive manufacturing in industry and academia. *CIRP Ann.* **2017**, *66*, 561–583. [CrossRef]
16. Marimuthu, S.; Triantaphyllou, A.; Antar, M.; Wimpenny, D.; Morton, H.; Beard, M. Laser polishing of selective laser melted components. *Int. J. Mach. Tools Manuf.* **2015**, *95*, 97–104. [CrossRef]
17. Lambarri, J.; Leunda, J.; Soriano, C.; Sanz, C. Laser Surface Smoothing of Nickel-based Superalloys. *Phys. Procedia* **2013**, *41*, 255–265. [CrossRef]
18. Pfefferkorn, F.E.; Duffie, N.A.; Li, X.; Vadali, M.; Ma, C. Improving surface finish in pulsed laser micro polishing using thermocapillary flow. *CIRP Ann.* **2013**, *62*, 203–206. [CrossRef]
19. Xu, Z.; Ouyang, W.; Jia, S.; Jiao, J.; Zhang, M.; Zhang, W. Cracks Repairing by Using Laser Additive and Subtractive Hybrid Manufacturing Technology. *J. Manuf. Sci. Eng.* **2020**, *142*, 031006. [CrossRef]
20. Zhou, J.; Han, X.; Li, H.; Liu, S.; Shen, S.; Zhou, X.; Zhang, D. In-Situ Laser Polishing Additive Manufactured AlSi10Mg: Effect of Laser Polishing Strategy on Surface Morphology, Roughness and Microhardness. *Materials* **2021**, *14*, 393. [CrossRef]
21. Xiao, H.; Chen, Y.; Liu, M.; Zhou, Y.; Du, C.; Zhang, W. Influence of Laser Additive Manufacturing and Laser Polishing on Microstructures and Mechanical Properties of High-Strength Maraging Steel Metal Materials. *Appl. Sci.* **2022**, *12*, 10340. [CrossRef]
22. Chen, L.; Richter, B.; Zhang, X.; Ren, X.; Pfefferkorn, F.E. Modification of surface characteristics and electrochemical corrosion behavior of laser powder bed fused stainless-steel 316L after laser polishing. *Addit. Manuf.* **2020**, *32*, 101013. [CrossRef]
23. Jaritngam, P.; Tangwarodomnukun, V.; Qi, H.; Dumkum, C. Surface and subsurface characteristics of laser polished Ti6Al4V titanium alloy. *Opt. Laser Technol.* **2020**, *126*, 106102. [CrossRef]
24. Hafiz, A.M.K.; Bordatchev, E.V.; Tutunea-Fatan, R.O. Experimental analysis of applicability of a picosecond laser for micro-polishing of micromilled Inconel 718 superalloy. *Int. J. Adv. Manuf. Technol.* **2013**, *70*, 1963–1978. [CrossRef]
25. Bhaduri, D.; Penchev, P.; Batal, A.; Dimov, S.; Soo, S.L.; Sten, S.; Harrysson, U.; Zhang, Z.; Dong, H. Laser polishing of 3D printed mesoscale components. *Appl. Surf. Sci.* **2017**, *405*, 29–46. [CrossRef]
26. Zhihao, F.; Libin, L.; Longfei, C.; Yingchun, G. Laser Polishing of Additive Manufactured Superalloy. *Procedia CIRP* **2018**, *71*, 150–154. [CrossRef]
27. Chen, L.; Richter, B.; Zhang, X.; Bertsch, K.B.; Thoma, D.J.; Pfefferkorn, F.E. Effect of laser polishing on the microstructure and mechanical properties of stainless steel 316L fabricated by laser powder bed fusion. *Mater. Sci. Eng. A Struct. Mater. Prop. Microstruct. Process.* **2021**, *802*, 140579. [CrossRef]

28. Zietala, M.; Durejko, T.; Polanski, M.; Kunce, I.; Plocinski, T.; Zielinski, W.; Lazinska, M.; Stepniowski, W.; Czujko, T.; Kurzydlowski, K.J.; et al. The microstructure, mechanical properties and corrosion resistance of 316 L stainless steel fabricated using laser engineered net shaping. *Mater. Sci. Eng. A Struct. Mater. Prop. Microstruct. Process.* **2016**, *677*, 1. [CrossRef]
29. Walter, R.; Kannan, M.B. Influence of surface roughness on the corrosion behaviour of magnesium alloy. *Mater. Eng.* **2011**, *32*, 2350–2354. [CrossRef]
30. Xu, Z.; Ouyang, W.; Liu, Y.; Jiao, J.; Liu, Y.; Zhang, W. Effects of laser polishing on surface morphology and mechanical properties of additive manufactured TiAl components. *J. Manuf. Process.* **2021**, *65*, 51–59. [CrossRef]
31. Ribeiro, A.M.; Alves, A.C.; Rocha, L.A.; Silva, F.S.; Toptan, F. Synergism between corrosion and wear on CoCrMo−Al_2O_3 biocomposites in a physiological solution. *Tribol. Int.* **2015**, *91*, 198–205. [CrossRef]
32. Sulima, I.; Kowalik, R.; Hyjek, P. The corrosion and mechanical properties of spark plasma sintered composites reinforced with titanium diboride. *J. Alloys Compd.* **2016**, *688*, 1195–1205. [CrossRef]
33. Cao, M.; Liu, L.; Yu, Z.; Fan, L.; Li, Y.; Wang, F. Electrochemical corrosion behavior of 2A02 Al alloy under an accelerated simulation marine atmospheric environment. *J. Mater. Sci. Technol.* **2019**, *35*, 651–659. [CrossRef]
34. Meng, L.J.; Long, J.Z.; Yang, H.; Shen, W.J.; Li, C.B.; Yang, C.; Wang, M.; Li, J.M. Femtosecond Laser Treatment for Improving the Corrosion Resistance of Selective Laser Melted 17-4PH Stainless Steel. *Micromachines* **2022**, *13*, 1089. [CrossRef]
35. Gu, D.D.; Zhang, H.; Dai, D.H.; Ma, C.L.; Zhang, H.M.; Li, Y.X.; Li, S.H. Anisotropic corrosion behavior of Sc and Zr modified Al-Mg alloy produced by selective laser melting. *Corros. Sci.* **2020**, *170*, 108657. [CrossRef]
36. Prabhakaran, S.; Kulkarni, A.; Vasanth, G.; Kalainathan, S.; Shukla, P.; Vasudevan, V.K. Laser shock peening without coating induced residual stress distribution, wettability characteristics and enhanced pitting corrosion resistance of austenitic stainless steel. *Appl. Surf. Sci.* **2018**, *428*, 17–30. [CrossRef]
37. Hong, T.; Nagumo, M. Effect of Surface roughness on early stages of pitting corrosion of Type 301 stainless steel. *Corros. Sci.* **1997**, *39*, 1665–1672. [CrossRef]
38. Burstein, G.T.; Pistorius, P.C. Surface Roughness and the Metastable Pitting of Stainless Steel in Chloride Solutions. *Corrosion* **1995**, *51*, 380–385. [CrossRef]

micromachines

MDPI

Article

Influence of Aging Treatment Regimes on Microstructure and Mechanical Properties of Selective Laser Melted 17-4 PH Steel

Dongdong Dong [1,2], Jiang Wang [3,*], Chaoyue Chen [3], Xuchang Tang [3], Yun Ye [2,*], Zhongming Ren [3], Shuo Yin [4], Zhenyu Yuan [5], Min Liu [2] and Kesong Zhou [1,2]

1 School of Materials and Energy, Guangdong University of Technology, Guangzhou 510006, China; kszhou2004@163.com (K.Z.)
2 National Engineering Laboratory of Modern Materials Surface Engineering Technology, Guangdong Provincial Key Laboratory of Modern Surface Engineering Technology, Institute of New Materials, Guangdong Academy of Sciences, Guangzhou 510651, China
3 State Key Laboratory of Advanced Special Steels, School of Materials Science and Engineering, Shanghai University, Shanghai 200444, China; cchen1@shu.edu.cn (C.C.)
4 Department of Mechanical and Manufacturing Engineering, Trinity College Dublin, University of Dublin, Parsons Building, D02 PN40 Dublin, Ireland
5 Wuhu Tianhang Equipment Technology Co., Ltd., Wuhu 241000, China
* Correspondence: jiangwang@i.shu.edu.cn (J.W.); yeyun@gdinm.com (Y.Y.)

Abstract: Aging is indispensable for balancing the strength and ductility of selective laser melted (SLM) precipitation hardening steels. This work investigated the influence of aging temperature and time on the microstructure and mechanical properties of SLM 17-4 PH steel. The 17-4 PH steel was fabricated by SLM under a protective argon atmosphere (99.99 vol.%), then the microstructure and phase composition after different aging treatments were characterized via different advanced material characterization techniques, and the mechanical properties were systematically compared. Coarse martensite laths were observed in the aged samples compared with the as-built ones, regardless of the aging time and temperature. Increasing the aging temperature resulted in a larger grain size of the martensite lath and precipitation. The aging treatment induced the formation of the austenite phase with a face-centered cubic (FCC) structure. With prolonged aging treatment, the volume fraction of the austenite phase increased, which agreed with the EBSD phase mappings. The ultimate tensile strength (UTS) and yield strength gradually increased with increasing aging times at 482 °C. The UTS reached its peak value after aging for 3 h at 482 °C, which was similar to the trend of microhardness (i.e., UTS = 1353.4 MPa). However, the ductility of the SLM 17-4 PH steel decreased rapidly after aging treatment. This work reveals the influence of heat treatment on SLM 17-4 steel and proposes an optimal heat-treatment regime for the SLM high-performance steels.

Keywords: selective laser melting; 17-4 PH steel; aging treatment regimes; microstructural evolution; mechanical properties

Citation: Dong, D.; Wang, J.; Chen, C.; Tang, X.; Ye, Y.; Ren, Z.; Yin, S.; Yuan, Z.; Liu, M.; Zhou, K. Influence of Aging Treatment Regimes on Microstructure and Mechanical Properties of Selective Laser Melted 17-4 PH Steel. *Micromachines* **2023**, *14*, 871. https://doi.org/10.3390/mi14040871

Academic Editor: Francesco Ruffino

Received: 7 March 2023
Revised: 2 April 2023
Accepted: 5 April 2023
Published: 18 April 2023

1. Introduction

As a typical additive manufacturing technique, selective laser melting (SLM) has gained great attention due to its unique advantages in fabricating sophisticated metallic parts [1]. Based on the powder bed fusion principle, a high-energy laser beam is used in the SLM technique to melt the powder feedstocks in a layer-by-layer sequence. Unlike conventional manufacturing methods, due to the micro-sized laser spot and fine particles, the SLM technique can fabricate metallic components with extremely high precision and geometrical complexity [2]. Additionally, the high-energy laser beam induces an ultra-high cooling rate during rapid solidification compared to conventional manufacturing techniques [3]. The laser absorptivity of the materials determines the quality of the components to a large extent [4]. Thus, the SLM-produced metallic components are commonly featured as equivalent or better mechanical properties [5]. To date, the SLM technique has been applied to the manufacturing of various kinds of metallic and composite materials, such as Ti alloys [6], Ni-based superalloys [7], Al-based alloys [8,9], Mg-based alloys [10,11], and Fe-based alloys [12].

Among all the steels, 17-4 PH (precipitation hardened) steel possesses an outstanding combination of high strength, excellent corrosion resistance, and favorable toughness [13–15]. Thus, 17-4 PH steel has been widely used in the tooling, molding, aerospace, and automotive industries [16–18]. Moreover, 17-4 PH steel is an excellent candidate for the SLM process due to its good weldability and wettability. Recently, some studies reported that the 17-4 PH steel fabricated by SLM achieved excellent mechanical properties by using optimal processing parameters [19,20]. Averyanova et al. [21] investigated the effect of the main processing parameters of SLM-fabricated 17-4 PH steel under the guidance of a fractional factorial approach. Although a complex function considering physical and geometrical parameters has been proposed, it merely describes the shapes of the fused tracks, while the mechanical properties based on the process parameters window were not reported. To further investigate the impacts of laser mode on microstructure features and mechanical properties, Ozsoy et al. [22] studied the impact of unique factors associated with pulsed laser processing using a design of experiments method. They successfully obtained SLM 17-4 PH steel samples with good mechanical performances (ultimate tensile strength = 899 MPa, elongation = 22.3%). However, the mechanical properties of 17-4 PH steel are still not adequate for some load-bearing applications.

Generally, the cyclic heating of deposited layers by subsequent laser movement in adjacent tracks and layers may trigger the precipitation of nano-scaled particles within the martensite matrix during the SLM process [23]. SLM-manufactured 17-4 PH steels are characterized by ultra-fine grains and a quantity of nanoparticles due to the intrinsic heat treatment (IHT) [24,25]. This IHT effect may also result in excellent strength for precipitation hardening steels such as 17-4 PH or maraging steel 300 (MS300) in the as-built condition [26]. However, limited improvement in the IHT effect means that SLM-manufactured 17-4 PH steels in as-built conditions cannot reach the standard of mechanical properties due to insufficient precipitation time.

Thus, post heat-treatments, such as solution and aging processes, are indispensable for 17-4 PH steels. The highly dispersed nano-scale soft Cu-rich particles induced by the post-heat treatments significantly improve the steel strength via the Orowan strengthening mechanism [27]. Many studies have demonstrated the significance of aging treatments for conventionally manufactured 17-4 PH steels. As such, inspired by the traditional heat treatments, heat-treated 17-4 PH steels are gradually becoming popular within the SLM community. Rafi et al. [28] reported that post-heat treatments such as solution and aging processes ameliorate the mechanical properties of SLM 17-4 PH steel. However, suitable post-heat treatments for SLM 17-4 PH steel samples were not summarized. Pasebani et al. [29] found that a high-temperature solution treatment would deteriorate the mechanical performance due to the coarsened martensite microstructure.

Most studies have focused on the effects of powder preparation methods, volumetric energy densities, and solution treatments on the microstructure and mechanical performance of 17-4 PH steel. However, systematic investigations on the direct aging treatment parameters for SLM 17-4 PH steels have not been fully completed and clearly understood. To fill this research gap, microstructural evolutions of SLM 17-4 PH steels treated at different aging temperatures and times were carefully characterized to reveal the influence of aging. The corresponding mechanical properties were also studied and analyzed based on microstructure features, especially the precipitation behavior.

2. Experimental Details

In this work, spherical 17-4 PH steel powder (Figure 1a) ranging from 30 μm to 40 μm was used as the feedstocks for SLM fabrication. The chemical composition is presented in Table 1. The 17-4 PH steel samples were fabricated using an SLM apparatus of three-dimensional System ProX-200, which was equipped with a fiber laser with a spot diameter of 75 μm, a maximum power of 300 W, and a wavelength of 1070 nm. The SLM fabrication process was conducted in a sealed chamber filled with a protective argon atmosphere (99.99 vol.%). During the SLM, the oxygen content of the atmosphere was controlled to below 200 ppm to avoid oxidation. Prior to the SLM fabrication, the feedstock powders were dried in a vacuum oven at 180 °C for 5 h. The SLM samples were fabricated using the optimal processing parameters to obtain the best relative density (i.e., 99.8 ± 0.12%), as determined by previous researches: a laser power of 200 W, a laser scanning speed of 300 mm/s, a hatch distance of 70 μm, and a layer thickness of 30 μm. Cubic samples with dimensions of 10 mm × 10 mm × 10 mm were fabricated for microhardness measurements and microstructure characterizations. Tensile samples were produced to investigate of the mechanical properties. A scheme of the tensile test specimens is presented in Figure 1b.

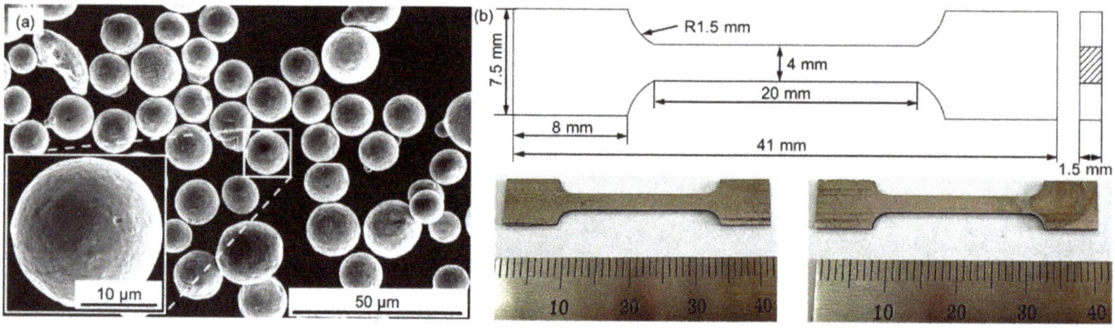

Figure 1. (**a**) Surface morphology of the 17-4PH steel powder; (**b**) scheme of the tensile test specimens used for the SLM fabrication.

Table 1. Chemical composition of the 17-4 PH steel powder (provided by EOS GmbH).

Elements	C	Cr	Ni	Cu	Mo	Nb	Mn	Si	P	S	Fe
Weight percentage (wt.%)	≤0.07	15.0–17.0	3.0–5.0	3.0–5.0	≤0.6	≤0.45	≤1.5	≤0.7	≤0.04	≤0.03	Bal.

Many researchers simply recommend an aging temperature and time of 482 °C and 1 h without providing any reason [28–30]. However, it is important to explore the effects of aging parameters on SLM 17-4PH steels. Therefore, different aging temperatures and times ranging from low to high levels were selected for this study. Following SLM manufacturing of the 17-4PH steel samples, direct aging treatments were carried out under various temperatures (from 382 °C to 682 °C) and duration times (from 0.5 to 5 h). These samples are denoted as X °C-Y h, indicating that the samples were aged at X °C for Y h (i.e., 382 °C-2 h representing 382 °C for 2 h).

To determine the phase composition, X-ray diffraction (XRD, Bruker D8 Advance) characterization was performed on the SLM-manufactured 17-4PH steel samples, targeting Cu Kα at a voltage of 40 kV and a current of 35 mA. The XRD spectra were collected by scanning 2θ angles ranging from 30° to 60° with a scanning step size of 0.02°. For microstructure observation, the SLM-manufactured 17-4PH steel samples were polished following the standard metallographic process and then etched with a solution of 5 g $FeCl_3$, 20 mL HCl, and 100 mL alcohol. The microstructure evolution of the SLM 17-4PH steel samples was investigated along the cross-sections of the building direction using optical microscopy (OM, OLYMPUS-DP71, Tokyo, Japan) and scanning electron microscopy (SEM, FEI Quenta 450, Portland, OR, USA). The electron-backscattered diffraction (EBSD, TSL-OIM) characterization was conducted on the SLM 17-4PH samples to determine the grain size and crystal orientation. The EBSD scanning was performed at a step size of 70 nm to ensure the adequate resolution of the grains.

Tensile tests of the SLM samples were conducted at a strain rate of 1 mm/min via a universal testing system (INSTRON 5982, Boston, MA, USA). A total of 4 tensile samples were tested in each state of the 17-4 PH steel. The maximum force applied by the universal testing machine Instron 5982 was 100 kN. All sample surfaces before the tests were ground (Ra $\leq$ 3.2 μm) to avoid measurement errors caused by poor or uneven surfaces. The Vickers microhardness of the SLM-fabricated 17-4 PH steel samples was tested at a load of 500 gf and a dwelling time of 15 s on a microhardness tester (Struers DuranScan 70 G5, Kuchl, Austria). The microhardness values of each sample were averaged over 10 tests. To ensure repitablity and accurancy of the test results, the SLM 17-4PH steel samples before and after different heat treatments were divided into nine groups with five cubic samples for phase detection, microstructural observations, and microhardness measurements, and four tensile samples for mechanical property tests. Please note that only typical experimental results were reported within this study.

3. Results and Discussion

3.1. Microstructural Evolution in As-Built Condition

Figure 2 presents the microstructure features and phase distribution of the as-built 17-4 PH steel along the building direction. The as-built samples were crack-free, and only a few tiny pores (white arrow) were observed in Figure 2a. Block structures (red arrow) consisting of numerous equiaxed grains (blue arrow) were the main features of the microstructure of as-built samples. Combined with the orientation map and phase distribution map (Figure 2b,c), each packet (purple arrows in Figure 2b) included several blocks that were composed of martensite laths with different misorientations. Moreover, a typical epitaxial growth of the coarse columnar structures (black arrows in Figure 2b) dominated the microstructure of the melt pool boundary of the as-built samples. This solidification mode was driven by the ratio of the thermal gradient (G) and cooling rate (R). Specifically, a large ΔG and a low R existed at the molten pool boundaries near the solidified components, leading to the growth of columnar dendrites perpendicular to the boundaries of the molten pool.

Interestingly, a quantity of ultra-fine austenite phase (mainly equiaxed grains in Figure 2c) also formed at the bottom of the molten pools (marked by pink arrows). This phenomenon was most likely due to the intrinsic heat treatment (IHT) of the SLM process. The repetitive re-heating and re-cooling induced by the laser scanning resulted in an IHT effect at temperatures between 300–500 °C, also known as a self-tempering effect [31]. This effect not only promoted the precipitation process of the nano Cu-rich particles [32], but also triggered the generation of the reverted γ-Fe phase (i.e., 3.7 vol.% within the as-built samples). Furthermore, a large fraction of high-angle grain boundaries (HAGBs) (i.e., 66%) was detected in the as-built samples, as presented in Figure 2d, resulting in a large average grain size of the as-built samples (around 10.19 μm). The low-angle grain boundaries (LAGBs) of 34% also indicated a large quantity of dislocations in the as-built 17-4PH steel samples.

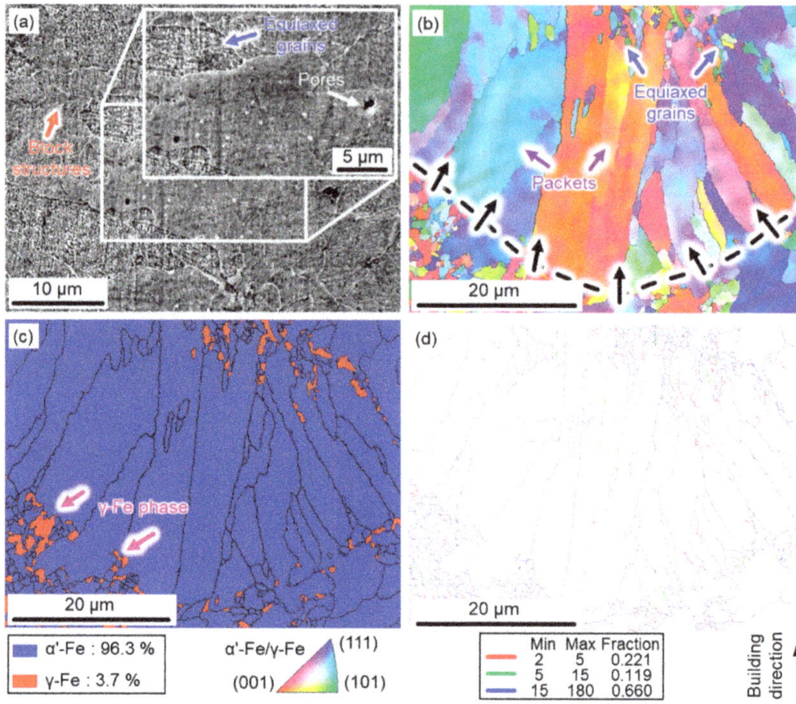

Figure 2. Microstructures of the SLM 17-4 PH steel under as-built conditions: (**a**) SEM; (**b**) IPF map; (**c**) phase distribution; (**d**) grain boundary map.

3.2. Microstructure and Phase Composition in Aging Conditions

Direct aging treatment is an effective method to improve the mechanical properties of martensitic precipitated hardening stainless steel, particularly with regard to temperature and time. Figure 3 presents the effects of aging temperature on the microstructure of the SLM 17-4 PH steel samples. Overall, columnar dendrites dominated the microstructure of the lath-like martensite matrix after aging at different temperatures. Comparatively, the dendrites in the samples aged at low temperatures (Figure 3a,b) were finer than those aged at high temperatures (i.e., Figure 3c,d). The primary dendrites (marked by red arrows in Figure 3c) and secondary dendrites (marked by blue arrows in Figure 3d) were formed with the increasing aging temperature. The essence of aging treatment is a type of tempering process. Therefore, when the 17-4 PH steel was aged at a high temperature (e.g., >500 °C), the lath-like martensite rapidly coarsened, further negatively impacting its mechanical properties.

Moreover, the critical effect of aging treatment is the generation of dispersive nano-Cu particles within the martensite matrix to enhance its material properties [33]. Nanoparticles were scarcely observed within the aged samples at 382 °C for 1 h, while the quantity of precipitated nano-Cu particles (marked by white arrows) swiftly increased with the rising temperature (from 482 °C to 682 °C). Particularly, numerous nano-Cu particles were detected in the aged samples treated at 682 °C for 1 h (Figure 3d). Overall, although the lath width treated at 382 °C was not obviously coarsened, there was no effective precipitation hardening effect (i.e., very few nano precipitates), indicating an under-aging state. After aging above 582 °C, large quantities of nanoparticles were evenly distributed in the matrix, but the lath width was clearly coarsened, biasing towards an over-aging state. Therefore, the 17-4 PH steel, after holding at 482 °C for different times, will be explored in the following section, taking into account the microstructure and the number of precipitated nanoparticles.

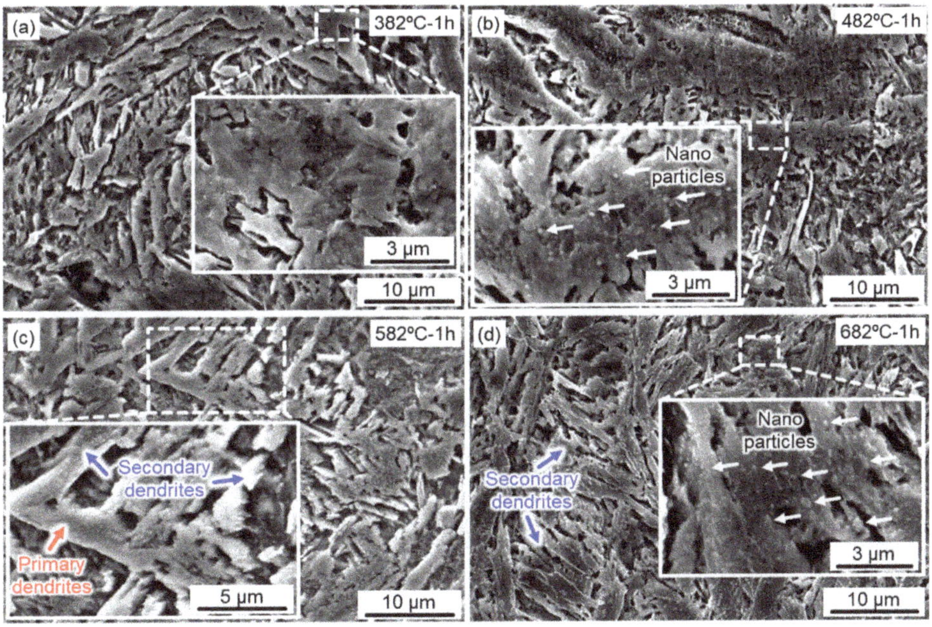

Figure 3. Microstructural evolution of the aged SLM 17-4 PH steel at various aging temperatures for 1 h: (**a**) 382 °C; (**b**) 482 °C; (**c**) 582 °C; (**d**) 682 °C.

To discuss the microstructural evolution during the aging treatment, high-magnification SEM images of the aged samples under different holding times are displayed in Figure 4. Typical lath-like martensite with coarsened secondary dendrites was clearly detected (marked with blue arrows), particularly for long holding times (Figure 4c,d). The strength and hardness of under-aged 17-4 PH steel were restricted despite the fact that the microstructure coarsening caused by short-term aging treatment was not obvious. Long-term aging treatment was beneficial to the precipitation and growth of the nanoparticles, but the coarsened lath martensite greatly reduced its mechanical properties. Therefore, it is extremely significant to seek a desirable combination of the aging temperature and time.

To analyze in detail the reason for under-aging and over-aging phenomena due to the synergic effects of temperature and time, EBSD characterizations were conducted on SLM 17-4 PH steel after different aging treatments. Unlike the coarse packets observed in the as-built samples (Figure 2b), columnar grains with a specific growth direction were replaced by the fine and homogenous martensite laths (Figure 5a). According to the statistics, the average grain size of the SLM 17-4 PH samples was reduced from 10.186 μm (as-builts state) to 3.479 μm (482 °C-1 h aged state), 4.599 μm (482 °C-3 h aged state), and 3.399 μm (582 °C-1 h aged state). Only some molten pools (marked by blue arrows) could be identified within the aged samples treated for 1 h, as displayed in Figure 5a,g, whereas they were difficult to detect in the aged sample designated 482 °C-3 h (Figure 5d). As depicted in Figure 5b,e,h, a similar phenomenon could also be found in the grain boundary distribution statistics. The fraction of the HAGBs within the 482 °C-3 h aged samples (i.e., 76.9%) was higher than that within the short-term aged samples (i.e., 65.2% in the 482 °C-1 h aged samples and 73.7% in the 582 °C-1 h aged samples). LAGBs hinder dislocation movement more significantly than HAGBs [34]. Thus, a higher fraction of LAGBs in the as-built samples (Figure 2d) contributed more strongly to the strengthening mechanism in terms of dislocation movement. However, long maintaining times and high treating temperatures were powerful factors to facilitate dynamic recovery (DRC) and recrystallization (DRX) via the aging processes, which further led to the annihilation of a quantity of subgrain boundaries and dislocations. As such, LAGBs greatly decreased from 34% in the as-built state (Figure 2d) to 23.1% in the 482 °C-3 h aged state (Figure 5e). Generally, as for the 17-4 PH steel, precipitation hardening could become a major strengthening mechanism instead of the working hardening triggered by high-density dislocation walls.

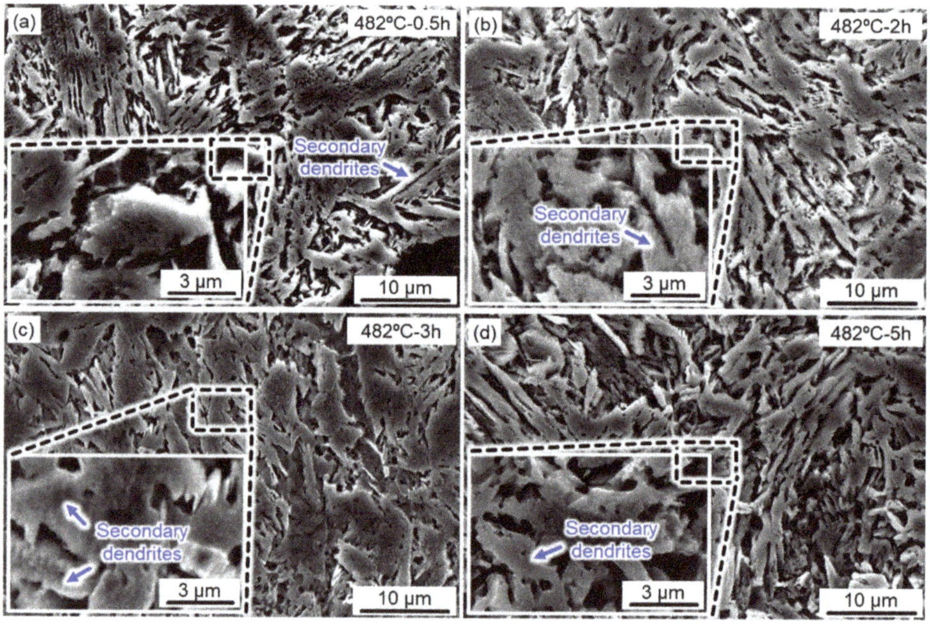

Figure 4. Microstructure features of the aged SLM 17-4 steel at 482 °C for different holding times: (**a**) 0.5 h; (**b**) 2 h; (**c**) 3 h; (**d**) 5 h.

Furthermore, phase mapping was also conducted on the aged samples to reveal the distribution and evolution of the reverted austenite (γ-Fe phase) within the 17-4 PH steel. After aging treatment at 482 °C for 1 h, the volume fraction of the austenite phase increased to 4.5% (Figure 5c) from 3.7% in the as-built sample (Figure 2c). With the prolongation of aging time to 3 h, the austenite phase further increased to 12.0% (Figure 5f). A high aging temperature of 582 °C for 1 h also resulted in an increase of the austenite phase to 8.2% (Figure 5i), compared with the aged sample at 482 °C for 1 h. Interestingly, the reverted austenite (i.e., γ-Fe phase) was concentratedly formed at the interface of the molten pools (marked by red arrows) after the aging treatments, as presented in Figure 5c,i. This was most likely due to the aging treatments contributing to the promotion of growth of the in situ generated γ-Fe phase around the molten pools. Notably, compared to other aged samples (Figure 5c,i), aging time was a critical factor to boost the formation of the γ-Fe phase (i.e., 12.0% in Figure 5f), leading to good ductility. Therefore, balancing the aging temperature and holding time was important to obtain excellent mechanical properties in the 17-4 PH steel.

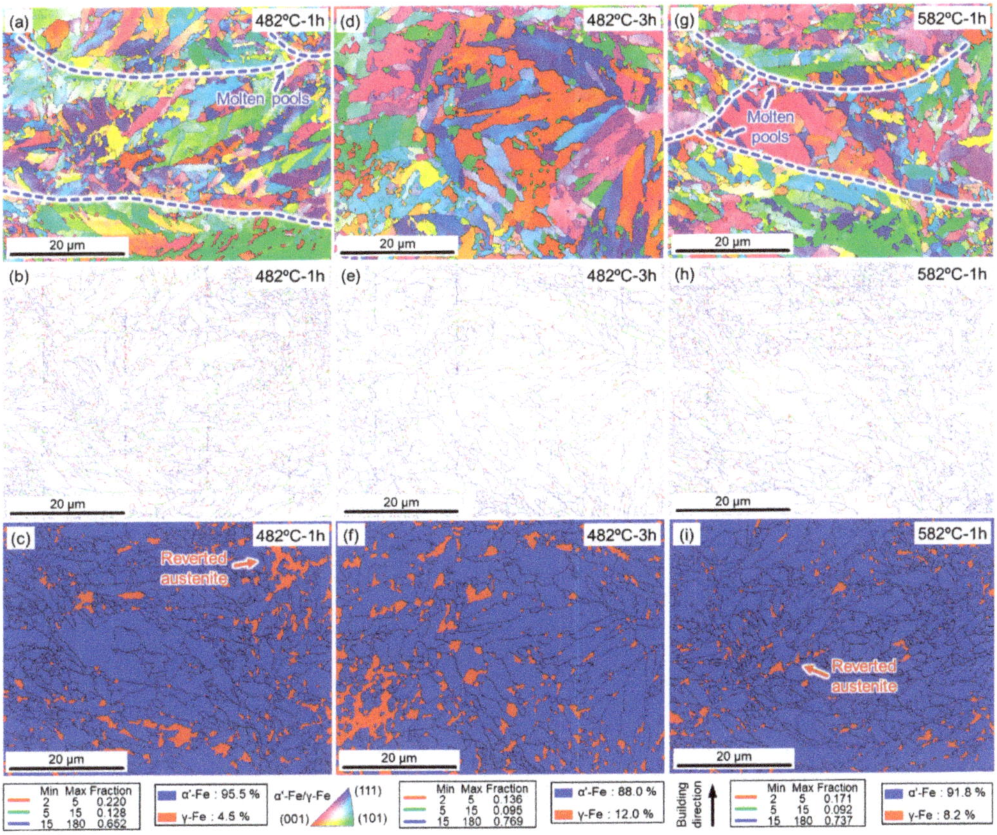

Figure 5. EBSD characterizations of the SLM 17-4 PH steel samples under different conditions: (**a**–**c**) aged samples at 482 °C for 1 h; (**d**–**f**) aged at 482 °C for 3 h; (**g**–**i**) aged at 582 °C for 1 h. IPF images are shown in the first row, grain boundary distribution in the second row, and phase mapping in the third row.

To further identify the phase distribution, XRD analysis was conducted on the SLM 17-4 PH steel under different aging states. As shown in Figure 6, the dominant phase compositions of the aged samples were α'-Fe phase (JCPDS # 06-0696) and γ-Fe phase (JCPDS # 23-0298). Figure 2c shows that the SLM 17-4 PH samples in the as-built state were mainly composed of the α'-Fe phase with a BCC structure and a small amount of γ-Fe phase with a FCC structure. Notably, the γ-Fe phase within the as-built samples did not exclusively include the retained austenite (i.e., γ-Fe phase) due to an incomplete transformation of austenite to martensite at a high temperature. The samples also included reverted austenite (named γ'-Fe phase to show this distinction) caused by the IHT effect during the SLM. After being processed by the aging treatment at various temperatures (Figure 6a) and times (Figure 6b), the reverted austenite with an FCC structure was rapidly formed alongside numerous nano-Cu particles. In short, phase transition within the aged samples can be described as follows:

$$\alpha' + \gamma + \gamma' \xrightarrow{\text{aging treatment}} \alpha' + \gamma' + \text{nano} - \text{Cu particles}$$

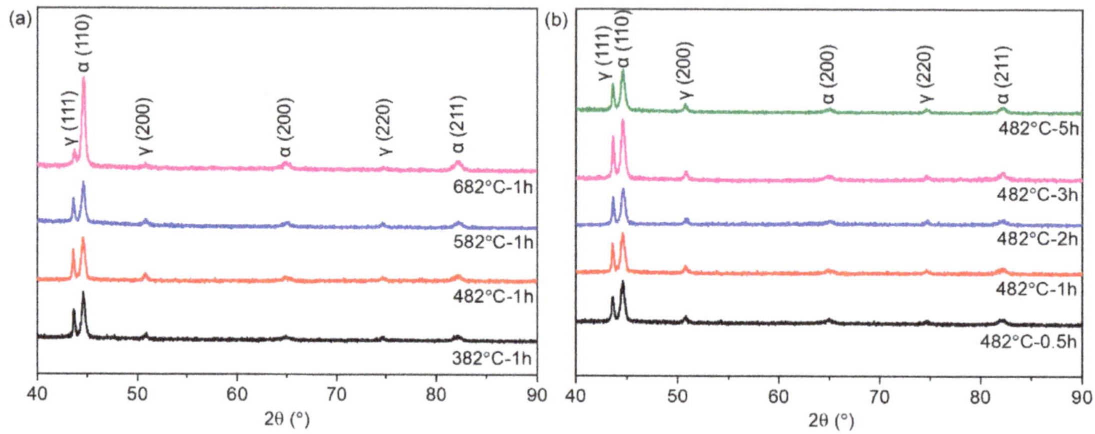

Figure 6. XRD spectra of the SLM-manufactured 17-4 PH steel samples under different (**a**) aging temperatures and (**b**) aging times.

The types of austenite (i.e., retained austenite and reverted austenite) could not be precisely distinguished. Thus, only 'γ-Fe phase' will be used throughout the remainder of this manuscript to improve readability. To quantitatively determine the amount of γ-Fe phase in the aged samples, a Reference Intensity Ratio (RIR) and Rietveld Refinement analysis were conducted on the XRD spectrum of the SLM 17-4 PH steel samples under different aging conditions. The 'error' in Table 2 represents the difference between the fitted curves and the XRD spectra; i.e., fitting curve error. As displayed in Table 2, the volume fraction of the γ-Fe phase increased with further aging treatment, which agreed with the EBSD phase mapping results (i.e., Figure 5c,f,i). This increase in austenite content at higher aging temperatures and longer duration complied with the phenomena reported in the previous section. Notably, the aging time was more vital than the holding temperature in determining the outcome of long-term aging (i.e., 19.6%). The formation of austenite grains through reversion from the martensite phase was most likely to occur at the boundaries of martensite laths during elongated aging times and elevated temperatures. Reverted austenite is beneficial to the chemical stabilization process by diffusing alloying element constituents and promoting ductility [35]. Nonetheless, the high-content γ-Fe phase was accompanied by Cu-rich dispersoids (above 550 °C [17]), reducing its tensile strength.

Table 2. Volume fractions of the α'-Fe and γ-Fe phases in the SLM 17-4 PH steel samples under different states.

Samples	α'-Fe Phase (%)	γ-Fe Phase (%)	Fitting Curve Error
As-built	96.3	3.7	/
Aged at 382 °C for 1 h	95.1	4.9	3.5
Aged at 482 °C for 1 h	93.8	6.2	4.2
Aged at 582 °C for 1 h	89.7	10.3	3.3
Aged at 682 °C for 1 h	80.4	19.6	2.1
Aged at 482 °C for 0.5 h	94.6	5.4	2.2
Aged at 482 °C for 2 h	88.9	11.1	1.8
Aged at 482 °C for 3 h	86.4	13.6	3.5
Aged at 482 °C for 5 h	83.7	16.3	2.6

In summary, the microstructure of the aged 17-4 PH steel frequently exhibited three states, owing to the different treatment temperatures and holding times, as shown in Figure 7. As depicted in the as-built samples (Figure 7a), only a small amount of γ-Fe phase (i.e., primarily in the form of retained austenite) was generated within the interface of the molten pools. A quantity of ultra-fine columnar grains (blue wireframes) and equiaxed grains (yellow hexagons) constituted the martensite packets (i.e., HAGBs) with numerous dislocations (i.e., LAGBs). Only a small amount of γ/γ'-Fe phase (pink ellipses) concentratedly grew at the interface of the molten pools due to the IHT effect. After an aging treatment with a low temperature or a short holding time, the first type of aging state, an under-aged state, is presented in Figure 7b. A few nano-Cu particles (green spots) were in situ precipitated within the martensite matrix, while coarse microstructures (i.e., coarsened columnar grains and elongated equiaxed grains) were also inevitably formed owing to the DRC and DRX process. Although in situ-formed nano-Cu particles boosted the hardness and tensile strength of the SLM 17-4 PH steel, the mediocre mechanical properties were not suitable for some load-bearing applications. Thus, to produce more nano-Cu particles that further improved the mechanical properties of the SLM 17-4 PH steel, a high aging temperature and long holding time were conducted on the as-built samples. As presented in Figure 7c, numerous dislocations were eliminated due to the DRC/X effects, thus leading to the coarse laths-like martensite and elongated equiaxed grains rather than packets. Not only quantities of nano-Cu particles, but also γ-Fe phase (mainly reverted austenite) were formed within the matrix. Thus, the mechanical properties were greatly improved due to the precipitation hardening aroused by the Cu-rich nanoparticles within the adequate-aged state instead of the dislocation strengthening within the as-builts state. However, with the further increase in the aging temperature or holding time, as displayed in Figure 7d, nano-Cu particles (marked by green spots) and microstructures were extremely coarsened, resulting in a great reduction of the mechanical performance; i.e., an over-aged state. The strength and hardness of the over-aged samples was significantly weakened due to the unfavorable microstructural features. In short, different aging conditions caused distinctive mechanical properties. Understanding the microstructural evolution mechanisms and seeking a series of optimal aging parameters are essential in this work.

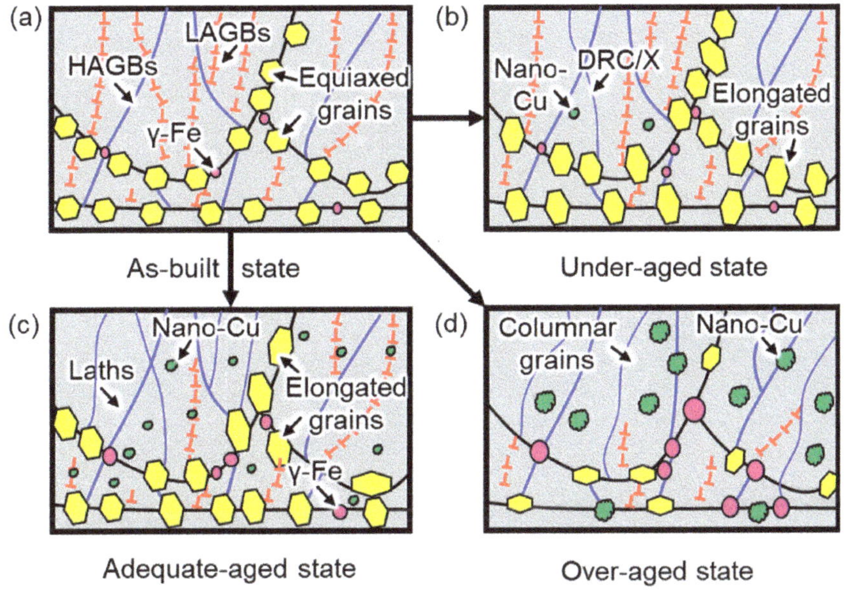

Figure 7. Typical microstructural evolution of the different aged states: (**a**) as-built; (**b**) under-aged; (**c**) adequate-aged; (**d**) over-aged.

3.3. Mechanical Properties

Figure 8 shows the microhardness variation of the SLM-manufactured 17-4 PH steel samples after different aging temperatures and holding times. As displayed in Figure 8a, with the prolongation of aging temperatures, the microhardness first increased rapidly and then slowly decreased from an under-aged state to an over-aged state. The microhardness peak (430 ± 8.5 $HV_{0.5}$) appeared in the 482 °C-1 h aged samples. The relationship between the microhardness (HV, $HV_{0.5}$) and the aging temperature (T, °C) can be depicted as follows:

$$HV = 337.88 + 0.33T - 3.09 \times 10^{-4}T^2$$

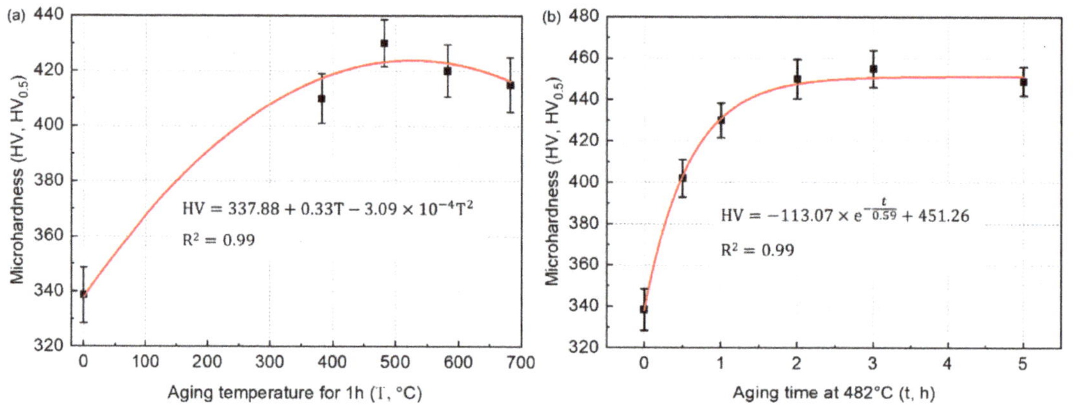

Figure 8. Microhardness variation of the SLM-manufactured 17-4 PH steel samples after aging treatments under different (**a**) temperatures and (**b**) holding times.

As for the effect of the aging time on the microhardness of the aged samples (Figure 8b), the microhardness also first increased, then remained roughly unchanged. The maximum value (455 ± 8.9 $HV_{0.5}$) occured within the 482 °C-3 h aged samples. The function of microhardness (HV, $HV_{0.5}$) vs. aging time (t, h) can be concluded as follows:

$$HV = -113.07 \times e^{-\frac{t}{0.59}} + 451.26$$

Thus, the optimal aging parameters suitable for SLM 17-4 PH steel can be selected as 482 °C for 3 h. Please note that the aforementioed equations only display the propbable relationship between the microhardness and aging temperature alongside the aging time. These functions cannot show the strict quantitative relationship between these factors due to limited testing results.

Figure 9a,b illustrates the strain-stress curves of the SLM 17-4 PH steel samples under different aging conditions, and the variations of tensile performances are summarized in Figure 9c,d, including ultimate tensile strength (UTS), yield strength (YS), and elongation to break (EL). As can be seen in Figure 9a,c, both the UTS and YS gradually increased withthe prolonged aging temperature at 1 h aging time. The UTS reached its peak value at the aging temperature of 482 °C, which was similar to the trend in microhardness. The ductility decreased rapidly (i.e., EL = 8.8%) when the aging treatment was conducted on the SLM 17-4 PH steel at 482 °C, despite having the highest strength (i.e., YS = 939.7 MPa and UTS = 1312.8 MPa) among the tested steel samples. Thus, the aging temperature should preferably be 482 °C if the strength is the optimization goal.

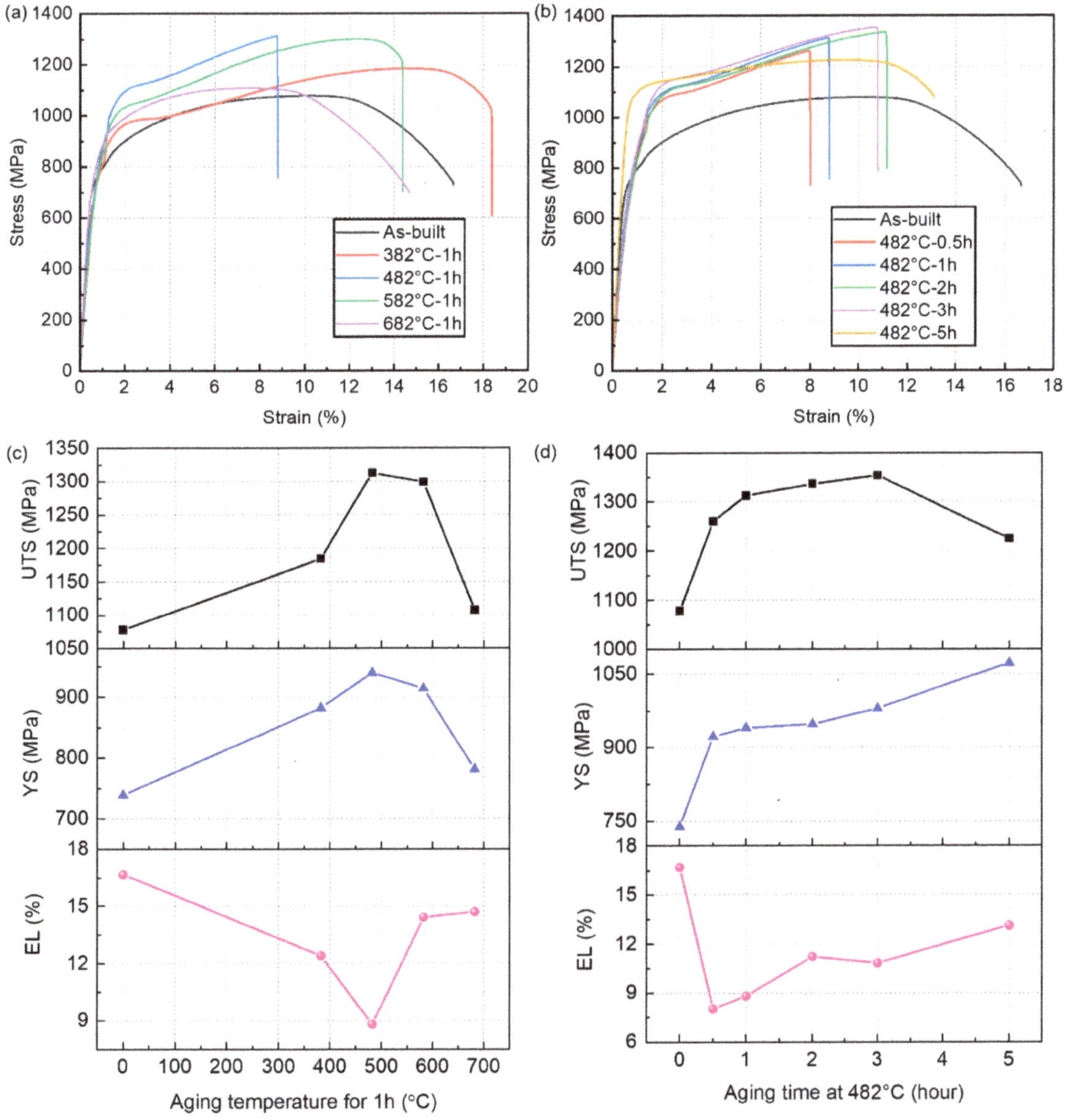

Figure 9. Mechanical properties of the SLM 17-4 PH steel samples under different post-aging treatment conditions: (**a**,**b**) strain-stress curves and (**c**,**d**) key indicator statistics. First row: mechanical properties at different aging temperatures for 1 h. Second row: mechanical properties at different aging times at 482 °C.

The results of further optimization for the aging temperature of 482 °C are shown in Figure 9b,d. Strength values, including the YS and UTS, increased with the prolongation of the aging time from 0.5 h to 5 h (Figure 9b). The highest strength I when the aging time was 3 h, indicating an adequate aging state that offered an important positive effect. Although EL was not the best among these aged samples, ultra-high UTS (i.e., 1353.4 MPa) with a desirable ductility (i.e., EL = 10.8%) displayed an excellent combination of strength and toughness under this aged state, as the statistics show in Figure 9d. Interestingly, YS and EL exhibited a slight elevation compared to the other aged states, while an extreme reduction of the UTS was shown. These effects could be attributed to the over-aging treatment, resulting in an obvious coarsening phenomenon of the nanoparticles and the microstructure. Thus, the fracture would occur immediately once the yield point was exceeded. In summary, the comprehensive properties of the 482 °C-3 h aged samples can be considered the best among these aged samples.

Moreover, the elongation value exhibited an inverse tendency compared with the tensile strength. Based on the above mentioned results, increases in aging time and temperature both initially increased the tensile strength of SLM 17-4 PH steel samples, followed by a subsequent decrease. The microstructure observation in Figure 5 shows that shorter aging times and lower aging temperatures both resulted in fewer nanoprecipitates. Thus, it is reasonable to conclude that higher aging temperatures and longer aging times resulted in a more significant strengthening effect, leading to higher strength values. On the contrary, further increasing the aging temperature and time can result in an increased fraction of the austenite phase and a coarsened microstructure, reducing the strengthening effect and lowering the strength values.

Figure 10 presents the fracture surfaces of the SLM 17-4 PH steel samples under different conditions, revealing the typical effect of aging. Compared with other aged states, only the as-built samples had a clear necking phenomenon (red arrows), as displayed in Figure 10a. Partially unmelted powder particles (blue arrows) and cracks (green arrows) could be also detected. The uniform distribution of the numerous micro-sized dimples (pink arrows) found using a high-magnification SEM shows that a ductile fracture behavior dominated the as-built samples. After an under-aging treatment (e.g., 482 °C for 1 h), Figure 10b shows the presence of not only small dimples but also many river patterns (purple arrows), indicating a mixed fracture mode of brittleness/ductility after the aging treatment. With an increase in the holding time, small dimples and obvious cleavage facets were observed on the fracture surface of the 482 °C-3 h aged samples, as displayed in Figure 10c. This phenomenon indicates that a large quantity of precipitated nano-Cu particles resulted in a brittle fracture, which is consistent with the trend in strength/ductility of tensile performance. When the aging time was further increased, an over-aged state was characterized, as shown in Figure 10d. Notably, tear ridges were formed along the interface of the martensite laths, and many dimples suggested poor ductility. As for the precipitation hardening alloys, the existence of secondary nanoparticles and precipitates could initiate the nucleation of micro-sized voids during the tensile fracture. Thus, the growth and coalescence of voids would result in the observed dimple features on the fracture surface in the tensile loading process. With a prolonged aging time, more prominent precipitation of secondary nanoparticles could promote the generation of more voids, leading to an increased number of dimples with small diameters.

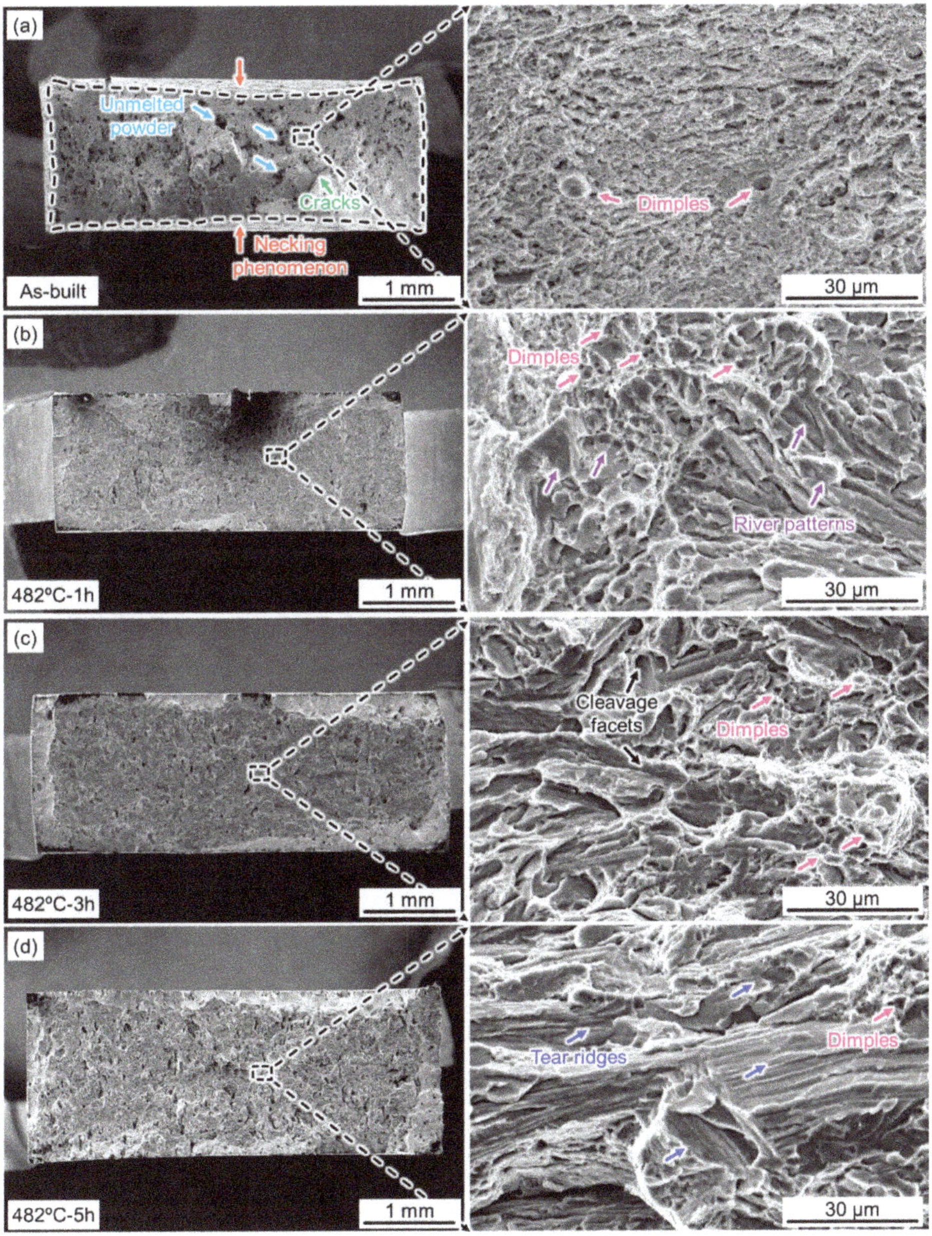

Figure 10. Fractographic features of the SLM 17-4 PH steel samples under different conditions: (**a**) as-built; (**b**) aged at 482 °C for 1 h; (**c**) aged at 482 °C for 3 h; (**d**) aged at 482 °C for 3 h.

4. Conclusions

Despite an ultra-high strength in the as-built condition, the post-aging treatment is still indispensable for SLM 17-4 PH steel to obtain balanced strength and ductility. This work investigated the effects of aging temperature and time on the microstructure and mechanical properties of SLM 17-4 PH steel. The main conclusions of this work are summarized as follows:

1. In the as-built conditions, the SLM 17-4 PH steel exhibited a dense microstructure with no significant pores and cracks. The microstructure of the as-built samples exhibited columnar grains growing along the building direction as a result of the largest temperature gradient in the SLM process. Ultra-fine equiaxed grains were decorated at the molten pool boundary, which was confirmed as γ-Fe phase.
2. Based on SEM and EBSD observations, coarse martensite laths were observed in the aged samples compared with the as-built ones regardless of the aging time and temperature. Increasing the aging temperature led to more significant grain growth of the martensite laths.
3. In the as-built condition, the 17-4 PH steel sampless were mainly composed of the martensite phase with a BCC structure and a small amount of austenite phase with a FCC structure. The aging treatment at various temperatures and times led to the formation of the reverted austenite phase with a FCC structure. As the aging time increased at 482 °C, the SLM 17-4 PH steel exhibited an increasing intensity of the austenite peak in comparison with the as-built samples. With further aging treatment, the increasing volume fraction of the austenite phase was found, which was consistent with the EBSD phase mapping results.
4. The mechanical properties showed that both the UTS and yield strength gradually increased as the aging time at 482 °C was prolonged. The UTS reached its peak value at an aging time of 3 h and a temperature of 482 °C, which was similar to the trend in microhardness. However, the ductility decreased rapidly with the aging treatment of SLM 17-4 PH steel.

Author Contributions: Conceptualization, C.C. and D.D.; methodology, D.D.; software, Y.Y. and S.Y.; validation, J.W., X.T. and Z.R.; formal analysis, D.D.; investigation, D.D.; resources, M.L. and K.Z.; writing—original draft preparation, D.D.; writing—review and editing, Z.Y., M.L. and K.Z.; visualization, M.L. and K.Z.; supervision, M.L. and K.Z.; project administration, M.L.; funding acquisition, C.C. and M.L. All authors have read and agreed to the published version of the manuscript.

Funding: This research was funded by GDAS' Project of Science and Technology Development (No. 2022GDASZH-2022010107), Guangdong Special Support Program (No. 2019BT02C629), Guangzhou Project of Science & Technology (No. 202007020008), the National Key Research and Development Program of China (No. 2022YFC2406000, 2019YFA0705300 and 2021YFB3702502), Natural Science Foundation of Shanghai (No. 23ZR1421500), the National Natural Science Foundation of China (No. 52001191, 52127807, 52271035), and the Shanghai Rising-Star Program (No. 20QA1403800). This research was also supported by the Independent Research Project of State Key Laboratory of Advanced Special Steel, Shanghai Key Laboratory of Advanced Ferrometallurgy, Shanghai University (SKLASS 2022-Z10) and the Science and Technology Commission of Shanghai Municipality (No.19DZ2270200).

Institutional Review Board Statement: Not applicable.

Informed Consent Statement: Not applicable.

Data Availability Statement: Not applicable.

Conflicts of Interest: The authors declare no conflict of interest.

References

1. Chen, Q.; Jing, Y.; Yin, J.; Li, Z.; Xiong, W.; Gong, P.; Zhang, L.; Li, L.; Pan, R.; Zhao, X.; et al. High Reflectivity and Thermal Conductivity Ag–Cu Multi-Material Structures Fabricated via Laser Powder Bed Fusion: Formation Mechanisms, Interfacial Characteristics, and Molten Pool Behavior. *Micromachines* **2023**, *14*, 362. [CrossRef] [PubMed]
2. Khorasani, M.; Gibson, I.; Ghasemi, A.H.; Hadavi, E.; Rolfe, B. Laser subtractive and laser powder bed fusion of metals: Review of process and production features. *Rapid Prototyp. J.* **2023**. [CrossRef]
3. Chen, C.; Xie, Y.; Yin, S.; Li, W.; Luo, X.; Xie, X.; Zhao, R.; Deng, C.; Wang, J.; Liao, H.; et al. Ductile and high strength Cu fabricated by solid-state cold spray additive manufacturing. *J. Mater. Sci. Technol.* **2023**, *134*, 234–243. [CrossRef]

4. Li, Z.; Li, H.; Yin, J.; Li, Y.; Nie, Z.; Li, X.; You, D.; Guan, K.; Duan, W.; Cao, L.; et al. A Review of Spatter in Laser Powder Bed Fusion Additive Manufacturing: In Situ Detection, Generation, Effects, and Countermeasures. *Micromachines* **2022**, *13*, 1366. [CrossRef] [PubMed]
5. Zhao, R.; Chen, C.; Shuai, S.; Hu, T.; Fautrelle, Y.; Liao, H.; Lu, J.; Wang, J.; Ren, Z. Enhanced mechanical properties of Ti6Al4V alloy fabricated by laser additive manufacturing under static magnetic field. *Mater. Res. Lett.* **2022**, *10*, 530–538. [CrossRef]
6. Chang, C.; Huang, J.; Yan, X.; Li, Q.; Liu, M.; Deng, S.; Gardan, J.; Bolot, R.; Chemkhi, M.; Liao, H. Microstructure and mechanical deformation behavior of selective laser melted Ti6Al4V ELI alloy porous structures. *Mater. Lett.* **2020**, *277*, 128366. [CrossRef]
7. Chen, L.Y.; Xu, J.Q.; Choi, H.; Pozuelo, M.; Ma, X.; Bhowmick, S.; Yang, J.M.; Mathaudhu, S.; Li, X.C. Processing and properties of magnesium containing a dense uniform dispersion of nanoparticles. *Nature* **2015**, *528*, 539–543. [CrossRef]
8. Yin, J.; Zhang, W.; Ke, L.; Wei, H.; Wang, D.; Yang, L.; Zhu, H.; Dong, P.; Wang, G.; Zeng, X. Vaporization of alloying elements and explosion behavior during laser powder bed fusion of Cu–10Zn alloy. *Int. J. Mach. Tools Manuf.* **2021**, *161*, 103686. [CrossRef]
9. Yi, J.; Chang, C.; Yan, X.; Xie, Y.; Liu, Y.; Liu, M.; Zhou, K. A novel hierarchical manufacturing method of the selective laser melted Al 7075 alloy. *Mater. Charact.* **2022**, *191*, 112124. [CrossRef]
10. Chang, C.; Liao, H.; Yi, L.; Dai, Y.; Cox, S.C.; Yan, M.; Liu, M.; Yan, X. Achieving ultra-high strength and ductility in Mg–9Al–1Zn–0.5Mn alloy via selective laser melting. *Adv. Powder Mater.* **2023**, *2*, 100097. [CrossRef]
11. Chang, C.; Yue, S.; Li, W.; Lu, L.; Yan, X. Study on microstructure and tribological behavior of the selective laser melted MgZnCa alloy. *Mater. Lett.* **2022**, *309*, 2–5. [CrossRef]
12. Chang, C.; Yan, X.; Deng, Z.; Lu, B.; Bolot, R.; Gardan, J.; Deng, S.; Chemkhi, M.; Liu, M.; Liao, H. Heat treatment induced microstructural evolution, oxidation behavior and tribological properties of Fe-12Cr-9Ni-2Al steel (CX steel) prepared using selective laser melting. *Surf. Coatings Technol.* **2022**, *429*, 127982. [CrossRef]
13. Viswanathan, U.K.; Nayar, P.K.K.; Krishnan, R. 17-4 PH stainless. *Mater. Sci. Technol.* **1989**, *5*, 346–349. [CrossRef]
14. Rack, H.J.; Kalish, D. Strength, Fracture Toughness, and Low Cycle Fatigue Behavior of 17-4 Ph Stainless Steel. *Met. Trans.* **1974**, *5*, 1595–1605. [CrossRef]
15. Murr, L.E.; Martinez, E.; Hernandez, J.; Collins, S.; Amato, K.N.; Gaytan, S.M.; Shindo, P.W. Microstructures and properties of 17-4 PH stainless steel fabricated by selective laser melting. *J. Mater. Res. Technol.* **2012**, *1*, 167–177. [CrossRef]
16. Dong, D.; Chang, C.; Wang, H.; Yan, X.; Ma, W.; Liu, M.; Deng, S.; Gardan, J.; Bolot, R.; Liao, H. Selective laser melting (SLM) of CX stainless steel: Theoretical calculation, process optimization and strengthening mechanism. *J. Mater. Sci. Technol.* **2021**, *73*, 151–164. [CrossRef]
17. Edition, S. *Metallurgy Second Edition Welding Metallurgy*; John Wiley & Sons, Inc.: Hoboken, NJ, USA, 2003; ISBN 3175723993.
18. Akita, M.; Uematsu, Y.; Kakiuchi, T.; Nakajima, M.; Kawaguchi, R. Defect-dominated fatigue behavior in type 630 stainless steel fabricated by selective laser melting. *Mater. Sci. Eng. A* **2016**, *666*, 19–26. [CrossRef]
19. Hu, Z.; Zhu, H.; Zhang, H.; Zeng, X. Experimental investigation on selective laser melting of 17-4PH stainless steel. *Opt. Laser Technol.* **2017**, *87*, 17–25. [CrossRef]
20. Rashid, R.; Masood, S.H.; Ruan, D.; Palanisamy, S.; Rahman Rashid, R.A.; Brandt, M. Effect of scan strategy on density and metallurgical properties of 17-4PH parts printed by Selective Laser Melting (SLM). *J. Mater. Process. Technol.* **2017**, *249*, 502–511. [CrossRef]
21. Averyanova, M.; Cicala, E.; Bertrand, P.; Grevey, D. Experimental design approach to optimize selective laser melting of martensitic 17-4 PH powder: Part i—Single laser tracks and first layer. *Rapid Prototyp. J.* **2012**, *18*, 28–37. [CrossRef]
22. Ozsoy, A.; Yasa, E.; Keles, M.; Tureyen, E.B. Pulsed-mode Selective Laser Melting of 17-4 PH stainless steel: Effect of laser parameters on density and mechanical properties. *J. Manuf. Process.* **2021**, *68*, 910–922. [CrossRef]
23. Chang, C.; Yan, X.; Deng, Z.; Chu, Q.; Deng, S.; Bolot, R.; Chemkhi, M.; Liu, M.; Liao, H.; Gardan, J. Effect of heat treatment on residual stress and wear resistance of CX stainless steel manufactured by Selective Laser Melting. *Procedia CIRP* **2021**, *104*, 738–743. [CrossRef]
24. Yan, X.; Chang, C.; Dong, D.; Gao, S.; Ma, W.; Liu, M.; Liao, H.; Yin, S. Microstructure and mechanical properties of pure copper manufactured by selective laser melting. *Mater. Sci. Eng. A* **2020**, *789*, 139615. [CrossRef]
25. Kürnsteiner, P.; Wilms, M.B.; Weisheit, A.; Gault, B.; Jägle, E.A.; Raabe, D. High-strength Damascus steel by additive manufacturing. *Nature* **2020**, *582*, 515–519. [CrossRef] [PubMed]
26. Kürnsteiner, P.; Wilms, M.B.; Weisheit, A.; Barriobero-Vila, P.; Jägle, E.A.; Raabe, D. Massive nanoprecipitation in an Fe-19Ni-xAl maraging steel triggered by the intrinsic heat treatment during laser metal deposition. *Acta Mater.* **2017**, *129*, 52–60. [CrossRef]
27. Nie, J.F. Effects of precipitate shape and orientation on dispersion strengthening in magnesium alloys. *Scr. Mater.* **2003**, *48*, 1009–1015. [CrossRef]
28. Rafi, H.K.; Pal, D.; Patil, N.; Starr, T.L.; Stucker, B.E. Microstructure and Mechanical Behavior of 17-4 Precipitation Hardenable Steel Processed by Selective Laser Melting. *J. Mater. Eng. Perform.* **2014**, *23*, 4421–4428. [CrossRef]
29. Pasebani, S.; Ghayoor, M.; Badwe, S.; Irrinki, H.; Atre, S.V. Effects of atomizing media and post processing on mechanical properties of 17-4 PH stainless steel manufactured via selective laser melting. *Addit. Manuf.* **2018**, *22*, 127–137. [CrossRef]
30. Hsu, T.H.; Chang, Y.J.; Huang, C.Y.; Yen, H.W.; Chen, C.P.; Jen, K.K.; Yeh, A.C. Microstructure and property of a selective laser melting process induced oxide dispersion strengthened 17-4 PH stainless steel. *J. Alloys Compd.* **2019**, *803*, 30–41. [CrossRef]
31. Wang, Y.; Sun, J.; Jiang, T.; Sun, Y.; Guo, S.; Liu, Y. A low-alloy high-carbon martensite steel with 2.6 GPa tensile strength and good ductility. *Acta Mater.* **2018**, *158*, 247–256. [CrossRef]

32. Sun, Y.; Hebert, R.J.; Aindow, M. Effect of heat treatments on microstructural evolution of additively manufactured and wrought 17-4 PH stainless steel. *Mater. Des.* **2018**, *156*, 429–440. [CrossRef]
33. Yadollahi, A.; Shamsaei, N.; Thompson, S.M.; Elwany, A.; Bian, L. Mechanical and microstructural properties of selective laser melted 17-4 PH stainless steel. *ASME Int. Mech. Eng. Congr. Expo. Proc.* **2015**, *57359*, V02AT02A014. [CrossRef]
34. Liu, Z.; Gao, Q.; Zhang, H.; Luo, S.; Zhang, X.; Li, W.; Jiang, Y.; Li, H. EBSD analysis and mechanical properties of alumina-forming austenitic steel during hot deformation and annealing. *Mater. Sci. Eng. A* **2019**, *755*, 106–115. [CrossRef]
35. Hadadzadeh, A.; Shahriari, A.; Amirkhiz, B.S.; Li, J.; Mohammadi, M. Additive manufacturing of an Fe–Cr–Ni–Al maraging stainless steel: Microstructure evolution, heat treatment, and strengthening mechanisms. *Mater. Sci. Eng. A* **2020**, *787*, 139470. [CrossRef]

micromachines

MDPI

Article

Effects of Building Directions on Microstructure, Impurity Elements and Mechanical Properties of NiTi Alloys Fabricated by Laser Powder Bed Fusion

Shuo Wang [1], Xiao Yang [1,*], Jieming Chen [1], Hengpei Pan [1], Xiaolong Zhang [2,*], Congyi Zhang [1], Chunhui Li [1], Pan Liu [1], Xinyao Zhang [1,3], Lingqing Gao [1,3] and Zhenzhong Wang [1]

1 Luoyang Ship Material Research Institute, Luoyang 471023, China; 18220580820@163.com (S.W.); jiemingch@163.com (J.C.); panhengpei@163.com (H.P.); congyi610@163.com (C.Z.); 18437955231@163.com (C.L.); liupan_lp@163.com (P.L.); 13525971194@163.com (X.Z.); gaolq725@163.com (L.G.); jindou8309@163.com (Z.W.)

2 Key Laboratory of Bionic Engineering, Ministry of Education, Jilin University, Changchun 130022, China

3 Henan Key Laboratory of Technology and Application Structural Materials for Ships and Marine Equipments, Luoyang 471023, China

* Correspondence: y19850419h@126.com (X.Y.); zxl727373312@126.com (X.Z.)

Abstract: For NiTi alloys prepared by the Laser Powder Bed Fusion (LPBF), changes in the building directions will directly change the preferred orientation and thus directly affect the smart properties, such as superelasticity, as well as change the distribution state of defects and impurity elements to affect the phase transformation behaviour, which in turn affects the smart properties at different temperatures. In this study, the relationship between impurity elements, the building directions, and functional properties; the effects of building directions on the crystallographic anisotropy; phase composition; superelastic properties; microhardness; geometrically necessary dislocation (GND) density; and impurity element content of NiTi SMAs fabricated by LPBF were systematically studied. Three building directions measured from the substrate, namely, 0°, 45° and 90°, were selected, and three sets of cylindrical samples were fabricated with the same process parameters. Along the building direction, a strong <100>//vertical direction (VD) texture was formed for all the samples. Because of the difference in transformation temperature, when tested at 15 °C, the sample with the 45° orientation possessed the highest strain recovery of 3.2%. When tested at the austenite phase transformation finish temperature (Af)+10 °C, the 90° sample had the highest strain recovery of 5.83% and a strain recovery rate of 83.3%. The sample with the 90° orientation presented the highest microhardness, which was attributed to its high dislocation density. Meanwhile, different building directions had an effect on the contents of O, C, and N impurity elements, which affected the transformation temperature by changing the Ni/Ti ratio. This study innovatively studied the impurity element content and GND densities of compressive samples with three building directions, providing theoretical guidance for LPBFed NiTi SMA structural parts.

Keywords: laser powder bed fusion (LPBF); NiTi SMAs; microstructure; mechanical response; building orientation; impurity elements

Citation: Wang, S.; Yang, X.; Chen, J.; Pan, H.; Zhang, X.; Zhang, C.; Li, C.; Liu, P.; Zhang, X.; Gao, L.; et al. Effects of Building Directions on Microstructure, Impurity Elements and Mechanical Properties of NiTi Alloys Fabricated by Laser Powder Bed Fusion. *Micromachines* **2023**, *14*, 1711. https://doi.org/10.3390/mi14091711

Academic Editor: Kunal Mitra

Received: 1 July 2023
Revised: 11 August 2023
Accepted: 12 August 2023
Published: 31 August 2023

1. Introduction

NiTi shape memory alloys (SMAs) have received widespread attention from the materials science and engineering fields because of their excellent superelasticity, shape memory effect, damping properties, and biocompatibility. NiTi SMAs have a wide range of applications in aerospace, biomedical devices, mechanical electronics, the construction industry, and daily life [1,2]. Normally, NiTi SMAs prepared by traditional vacuum arc induction melting and powder metallurgy methods have homogeneous structures. However, thermal processing methods, such as drawing, forging, rolling, extrusion, and welding, are prone to introducing impurities, such as oxygen, carbon, and nitrogen, making the phase transformation temperature difficult to regulate. Frenzel et al. [3] pointed out that for the ratio of Ni and Ti elements, even a slight change of 0.1 at.%, can change their martensite transformation start temperature (Ms) by up to ~10 °C, thus significantly affecting their microstructure and mechanical properties, which greatly limits the application of NiTi SMAs in complex working conditions [4,5].

Laser powder bed fusion (LPBF) has a wide range of applications and has been widely used in a variety of materials, such as steel, high-temperature alloys, titanium alloys and aluminium alloys [6–9]. LPBF technology can transform complex three-dimensional shapes into two-dimensional layers and fabricate the desired shapes in one process. Its tunability in the process can solve the limitation of the phase transformation temperature, which is difficult to regulate during the traditional preparation and processing of NiTi SMAs [10,11]. Furthermore, LPBF also has great potential for the manufacturing of gradient materials. For example, Song et al. [12] proposes an analytical approach to design stretching-dominated truss lattices with tailored elastic properties, including isotropic elasticity, tailored zero/negative Poisson's ratios, tailored Young's moduli ratios along specified directions, and the prototype was prepared using the Micro-LPBF technique, these techniques could be used to develop new engineering applications and promote the development of NiTi SMAs [13–15]. Therefore, the NiTi SMAs fabricated by LPBF have become a major topic in the research community recently [16–19]. In these studies, the mechanical properties of LPBFed NiTi SMAs showed an obvious orientation dependence. For example, the type of defects has a great influence on the tensile strength, and the building directions have a significant effect on the wear properties of LPBFed NiTi SMAs [20,21]. Dadbakhsh et al. [22] investigated the anisotropy of LPBFed NiTi SMAs (with different building directions) and clarified that the crystallographic textures have an important effect on the mechanical properties of LPBFed NiTi SMAs. Gu et al. [15] optimized the process parameters of LPBFed NiTi, an excellent shape recovery rate of 88.23% was achieved under the optimal parameters, and a shape-recovery rate of 96.7% was achieved under electrical actuation for a structure with a pre-compressed strain of 20%. Shi et al. [19] investigated the effects of crystallographic anisotropy on the microstructure, phase transformation and the tribological properties of NiTi shape memory alloys fabricated by LPBF and revealed how different LPBF-induced microstructures affect mechanical properties and wear properties. Most of the studies above focused on the effects of the building directions on the crystallographic texture, phase composition, and thermomechanical response. However, the effect of the building directions on the impurity elements and the magnitude of the dislocation densities in LPBFed NiTi SMAs has not been reported.

In this study, the microstructures, mechanical properties, and impurity element contents of LPBFed NiTi SMAs with three different building directions (0°, 45°, and 90°) were investigated. In addition, the relationship between the preferred orientations and the dislocation densities was analyzed. Furthermore, the influence of the defects in the samples with different building directions on the introduction of impurity elements was also investigated in detail.

2. Materials and Methods

2.1. NiTi Samples Fabrication by LPBF

$Ni_{50.8}Ti_{49.2}$ powder was prepared by Minatech Ltd. (Shenzhen, China) using the electrode induction-melting gas atomization (EIGA) technique. The main composition (wt.%) of the NiTi powder was determined to be 55.80 wt.% Ni, 0.0576 wt.% O, 0.0066 wt.% C, 0.0067 wt.% N, and balance Ti. Figure 1a shows the scanning electron microscopy (SEM) image of the NiTi powder, which exhibits a regular spherical shape with less satellite powder, and the size range of the powder particles is from 15 μm to 53 μm (D50 = 36.8 μm). As shown in Figure 1b, three sets of cylindrical specimens were prepared with a height of 10 mm and a diameter of 6 mm. Figure 1c shows the scanning strategy of 67° rotation angles between the adjacent layers. The LPBF processing was performed in a BLT (BLT

S210, Shaanxi, China) machine equipped with a 500 W ytterbium-doped laser under argon protection to keep the oxygen level below 100 ppm. Previous research has shown that NiTi alloys exhibits favorable superelasticity and low porosity when exposed to the energy density of about 72 J/mm^3. To control independent variables, the same process parameters and different building directions are applied. Table 1 shows the optimized parameters in the LPBF processing, and the energy density was calculated with E = P/vht [21].

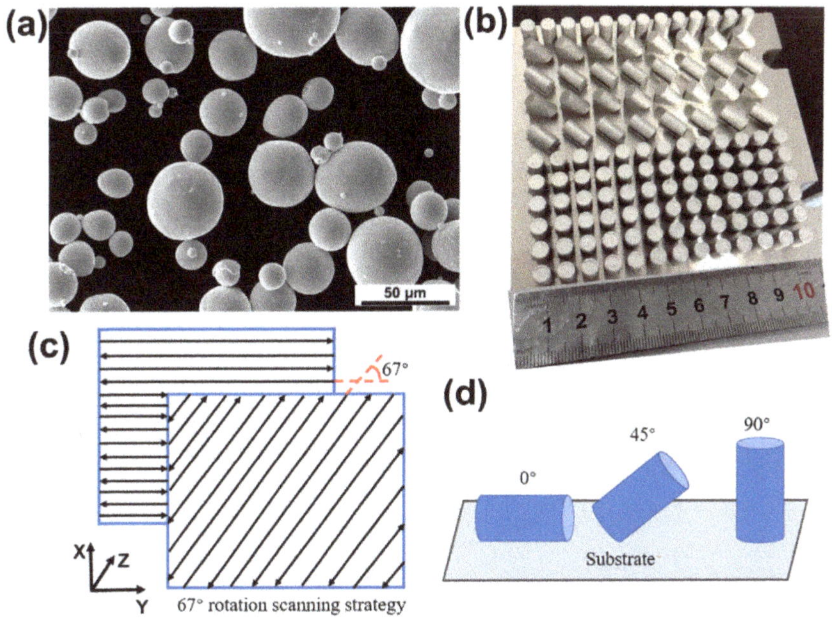

Figure 1. (**a**) Morphology of $Ni_{50.8}Ti_{49.2}$ powder, (**b**) actual view of LPBFed NiTi SMA samples, (**c**) scanning strategy implemented in this research and (**d**) schematic of LPBFed NiTi SMA samples.

Table 1. Processing parameters for LPBFed NiTi SMAs.

Sample	Laser Power P (W)	Scanning Speed v (mm/s)	Hatch Spacing h (μm)	Layer Thickness t (μm)	Energy Density E (J/mm^3)
0°	105	600	80	30	72.92
45°	105	600	80	30	72.92
90°	105	600	80	30	72.92

2.2. Microstructure and Property Characterization

ADSC250 differential scanning calorimeter (DSC, TA) was used to determine the phase transformation temperature of the samples. The weight of the DSC specimens was 5~20 mg, and the heating/cooling rate was 15 °C/min from −80 °C to 70 °C. The LPBFed NiTi SMAs were ground, polished, and etched using a mixture of 70 vol% H_2O + 20 vol% HNO_3 + 10 vol% HF solution. The metallography of the samples was observed by an optical microscope (OM, Zeiss, Oberkochen, Germany). The atomic ratios of Ni and Ti were analyzed by scanning electron microscopy (SEM, FEI Scios 2, Waltham, MA, USA) equipped with an EDAX X-ray energy spectrometer (EDS). Five random points were selected to test each sample to obtain the average composition. The tested samples were polished to 2.5 μm with 180~2000 grit silicon carbide sandpaper, followed by electrolytic polishing in a HNO_3/CH_3OH = 1:10 (vol%) solution at 20 V for 15 s. The crystallographic orientation, phase composition and dislocation density were analyzed by SEM equipped with an EDAX electron backscatter diffraction (EBSD) system.

Transmission electron microscopy (TEM) samples were ground to a thickness of 50 μm and then electropolished using a twin-jet thinning electropolishing device and an electrolyte consisting of 4% perchloric acid and 96% ethanol (vol%) at −20 °C. TEM observation and electron diffraction analysis were performed in a JEM2100 (JEOL, Tokyo, Japan) electron microscope at 200 kV.

Sample impurity elements were collected by an ONH836 gas analyser and a CS800 carbon and sulphur analyser. Compression tests were performed in an INSTRON 8862 mechanical testing machine with a strain rate of 5×10^{-4}/s. The microhardness was tested on a Wilson VH3300 (Buehler, Lake Bluff, IL, USA) microhardness tester. Among them, five positions were collected for each sample to obtain the average microhardness value.

3. Results and Discussion

3.1. Defects and Impurities

As shown in Figure 2, the OM images of the 0°, 45° and 90° samples demonstrate that the porosity defects represented by the black dots are distributed inside the grains and at the grain boundaries. It is noteworthy that the unmelted defects are concentrated at the melt pool boundaries. Specifically, the 0° samples have more unmelted defects but smaller sizes and fewer porosity defects. The 90° samples have more serious porosity defects and fewer unmelted defects, while the number and size of unmelted defects and porosity defects of the 45° sample are moderate. Actually, the difference in the building direction changes the defect distribution of the samples. During LPBF processing, different thermal histories result in different grain solidification directions and differences in convection, leading to differences in the distribution of unmelted defects and differences in microporous defects [21]. In addition, the spattering of the molten metal varies depending on the direction of solidification, leading to differences in the distribution of the defects [23–26].

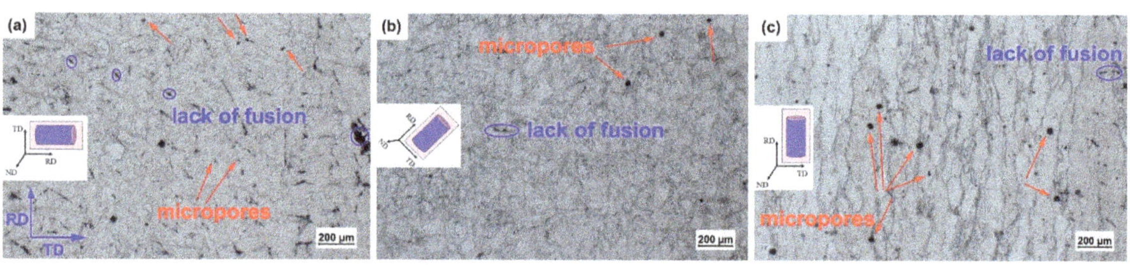

Figure 2. Optical microstructure of (**a**) 0°, (**b**) 45° and (**c**) 90° samples (the rolling direction (RD) and transverse direction (TD) planes are the observation planes).

Slight changes in the Ni/Ti ratios can cause drastic changes in the phase transformation temperatures [27,28]. Among them, the introduction of impurity elements could change the phase composition of the samples by forming the second phases, i.e., $Ti_4Ni_2O_X$, TiN, TiO_2 and TiC, which will change the transformation temperatures. The functional properties of the samples will change further [29,30]. Chemical element analysis was performed on the virgin powder and the samples. The results are shown in Figure 3, and the LPBF samples have a slightly higher oxygen content than the virgin powder. Among them, the 0° sample has the highest oxygen content, while the 45° sample has the lowest. The results are consistent with the analysis of the OM images, where the 0° sample has the highest number of defects, while the 45° sample has the lowest number of defects. In addition, the carbon and nitrogen contents of the as-fabricated samples are lower than those of the virgin powder, which indicates that the high temperature of the laser cladding process acts on the powder, resulting in the loss of carbon or nitrogen.

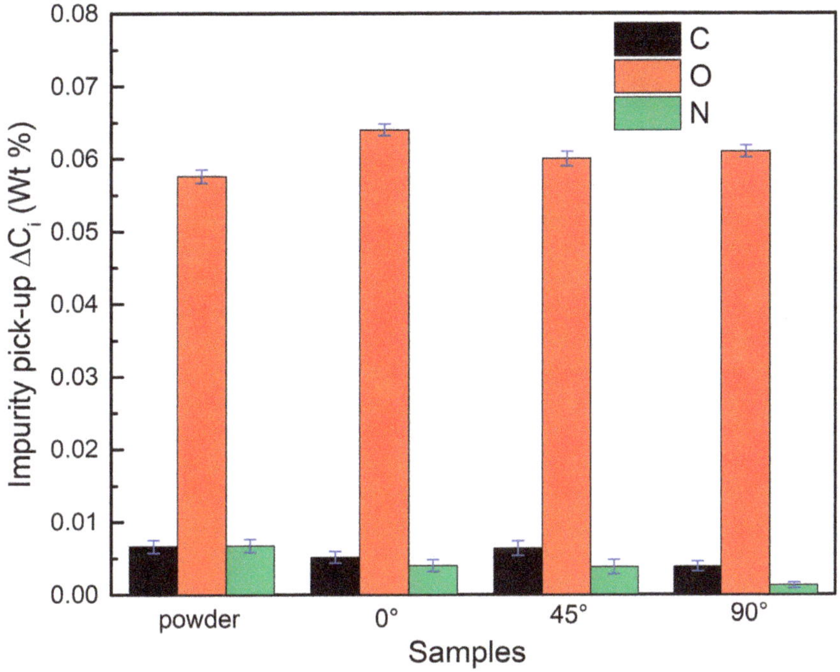

Figure 3. The contents of carbon, oxygen and nitrogen in the virgin powder and 0°, 45° and 90° samples.

3.2. *Microstructure Analysis*

LPBF has the characteristics of directional temperature gradients and interlayer remelting. The grains could grow along the highest thermal gradient, and the competition between the grains for growth will also lead to the preferential growth of favourable crystallographic planes and favourable crystallographic directions during the processing of LPBF. Obviously, the building directions will significantly change the preferred orientation of the samples. Therefore, the texture of the LPBFed NiTi SMAs will be changed by the difference in the building directions in this work [31–34].

To clarify the texture of the samples with different building directions, EBSD analysis was performed. As shown in Figure 4a, the inverse pole figure (IPF) of the 0° sample along the RD-TD direction shows a clear solidification texture of <100>//TD along the VD. According to the pole figure of the 45° sample (Figure 4b), the {001} pole appears at 45° from the centre of the projection plane, indicating that the {001} texture is rotated by 45° compared to the 0° sample. Essentially, the {001} of the 45° sample is still growing parallel to the VD. As shown in Figure 4c, the 90° sample has a strong solidification texture of <100>//RD//VD. Interestingly, the pole figure of the 90° sample overlaps with that of the 0° sample by rotating the pole figure of the 90° sample by 90°, which indicates that the 90° sample has the same preferred orientation as the 0° sample. According to the analysis of the EBSD results, the preferred orientation of the LPBFed NiTi SMAs can be obviously changed by adjusting the building direction; meanwhile, a <100>//VD was formed for all samples with different building directions.

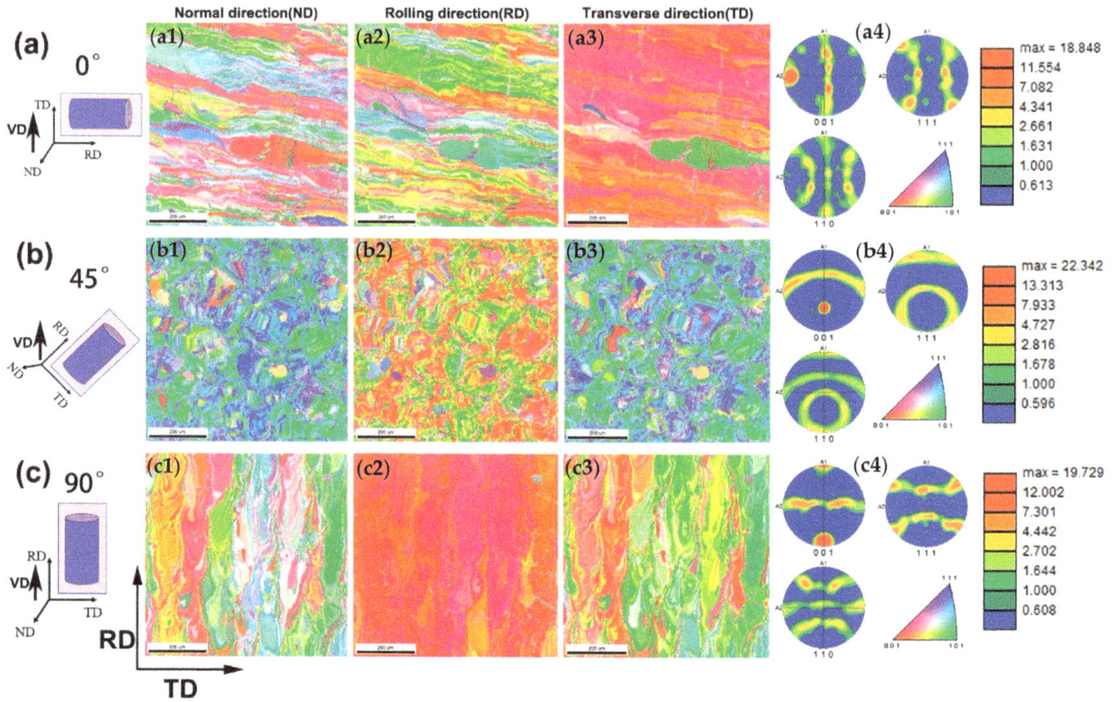

Figure 4. Schematic of the EBSD collection position for the (**a**) 0°, (**b**) 45° and (**c**) 90° samples: (**a1–c1**) are IPF orientation maps of LPBFed NiTi SMAs along the normal direction; (**a2–c2**) are IPF orientation maps of LPBFed NiTi SMAs along the rolling direction; (**a3–c3**) are IPF orientation maps of LPBFed NiTi SMAs along the transverse direction; and (**a4–c4**) are corresponding {001}, {110} and {111} pole figures of the 0°, 45° and 90° samples.

Figure 5 shows the phase composition, kernel average misorientation (KAM) and grain boundary distribution of samples with different building directions. As shown in Figure 5a,d,g, the martensite phase (shown in green) content of the 0°, 45° and 90° samples are 5.5%, 0.7% and 3.1%, respectively. Obviously, the martensite phase content of the 45° sample is lower than that of the others, which indicates that the martensite phase transformation temperature of the 45° sample is lower than that of the others. The KAM value is the statistic for calculating the overall average misorientation of the pixel points collected, as shown in Figure 5b,e,h. The KAM values of the three samples are basically the same, i.e., 0.73, 0.72 and 0.73, respectively. Moreover, the grain boundary distribution of different samples is shown in Figure 5c,f,i. Generally, the lattice distortion is stronger in the area of the martensitic phase. the area perpendicular to the melt pool that is parallel to the building direction. Meanwhile, as shown in the KAM maps, there is also a large lattice distortion at the high-angle grain boundary, which indicates that the grains at this location are small, and more nucleation sites accompanied by large heat flow gradients during the LPBF process have led to the formation of more fine grains.

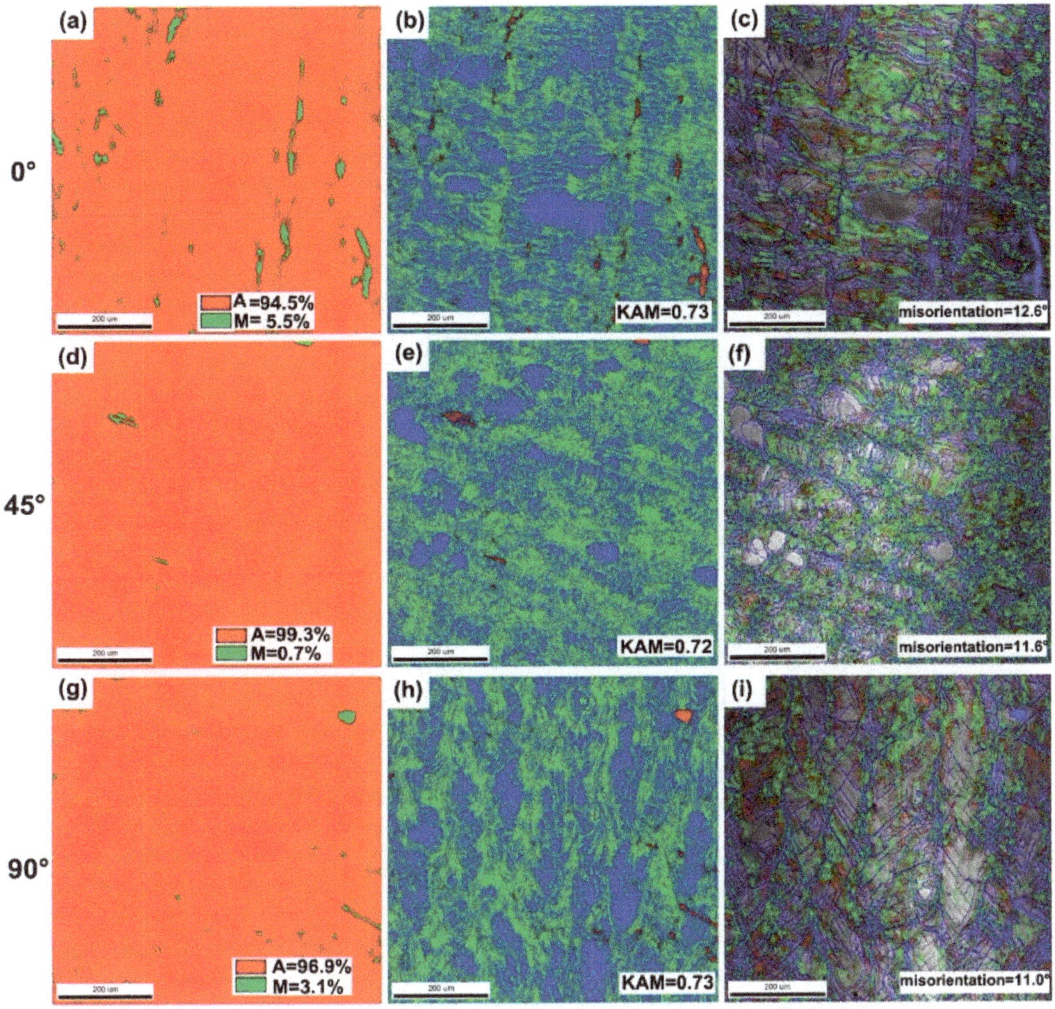

Figure 5. (**a–c**) EBSD analyses of the 0° sample. (**d–f**) EBSD analyses of the 45°sample. (**g–i**) EBSD analyses of the 90°sample. (**a,d,g**) Phase composition analyses. (**b,e,h**) KAM analyses. (**c,f,i**) Grain boundary distribution analyses.

Figure 6 shows the geometrically necessary dislocation (GND) densities of the 0°, 45° and 90° samples. The specific values calculated by EBSD are shown in Figure 6d. Among them, the GND densities of the 0°, 45° and 90° samples are $4.45 \times 10^{13}/m^2$, $3.89 \times 10^{13}/m^2$ and $3.84 \times 10^{13}/m^2$, respectively. In contrast, the 45° sample has the highest GND density, while the densities of the 0° and 90° samples were relatively low. Since the martensitic phase contains more dislocations and twin substructures, which lead to an increase in the overall dislocation density value, the limit of the misorientation value for calculating the GND density is set to 1.5°, excluding the influence of the martensitic phase.

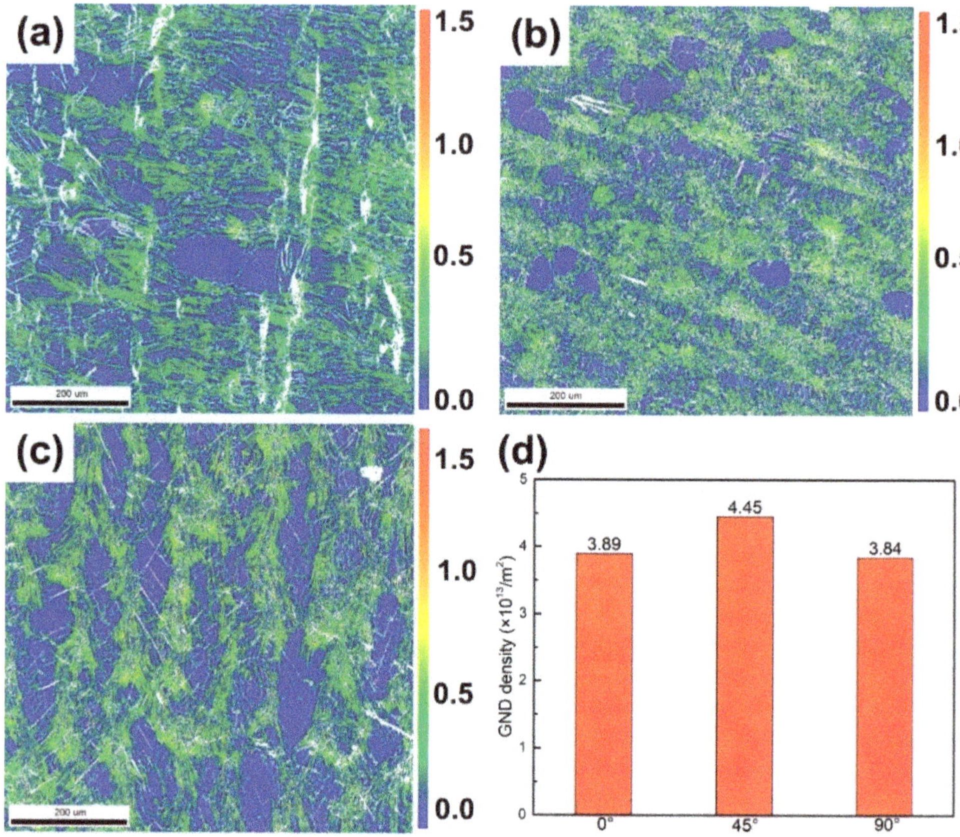

Figure 6. (**a**) GND distribution of the 0° sample, (**b**) 45° sample and (**c**) 90° sample and their corresponding (**d**) GND density values.

To further analyze the effect of the building directions on the samples, TEM was used to observe the phase composition and phase distribution of different samples. As shown in Figure 7, a large number of Ti_2Ni phases or $Ti_4Ni_2O_X$ phases with sizes ranging from 50 to 100 nm can be observed. In addition, the lath-shaped B19′ martensite phase with sizes ranging from 100 to 300 nm can also be observed. Selected area electron diffraction was applied to identify the $Ti_2Ni/Ti_4Ni_2O_X$ phase and the B19′ martensite phase. Similarly, Figure 8 shows the distribution of dislocations, Ti_2Ni phase and B19′ martensite phase of all the samples. Among them, the size of the $Ti_2Ni/Ti_4Ni_2O_X$ phase in the 0° and 90° samples is larger than that in the 45° sample. In addition, according to the TEM bright-field morphologies, the dislocation density of the 45° sample is higher than that of the 0° and 90° samples, while more martensite phases were observed in the 0° and 90° samples. Based on the phenomenon reflected by the TEM morphology above, we found that the statistical results of the GND density and martensite phase content are consistent with the EBSD analysis. TEM analysis proved that larger sized $Ti_2Ni/Ti_4Ni_2O_X$ phases were associated with higher oxygen content in the 0° and 90° samples and that samples with different building directions contained both $Ti_2Ni/Ti_4Ni_2O_X$ phases and martensite phase.

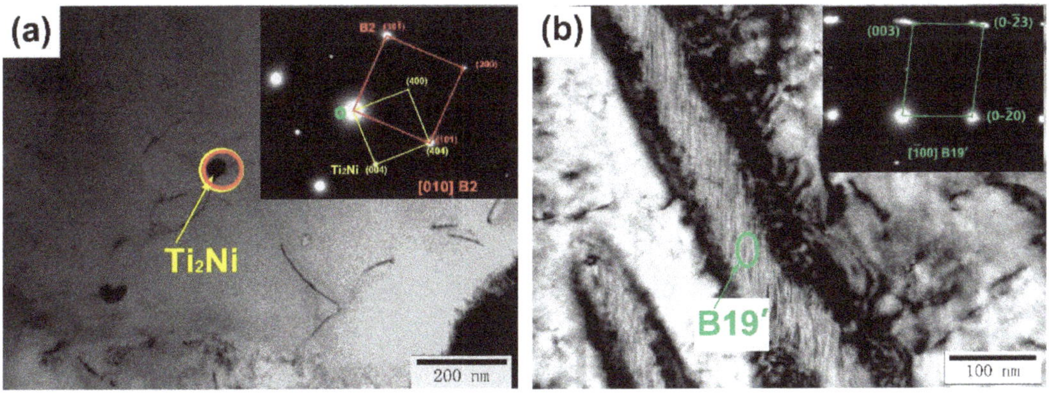

Figure 7. (**a**) Bright field image and selected area electron diffraction in the corresponding circle region for the 0° sample (matrix and Ti_2Ni phase). (**b**) Bright field image and selected electron diffraction in the corresponding elliptic region for the 0° sample.

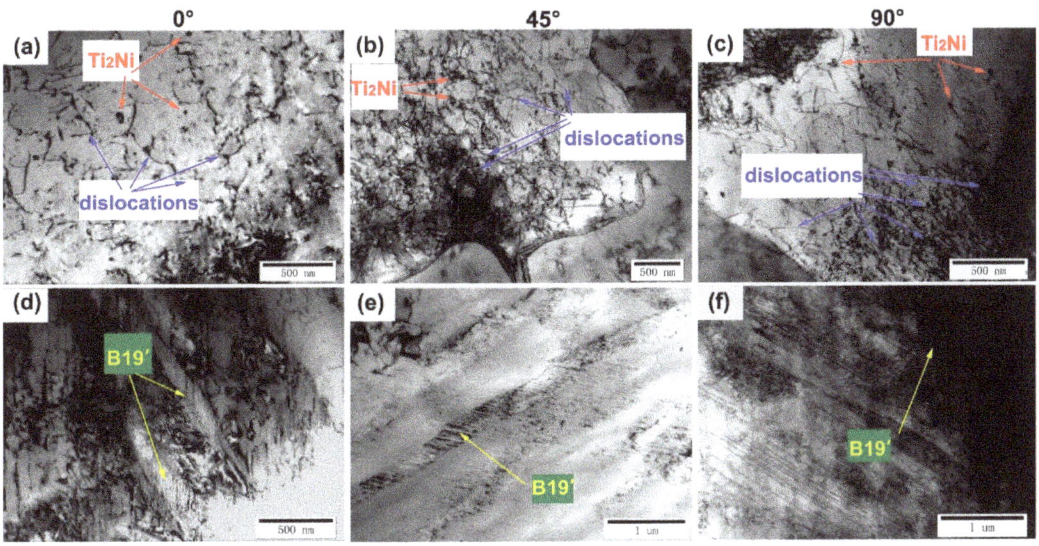

Figure 8. (**a**,**d**) TEM morphology of 0° samples. (**b**,**e**) TEM morphology of 45° samples. (**c**,**f**) TEM morphology of 90° samples.

3.3. Phase Transformation Analysis

The reversible phase transformation between the austenite phase and martensite phase is the theoretical basis for the shape memory effect and superelasticity of NiTi alloys. Among them, the martensite start temperature (Ms), martensite finish temperature (Mf), austenite start temperature (As) and austenite finish temperature (Af) are the important phase transformation temperatures during the phase transformation process. Figure 9a shows the result of the DSC measurement of the three samples. The corresponding phase transformation temperatures obtained by the tangent method are listed in Table 2. The Ms values of the 0°, 45° and 90° samples are 22.1 °C, 11.4 °C and 13 °C, respectively. Obviously, the Ms of the 0° sample was the highest, while that of the 45° sample was the lowest. Correspondingly, the order of the martensite phase content in all samples at ambient temperature (15 °C) was 0° sample > 90° sample > 45° sample, which is consistent with the result of the EBSD analysis. For each sample, five EDS results were collected to calculate the average value, and the Ni content results are shown in Figure 9b. It can be seen that the higher the Ni content of the sample, the lower the phase transformation temperature, which is consistent with the DSC results.

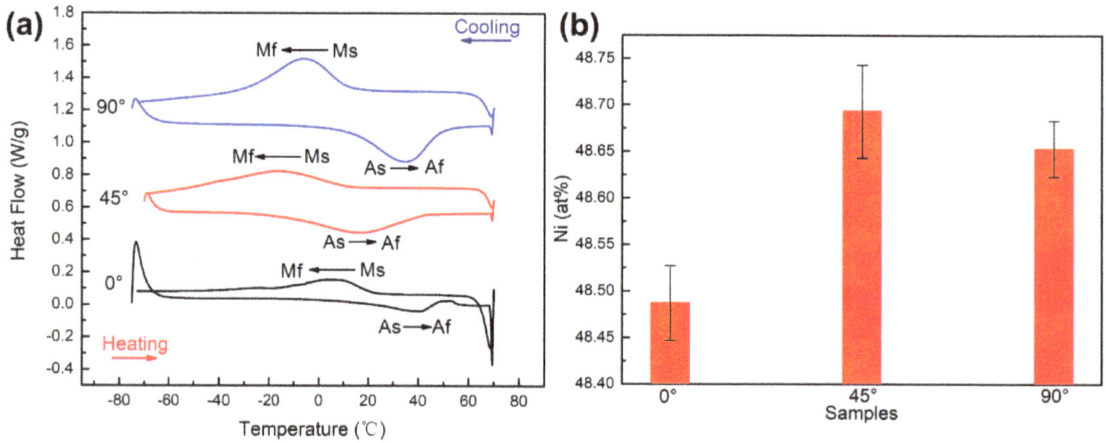

Figure 9. (**a**) DSC curves of 0°, 45° and 90° samples. (**b**) Corresponding Ni content of 0°, 45° and 90° samples.

Table 2. Phase transformation temperatures of the 0°, 45° and 90° samples.

Sample	0°	45°	90°
Mf (°C)	−17.5	−54.2	−30.7
Ms (°C)	22.1	11.4	13
As (°C)	13.9	−15.9	11.1
Af (°C)	52.3	43.7	51

The evaporation of Ni, the introduction of impurity elements, and the difference in thermal conductivity are the reasons for the difference in the phase transformation temperature. During the processing of LPBF, a higher energy input could lead to the evaporation of Ni, which in turn changes the Ni/Ti ratio in the NiTi alloys [30,35]. Nevertheless, the LPBF processing parameters were the same except for the building direction in this work, so the evaporation of Ni in different samples did not have a significant effect on the phase transformation temperature. Second, impurities are introduced, i.e., C, O, N, etc. Walker et al. [36] pointed out that the introduction of impurities will react with the matrix elements to form compounds, i.e., TiC, $Ti_4Ni_2O_X$, TiO_2, etc., which will affect the Ni/Ti ratios and further the phase transformation temperature. As shown in Figure 3, the results of the chemical test demonstrate that the content of oxygen is 0° sample > 90° sample > 45° sample. Obviously, the loss of Ni was greatest in the 0° sample, which led to a decrease in the Ni/Ti ratio and an increase in the phase transformation temperature. Last, the difference in thermal conductivity. As described above, the raw powder and LPBF parameters used in this work are the same. Therefore, it is considered that thermal conductivity has no significant effect on the phase transformation temperature. In conclusion, the difference in the content of impurity elements caused by different building directions is the most important factor causing the difference in phase transformation temperatures.

3.4. Mechanical Properties Analysis

Figure 10 shows the microhardness test results for the three samples, with the 45° sample having the highest microhardness and the 0° sample having the lowest microhardness.

Generally, dislocation densities are an important factor for the microhardness of metallic materials. The increase in the dislocation density could lead to an increase in the matrix strength and hardness of the material. As described in Section 3.2, the 45° sample has the highest dislocation density, while the dislocation densities of the 0° and 90° samples were relatively low; therefore, as a result, the highest microhardness values were obtained for the 45° samples. In addition to the dislocation density, phase transformation is also an important factor affecting the microhardness. Stress-induced martensite phase transformation and martensite reorientation may occur during microhardness testing, which can lead to a change in the microhardness values [37]. Based on the phase transformation temperature (Table 2), we can see that the 0° sample has the highest phase transformation temperatures, followed by the 90° sample, while the 45° sample has the lowest. However, as shown in Figure 10, the trend of the microhardness values of the 0° sample, 45° sample and 90° sample are opposite to the phase transformation temperatures: 45° sample > 90° sample > 0° sample. A high phase transformation temperature is prone to induce the stress-induced martensite phase transformation, and martensite reorientation may occur with less difficulty. Therefore, the microhardness value of the 0° sample was the lowest.

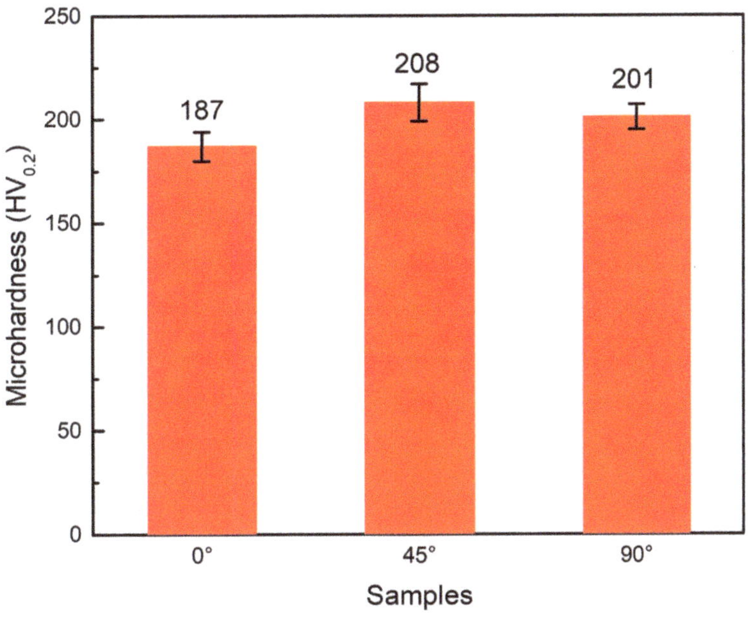

Figure 10. Microhardness of 0°, 45° and 90° samples.

Figure 11 shows the compression stress–strain curves. The recoverable strains and recovery ratios are shown in Table 3. As shown in Figure 11a, all the samples were subjected to a strain of 5% at 15 °C. Table 3 shows that the 45° sample has the highest recoverable strain of 3.2%, while the lowest recoverable strain is 1.8% for the 0° sample under the conditions above. As described in the phase transformation temperature section, the Ms of the 45° sample is the lowest compared with the others. Therefore, during the superelasticity test, the 45° sample with the highest proportion of the austenite phase shows the greatest driving force of recovery and the highest recoverable strain.

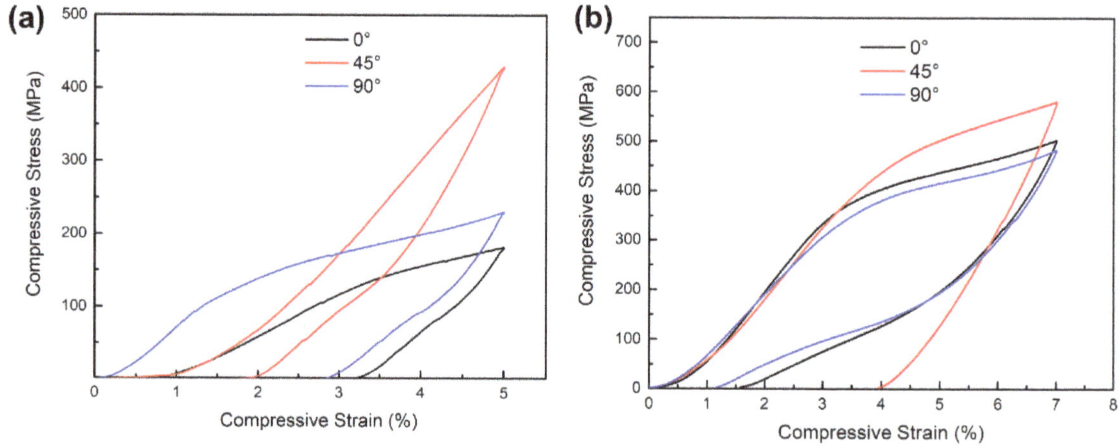

Figure 11. Compressive engineering stress-engineering strain curves for 0°, 45° and 90° samples (**a**) tested at ambient temperature (≈15 °C) and (**b**) tested at Af + 10 °C.

Table 3. Summary of compressive superelastic properties for 0°, 45° and 90° samples.

Test Methods	Tested at 15 °C; 5% Compressive Strain			Tested at Af + 10 °C; 7% Compressive Strain		
Samples	**0°**	**45°**	**90°**	**0°**	**45°**	**90°**
Recoverable strain (%)	1.8	3.2	2.2	5.42	3.06	5.83
Irrecoverable strain (%)	3.2	1.8	2.8	1.58	3.94	1.17
Recovery ratio (%)	36	64	44	77.4	43.7	83.3

As shown in Figure 11b, the superelasticity was tested at the temperature of Af + 10 °C. The recoverable strains of the 0° and 90° samples were improved to 5.42% and 5.83%, respectively. Interestingly, the 90° sample has the best recoverable strain, while the recoverable strain of the 45° sample is the lowest. As discussed above, samples with different building directions have different textures along the loading direction. According to the result of the EBSD analysis, the loading direction of the 90° sample coincides with the <100> orientation. Generally, <100> is a hard orientation, which is not conducive to plastic deformation. Moreover, the 90° sample is more likely to drive stress-induced martensitic phase transformation behaviour without causing dislocation accumulation [38]. Therefore, the 90° sample has the best superelasticity compared to the others at the temperature of Af + 10 °C. The 45° sample has a <110> texture along the loading direction with a relatively large Schmidt factor, which is prone to plastic deformation. Therefore, the plastic deformation will remain in the matrix after unloading and cannot be recovered. Meanwhile, the low Ms of the 45° sample requires a larger stress to drive the martensitic phase transformation, resulting in plastic deformation without a significant stress-induced martensitic phase transformation platform.

Therefore, the superelasticity of the 45° sample is worst compared with the others at the temperature of Af + 10 °C.

A summary of the research work on LPBFed NiTi SMAs with superior superelastic properties (tested at Af + 10 °C) was performed, and the results are shown in Table 4. It can be seen that at smaller deformations, the samples generally have higher strain recovery rates, while larger strain is prone to have higher strain recovery, but the recovery rates are not ideal. The superelastic properties in this work are similar to those in similar studies.

Table 4. Comparison of the optimal superelastic properties (tested at Af + 10 °C) in various LPBFed NiTi.

Reference (Tested at Af + 10 °C)	Recoverable Strain (%)	Recovery Ratio (%)
[39]	5.62	98
[40]	5.5	94.8
[41]	5.32	93.2
This work	5.83	83.3

4. Conclusions

In this study, LPBF-NiTi samples with different building directions (0°, 45° and 90°) were tested and analyzed for microstructure, defect distribution, impurity element, crystallographic orientation, dislocation density, phase transformation behaviour and mechanical properties using the same LPBF processing parameters, and the following conclusions were obtained:

(1) The samples with different building directions exhibited distinct defect distributions. The 0° and 90° samples, which had more unmelted defects, demonstrated a higher oxygen content. The introduction of oxygen impurities altered the Ni/Ti ratios in the matrix and significantly raised the phase transformation temperature. The order of martensitic phase transformation temperature was as follows: 0° sample > 90° sample > 45° sample. This order is in line with the oxygen content present in each sample. Consequently, the quantity of martensite phase in samples with different building directions also followed the same trend: 0° sample > 90° sample > 45° sample.

(2) The increase in dislocation density enhanced the microhardness of the matrix, while the increase in phase transformation temperature facilitated the occurrence of stress-induced martensitic phase transformation. This phenomenon was also the primary factor responsible for altering the hardness of LPBFed NiTi SMAs.

(3) LPBFed NiTi SMAs with different building directions had different preferred orientations along the loading direction, but all had textures of VD//<100>.

The stress-induced martensitic phase transformation was related to crystallographic orientation. Unfavourable orientations can result in plastic deformation and obscure the platform of the stress-induced martensitic phase transition. When the phase transformation temperature (Af) exceeded the ambient temperature, it significantly impacted the superelastic properties, with higher temperatures leading to poorer superelastic properties.

The main idea of this study was to change the introduction of impurity elements in NiTi alloys and alter their microstructures by adjusting the LPBF building direction, which in turn affects their phase transformation temperatures and superelasticity. In conclusion, for engineering applications, the 45° sample had a lower phase transformation temperature, which was more favourable for superelastic recovery under certain temperatures (As < testing temperature < Af), while the preferred orientation of the 90° sample ensured a greater superelastic recovery strain when tested at Af + 10 °C.

Author Contributions: S.W. conceived and designed the experiments; X.Y., S.W., C.Z. and X.Z. (Xiaolong Zhang), performed the experiments; X.Y., S.W., J.C., C.L. and H.P. analysed the data; X.Y., X.Z. (Xinyao Zhang), P.L., Z.W. and L.G. contributed reagents/materials/analysis tools; X.Y. wrote the paper. All authors have read and agreed to the published version of the manuscript.

Funding: This research received no external funding.

Data Availability Statement: The data presented in this study are available on request from the corresponding author. The data are not publicly available due to privacy.

Conflicts of Interest: The authors declare no conflict of interest.

References

1. Otsuka, K.; Ren, X. Physical metallurgy of Ti-Ni-based shape memory alloys. *Prog. Mater. Sci.* **2005**, *50*, 511–678. [CrossRef]
2. Frenzel, J.; Zhang, Z.; Somsen, C.; Neuking, K.; Eggeler, G. Influence of carbon on martensitic phase transformations in NiTi shape memory alloys. *Acta Mater.* **2007**, *55*, 1331–1341. [CrossRef]
3. Frenzel, J.; George, E.; Dlouhy, A.; Somsen, C.; Wagner, M.-X.; Eggeler, G. Influence of Ni on martensitic phase transformations in NiTi shape memory alloys. *Acta Mater.* **2010**, *58*, 3444–3458. [CrossRef]
4. Otubo, J.; Rigo, O.; Neto, C.M.; Mei, P. The effects of vacuum induction melting and electron beam melting techniques on the purity of NiTi shape memory alloys. *Mater. Sci. Eng. A* **2006**, *438*, 679–682. [CrossRef]
5. Resnina, N.; Palani, I.; Belyaev, S.; Prabu, S.M.; Liulchak, P.; Karaseva, U.; Manikandan, M.; Jayachandran, S.; Bryukhanova, V.; Sahu, A.; et al. Structure, martensitic transformations and mechanical behavior of NiTi shape memory alloy produced by wire arc additive manufacturing. *J. Alloys Compd.* **2020**, *851*, 156851. [CrossRef]
6. Khorasani, M.; Ghasemi, A.; Leary, M.; Downing, D.; Gibson, I.; Sharabian, E.G.; Veetil, J.K.; Brandt, M.; Bateman, S.; Rolfe, B. Benchmark models for conduction and keyhole modes in laser-based powder bed fusion of Inconel 718. *Opt. Laser Technol.* **2023**, *164*, 109509. [CrossRef]
7. Lavecchia, F.; Pellegrini, A.; Galantucci, L.M. Comparative study on the properties of 17-4 PH stainless steel parts made by metal fused filament fabrication process and atomic diffusion additive manufacturing. *Rapid Prototyp. J.* **2023**, *29*, 393–407. [CrossRef]
8. Xue, A.; Lin, X.; Wang, L.; Lu, X.; Yuan, L.; Ding, H.; Huang, W. Achieving fully-equiaxed fine β-grains in titanium alloy produced by additive manufacturing. *Mater. Reasearch Lett.* **2023**, *11*, 60–68. [CrossRef]
9. Qi, Y.; Hu, Z.; Zhang, H.; Nie, X.; Zhang, C.; Zhu, H. High strength Al-Li alloy development for laser powder bed fusion. *Addit. Manuf.* **2021**, *47*, 102249. [CrossRef]
10. Yu, H.; Yang, J.; Yin, J.; Wang, Z.; Zeng, X. Comparison on mechanical anisotropies of selective laser melted Ti-6Al-4V alloy and 304 stainless steel. *Mater. Sci. Eng. A* **2017**, *695*, 92–100. [CrossRef]
11. Shipley, H.; McDonnell, D.; Culleton, M.; Coull, R.; Lupoi, R.; O'Donnell, G.; Trimble, D. Optimisation of process parameters to address fundamental challenges during selective laser melting of Ti-6Al-4V: A review. *Int. J. Mach. Tools Manuf.* **2018**, *128*, 1–20. [CrossRef]
12. Ma, Q.; Zhang, L.; Ding, J.; Qu, S.; Fu, J.; Fu, M.W.; Song, X.; Wang, M.Y. Analytical design of stretching-dominated truss lattices with tailored elasticity from transversely isotropic base materials. *Mater. Design* **2023**, *230*, 111995. [CrossRef]
13. Yang, Y.; Zhan, J.; Sui, J.; Li, C.; Yang, K.; Castany, P.; Gloriant, T. Functionally graded NiTi alloy with exceptional strain-hardening effect fabricated by SLM method. *Scr. Mater.* **2020**, *188*, 130–134. [CrossRef]
14. Yuan, L.; Gu, D.; Lin, K.; Liu, H.; Sun, J.; Yang, J.; Liu, X.; Chen, W.; Song, Y. Electrically Actuated Shape Recovery of NiTi Components Processed by Laser Powder Bed Fusion after Regulating the Dimensional Accuracy and Phase Transformation Behavior. *Chin. J. Mech. Eng. Addi. Manuf. Frontiers.* **2022**, *4*, 100056. [CrossRef]
15. Qu, S.; Wang, L.; Ding, J.; Fu, J.; Gao, S.; Ma, Q.; Liu, H.; Fu, M.; Lu, Y.; Song, X. Superelastic NiTi Functional Components by High-PrecisionLaser Powder Bed Fusion Process: The Critical Roles of Energy Density and Minimal Feature Size. *Micromachines* **2023**, *14*, 1436. [CrossRef] [PubMed]
16. Safaei, K.; Nematollahi, M.; Bayati, P.; Kordizadeh, F.; Andani, M.T.; Abedi, H.; Poorganji, B.; Elahinia, M. On the crystallographic texture and torsional behavior of NiTi shape memory alloy processed by laser powder bed fusion: Effect of build orientation. *Addit. Manuf.* **2022**, *59*, 103184. [CrossRef]
17. Nematollahi, M.; Saghaian, S.E.; Safaei, K.; Bayati, P.; Bassani, P.; Biffi, C.; Tuissi, A.; Karaca, H.; Elahinia, M. Building orientation-structure-property in laser powder bed fusion of NiTi shape memory alloy. *J. Alloys Compd.* **2021**, *873*, 159791. [CrossRef]
18. Thijs, L.; Sistiaga, M.L.M.; Wauthle, R.; Xie, Q.; Kruth, J.-P.; Van Humbeeck, J. Strong morphological and crystallographic texture and resulting yield strength anisotropy in selective laser melted tantalum. *Acta Mater.* **2013**, *61*, 4657–4668. [CrossRef]
19. Kok, Y.; Tan, X.; Wang, P.; Nai, M.; Loh, N.; Liu, E.; Tor, S. Anisotropy and heterogeneity of microstructure and mechanical properties in metal additive manufacturing: A critical review. *Mater. Des.* **2017**, *139*, 565–586. [CrossRef]
20. Shi, G.; Li, L.; Yu, Z.; Sha, P.; Cao, Q.; Xu, Z.; Liu, Y.; Guo, Y.; Si, J.; Liu, J. Effect of crystallographic anisotropy on phase transformation and tribological properties of Ni-rich NiTi shape memory alloy fabricated by LPBF. *Opt. Laser Technol.* **2023**, *157*, 108731. [CrossRef]
21. Moghaddam, N.S.; Saghaian, S.E.; Amerinatanzi, A.; Ibrahim, H.; Li, P.; Toker, G.P.; Karaca, H.E.; Elahinia, M. Anisotropic tensile and actuation properties of NiTi fabricated with selective laser melting. *Mater. Sci. Eng. A* **2018**, *724*, 220–230. [CrossRef]
22. Chen, Q.; Jing, Y.; Yin, J.; Li, Z.; Xiong, W.; Gong, P.; Zhang, L.; Li, S.; Pan, R.; Zhao, X.; et al. High Reflectivity and Thermal Conductivity Ag-Cu Multi-Material Structures Fabricated via Laser Powder Bed Fusion: Formation Mechanisms, Interfacial Characteristics, and Molten Pool Behavior. *Micromachines* **2023**, *14*, 362. [CrossRef] [PubMed]
23. Li, Z.; Li, H.; Yin, J.; Li, Y.; Nie, Z.; Li, X.; You, D.; Guan, K.; Duan, W.; Cao, L.; et al. A Review of Spatter in Laser Powder Bed Fusion Additive Manufacturing: In Situ Detection, Generation, Effects, and Countermeasures. *Micromachines* **2022**, *13*, 1366. [CrossRef]
24. Yin, J.; Zhang, W.; Ke, L.; Wei, H.; Wang, D.; Yang, L.; Zhu, H.; Dong, P.; Wang, G.; Zeng, X. Vaporization of alloying elements and explosion behavior during laser powder bed fusion of Cu–10Zn alloy. *Int. J. Mach. Tools Manuf.* **2021**, *161*, 103686. [CrossRef]

25. Yang, G.; Xie, Y.; Zhao, S.; Qin, L.; Wang, X.; Wu, B. Quality Control: Internal Defects Formation Mechanism of Selective Laser Melting Based on Laser-powder-melt Pool Interaction: A Review. *Chin. J. Mech. Eng. Addit. Manuf. Front.* **2022**, *3*, 100037. [CrossRef]
26. Yin, J.; Wang, D.; Wei, H.; Yang, L.; Ke, L.; Hu, M.; Xiong, W.; Wang, G.; Zhu, H.; Zeng, X. Dual-beam laser-matter interaction at overlap region during multilaser powder bed fusion manufacturing. *Addit. Manuf.* **2021**, *46*, 102178. [CrossRef]
27. Dadbakhsh, S.; Vrancken, B.; Kruth, J.-P.; Luyten, J.; Van Humbeeck, J. Texture and anisotropy in selective laser melting of NiTi alloy. *Mater. Sci. Eng. A* **2016**, *650*, 225–232. [CrossRef]
28. Gu, D.; Hagedorn, Y.-C.; Meiners, W.; Meng, G.; Batista, R.J.S.; Wissenbach, K.; Poprawe, R. Densification behavior, microstructure evolution, and wear performance of selective laser melting processed commercially pure titanium. *Acta Mater.* **2012**, *60*, 3849–3860. [CrossRef]
29. Wen, S.; Liu, Y.; Zhou, Y.; Zhao, A.; Yan, C.; Shi, Y. Effect of Ni content on the transformation behavior and mechanical property of NiTi shape memory alloys fabricated by laser powder bed fusion. *Opt. Laser Technol.* **2021**, *134*, 106653. [CrossRef]
30. Wang, X.; Yu, J.; Liu, J.; Chen, L.; Yang, Q.; Wei, H.; Sun, J.; Wang, Z.; Zhang, Z.; Zhao, G.; et al. Effect of process parameters on the phase transformation behavior and tensile properties of NiTi shape memory alloys fabricated by selective laser melting. *Addit. Manuf.* **2020**, *36*, 101545. [CrossRef]
31. Mentz, J.; Bram, M.; Buchkremer, H.P.; Stöver, D. Improvement of Mechanical Properties of Powder Metallurgical NiTi Shape Memory Alloys. *Adv. Eng. Mater.* **2010**, *8*, 247–252. [CrossRef]
32. Haberland, C.; Elahinia, M.; Walker, J.M.; Meier, H.; Frenzel, J. On the development of high quality NiTi shape memory and pseudoelastic parts by additive manufacturing. *Smart Mater. Struct.* **2014**, *23*, 104002. [CrossRef]
33. Antonysamy, A.A.; Meyer, J.; Prangnell, P.B. Effect of build geometry on the β-grain structure and texture in additive manufacture of Ti-6Al-4V by selective electron beam melting. *Mater. Charact.* **2013**, *84*, 153–168. [CrossRef]
34. Holland, S.; Wang, X.; Chen, J.; Cai, W.; Yan, F.; Li, L. Multiscale characterization of microstructures and mechanical properties of Inconel 718 fabricated by selective laser melting. *J. Alloys Compd.* **2019**, *784*, 182–194. [CrossRef]
35. Goll, D.; Trauter, F.; Loeffler, R.; Gross, T.; Schneider, G. Additive Manufacturing of Textured FePrCuB Permanent Magnets. *Micromachines* **2021**, *12*, 1056. [CrossRef]
36. Safaei, K.; Andani, N.T.; Poorganji, B.; Andani, M.T.; Elahinia, M. Controlling texture of NiTi alloy processed by laser powder bed fusion:Smart build orientation and scanning strategy. *Addit. Manuf. Lett.* **2023**, *5*, 100126. [CrossRef]
37. Bormann, T.; Schumacher, R.; Müller, B.; Mertmann, M.; de Wild, M. Tailoring Selective Laser Melting Process Parameters for NiTi Implants. *J. Mater. Eng. Perform.* **2012**, *21*, 2519–2524. [CrossRef]
38. Walker, J.M.; Haberland, C.; Andani, M.T.; Karaca, H.E.; Dean, D.; Elahinia, M. Process development and characterization of additively manufactured nickel-titanium shape memory parts. *J. Intell. Mater. Syst. Struct.* **2016**, *27*, 2653–2660. [CrossRef]
39. Saedi, S.; Moghaddam, N.S.; Amerinatanzi, A.; Elahinia, M.; Karaca, H.E. On the effects of selective laser melting process parameters on microstructure and thermomechanical response of Ni-rich NiTi. *Acta Mater.* **2018**, *144*, 552–560. [CrossRef]
40. Saedi, S.; Turabi, A.S.; Andani, M.T.; Haberland, C.; Karaca, H.; Elahinia, M. The influence of heat treatment on the thermomechanical response of Ni-rich NiTi alloys manufactured by selective laser melting. *J. Alloys Compd.* **2016**, *677*, 204–210. [CrossRef]
41. Lu, H.; Liu, L.; Yang, C.; Luo, X.; Song, C.; Wang, Z.; Wang, J.; Su, Y.; Ding, Y.; Zhang, L.; et al. Simultaneous enhancement of mechanical and shape memory properties by heat-treatment homogenization of Ti2Ni precipitates in TiNi shape memory alloy fabricated by selective laser melting. *J. Mater. Sci. Technol.* **2021**, *101*, 205–216. [CrossRef]

micromachines

Article

The Influence of Laser Process Parameters on the Adhesion Strength between Electroless Copper and Carbon Fiber Composites Determined Using Response Surface Methodology

Xizhao Wang [1,2], Jianguo Liu [2], Haixing Liu [1], Zhicheng Zhou [2], Zhongli Qin [3,*] and Jiawen Cao [3]

[1] Institute of Laser and Intelligent Manufacturing Technology, South-Central Minzu University, Wuhan 430074, China; wangxizhao@scuec.edu.cn (X.W.); liuhaixing98@gmail.com (H.L.)
[2] Wuhan National Laboratory for Optoelectronics (WNLO), Huazhong University of Science and Technology (HUST), Wuhan 430074, China; liujg@mail.hust.edu.cn (J.L.); m202170333@hust.edu.cn (Z.Z.)
[3] School of Electronics and Information Engineering, Hubei University of Science and Technology, Xianning 437100, China; caojiawen@hbust.edu.cn
* Correspondence: qinzhongli@hbust.edu.cn; Tel.: +86-18062796898

Abstract: Laser process technology provides a feasible method for directly manufacturing surface-metallized carbon fiber composites (CFCs); however, the laser's process parameters strongly influence on the adhesion strength between electroless copper and CFCs. Here, a nanosecond ultraviolet laser was used to fabricate electroless copper on the surface of CFCs. In order to achieve good adhesion strength, four key process parameters, namely, the laser power, scanning line interval, scanning speed, and pulse frequency, were optimized experimentally using response surface methodology, and a central composite design was utilized to design the experiments. An analysis of variance was conducted to evaluate the adequacy and significance of the developed regression model. Also, the effect of the process parameters on the adhesion strength was determined. The numerical analysis indicated that the optimized laser power, scanning line interval, scanning speed, and pulse frequency were 5.5 W, 48.2 μm, 834.0 mm/s, and 69.5 kHz, respectively. A validation test confirmed that the predicted results were consistent with the actual values; thus, the developed mathematical model can adequately predict responses within the limits of the laser process parameters being used.

Keywords: carbon fiber composites; adhesion strength; response surface methodology; optimization; laser process parameters

Citation: Wang, X.; Liu, J.; Liu, H.; Zhou, Z.; Qin, Z.; Cao, J. The Influence of Laser Process Parameters on the Adhesion Strength between Electroless Copper and Carbon Fiber Composites Determined Using Response Surface Methodology. *Micromachines* **2023**, *14*, 2168. https://doi.org/10.3390/mi14122168

Academic Editor: Antonio Ancona

Received: 1 November 2023
Revised: 19 November 2023
Accepted: 21 November 2023
Published: 29 November 2023

1. Introduction

As appealing materials, carbon fiber composites (CFCs) are extensively applied in many fields, including the automobile and aircraft/aerospace industries, due to their unique properties: excellent mechanical performance, low density, high thermal stability, and so on [1–3]. However, they also have some obvious disadvantages, such as poor electrical conductivity and low erosion resistance, which limit their further application in the field of aircraft structure/manufacture, as materials with better electrical conductivity can effectively avoid damage from lightning strikes [4]. Therefore, there is an urgent need to establish a project that focuses on protecting the composite outer skin of aircraft against lightning damage. The protective layer formed via the metallization of CFC surfaces exhibits excellent electrical conductivity, making it one of the most effective solutions to solve this problem.

In recent years, there has been growing interest in the surface metallization of CFCs [5]. To make CFCs electrically conductive, metallic materials can be coated onto the polymer surface [6]. However, the intrinsic hydrophobicity of CFC surfaces leads to poor adhesion between the metal layer and CFCs, which severely limits their wide application. Therefore, obtaining a high-strength metal layer has become critical in the fabrication of conductive

CFCs. Various technologies have been developed to fabricate metal layers with high adhesion to CFC surfaces, including electroless plating, electrodeposition, spraying, and plasma etching [7–10]. For instance, Chen et al. deposited copper onto CFCs using an electroless method and obtained copper-coated CFCs with good adhesion and excellent conductivity [11]. Wang et al. reported a pulse-reverse electrodeposition process to fabricate Ni-coated CFCs, which exhibited good adhesion [12]. Archambault et al. successfully cold-sprayed copper onto CFCs to prepare Cu-coated CFCs with an adhesion strength of 2.6 MPa $\pm$ 0.8 MPa [13]. Prysiazhnyi et al. employed nitrogen plasma modification for surface treatment to enhance the adhesion between electroless copper and CFCs [14]. However, the above-mentioned approaches suffer from inherent disadvantages, e.g., a tedious operation process, high cost, corrosive or poisonous chemical reagents, poor adhesion strength, and so on, which have seriously limit their practical applications.

In comparison, laser process technology offers numerous many advantages, including noncontact operation, high precision, region selection, controllability, simplicity, and applicability to a wide range of materials [15–19], which make it a promising method for preparing metal-coated CFCs with high adhesion. For instance, Gustke et al. proposed a pulsed laser roughening method for enhancing the adhesion strength between a sprayed copper layer and CFCs. It was shown that the adhesion strength was increased by 200% [20]. Li et al. investigated the effects of the pretreatment of a picosecond infrared laser and an excimer ultraviolet laser on improving the shear strength between an aluminum alloy and CFCs, respectively. It was revealed that the shear strength was determined based on chemical bonding rather than mechanical interlocking [21]. Palavra et al. studied the effect of laser surface pretreatment parameters on the adhesion strength between a titanium layer and CFCs, including the laser power and pulse energy [22]. In summary, most researchers in this field have primarily focused on the impact of laser treatment on enhancing the adhesion strength between metal layers and CFCs. The influence of the key input parameters of the laser process, such as the laser power, scanning line interval, scanning speed, and pulse frequency, on the adhesion strength between metal layers and CFCs have not been taken into account. In addition, the interactive effects of process input parameters on adhesion strength have not been studied.

To understand their effects on adhesion strength, laser process parameters should be extensively analyzed and optimized. Response surface methodology (RSM), used for experimental design, is among the most effective optimization techniques [23], which can not only reveal the relationship between the input variables and output responses but also predict the optimal output responses through numerical optimization under certain conditions. And, the RSM is often employed in numerous fields to find the optimal parametric combination from a group of given variables for achieving the desired output responses [24–26].

Generally, the adhesion strength between electroless copper and CFCs is influenced by numerous factors. Thus, to obtain excellent adhesion strength with copper plating, it is necessary to investigate the effect of different laser process parameters on the adhesion strength. However, to the best of our knowledge, the influence of the key laser process parameters on the adhesion strength between electroless copper and CFCs is relatively poorly understood. The interactive effects of laser process parameters on adhesion strength have not been reported.

Central composite design (CCD) is the primary type of RSM, and it is effective for obtaining more information about experimental variables and experimental errors with the fewest experimental cycles. In this study, a standard RSM with a CCD was applied to develop a model for optimizing the laser process parameters to achieve good adhesion strength. The key laser process parameters considered were laser power, scanning line interval, scanning speed, and pulse frequency. An analysis of variance (ANOVA), main effect plots, contour plots, and the corresponding 3D response surface were employed to assess the effect of each factor on the adhesion strength between electroless copper and CFCs. Furthermore, a mathematical model was developed to predict the optimal adhesion strength.

2. Materials and Methods

2.1. Materials

Commercial CFC sheets, which had a thickness of 2.0 mm and consisted of carbon fiber and epoxy polymer, were purchased from Beijing Plastic Manufacturing Co., Ltd. (Beijing, China). Before laser ablation, the CFC sheet was cut into small pieces (30.0 mm × 30.0 mm). Next, the samples were rinsed ultrasonically with acetone for 15 min, followed by rinsing with deionized water for an additional 15 min. Pentahydrate copper sulfate ($CuSO_4{\cdot}5H_2O$) and EDTA disodium (Na_2EDTA) were purchased from Sinopharm Chemical Reagent Co., Ltd. (Shanghai, China). Formaldehyde (HCHO), sodium hydroxide (NaOH), and potassium sodium tartrate tetrahydrate ($C_4H_{12}KNaO_{10}$) were purchased from Aladdin Reagent (Minneapolis, Michigan, USA). All chemicals were analytical grade and used directly without further purification.

2.2. Fabrication of Electroless Copper on CFCs

Figure 1 illustrates the fabrication process of electroless copper on the CFC surface, including formation, coarsening, removal and electroless plating: (1) The CFC sheet could not initiate the process of electroless plating itself due to the lack of catalytically active centers (i.e., active seeds). Many precious metals, such as gold (Au), palladium (Pd), and silver (Ag), are usually used as active seeds. Among them, Pd is optimal due to its excellent catalytic activity. As shown in Figure 1, the CFC sheets were first immersed in a 1.0 g/L $PdCl_2$ aqueous solution for 10 min and then taken out and dried. Thus, the CFC sheets coated with $PdCl_2$ films were obtained. (2) The coated CFC sheets were coarsened using laser direct ablation equipment (Figure 2a). As shown in Figure 2b, the direct laser ablation system was mainly composed of a 355 nm nanosecond laser, an optical system, a multiaxis workbench system, and a controlling system. The maximum average laser power was about 20 W, the highest scanning speed was about 10,000 mm/s, the pulse width was about 16 ns, the value of the beam quality factor M^2 was less than 1.2, and the repetition rate ranged from 200 kHz to 2 MHz. A two-mirror galvanometric scanner with an F-theta objective lens was employed to focus and scan the laser beam in the x−y direction. The Gaussian-profile laser beam at $1/e^2$ of its maximum intensity had a focused spot diameter of about 15 μm. During laser direct ablation, the CFC sheet (Figure 3a)-coated $PdCl_2$ films were mounted on an x-y-z translation stage precisely controlled using a computer. A line-by-line scanning method in the x direction and then in the y direction was used to fabricate a rough surface (Figure 3b–f) on the CFC sheets for obtaining the electroless copper with high adhesion in an atmospheric environment. (3) The ablated CFC sheets were cleaned ultrasonically in deionized water to selectively remove the active seeds (i.e., $PdCl_2$) in the non-laser-irradiated zone, while keeping the active seeds in the laser-irradiated zone. (4) The cleaned CFC sheets were immersed into a commercial copper bath solution for selective copper plating. The obtained electroless coppers on the surfaces of CFC sheet are shown in Figure 4.

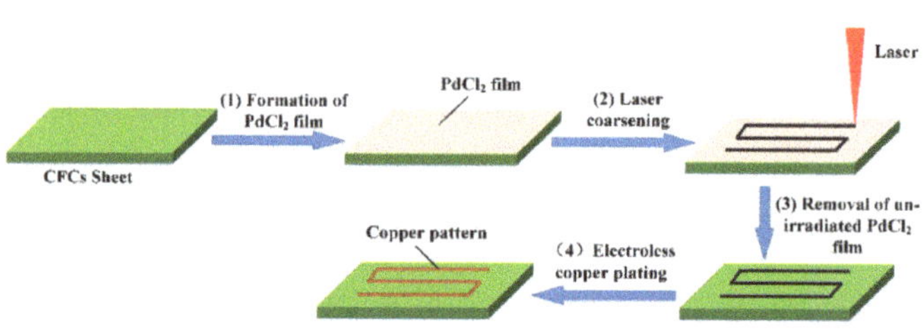

Figure 1. Schematic diagram of the fabrication process of laser-induced copper plating on CFC surface.

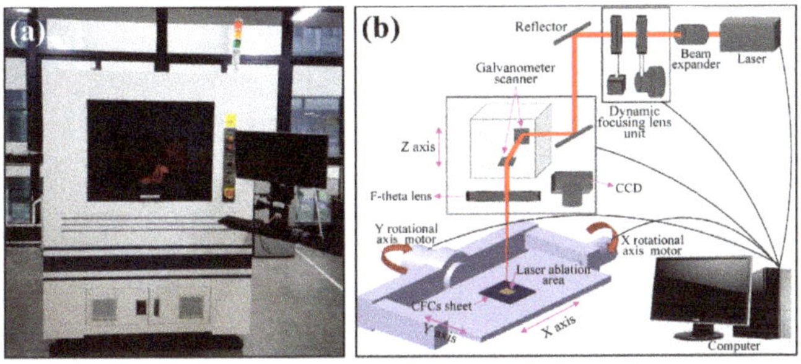

Figure 2. Nanosecond ultraviolet laser 3D process system: (**a**) equipment photograph and (**b**) schematic diagram.

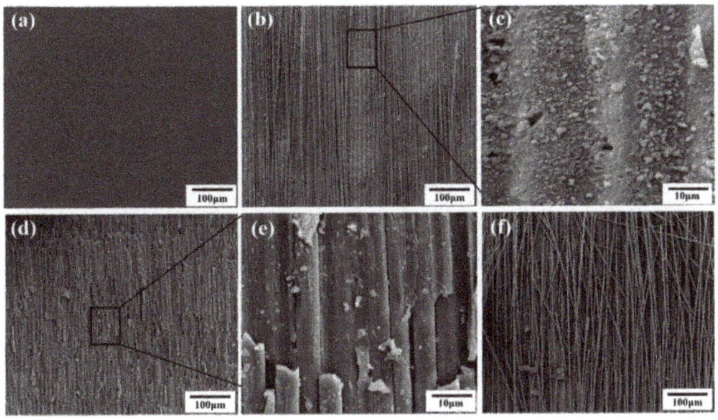

Figure 3. SEM images of pristine CFC surface (**a**) and laser-ablated CFC surface (**b–f**); (**c**,**e**) are the local magnifications of (**b**,**d**), respectively. Laser power for (**b**,**d**,**f**) is 2.5 W, 5 W and 7 W, respectively, while other laser parameters (U = 40 μm, V = 300 mm/s, f = 60 kHz) remained unchanged and were scanned only once with the laser.

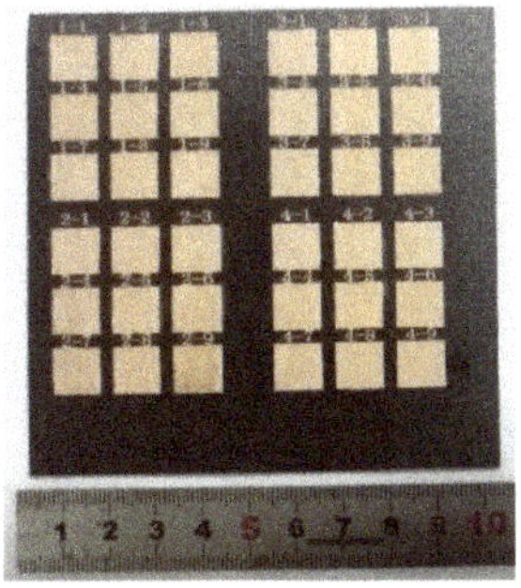

Figure 4. Specimen photograph of electroless copper pattern on CFC surface.

2.3. Measurements and Characterization

The morphological structures of the laser-ablated CFC surfaces were analyzed using a Nova NanoSEM 450 scanning electron microscope (SEM, FEI, Hillsboro, OR, USA). The adhesion strength between the prepared copper plating layer and CFC sheet was measured via the vertical pulling force method shown in Figure 5. First, a copper plating layer on the CFC surface with dimensions of 2.0 mm × 2.0 mm was prepared. Then, it was welded to a metal wire using lead–tin solder. Next, the pulling force of the dynamometer was gradually increased along the direction perpendicular to the CFC sheet, and the maximum tension pulling force (F_M) displayed on the dynamometer was recorded until the copper metal layer was pulled off. The adhesion strength (Y_M) of the copper plating layer on the CFC surface was calculated according to Equation (1) [27].

$$Y_M = \frac{F_M}{S} \quad (1)$$

where S is the surface area of the electroless copper. For each sample, five parallel tests of the adhesion strength Y_M were performed to obtain an average value.

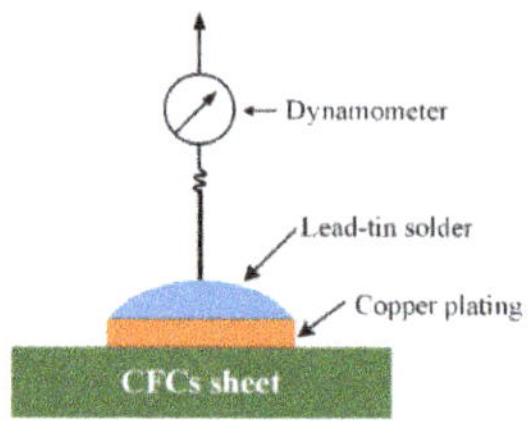

Figure 5. Schematic diagram of vertical pulling force measurement.

2.4. Modeling of Pulsed Laser Process

2.4.1. Response Surface Methodology

RSM is a collection of mathematical and statistical techniques, which is usually used to model and explore a problem in which a response is influenced by multiple variables. The RSM can predict the relationship between output responses and input independent variables in a specific range, including optimization methods to explore the optimum values of the input independent parameters that produce the desirable output responses [28]. Based on measurable, continuous, and controllable input independent variables (x_1, x_2, x_3, ..., x_k), and when the error is negligible, the linear output response Y_M can be described as follows:

$$Y_M = \varphi(x_1, x_2, x_3, \ldots, x_k) + \varepsilon \quad (2)$$

where φ represents the true response function, the accurate form of which is unknown and very complex; ε is a term associated with other sources of variability not taken into account in φ. Commonly, ε represents the effects of measurement error on response and background noise, as well as the influence of other variables, and so on. In general, ε is often applied as a statistical error, which is assumed to have normal distribution with a mean value of zero and a variance value of σ^2.

Generally, the second-order polynomial regression Equation (3) is used in RSM.

$$Y_M = \beta_0 + \sum_{i=1}^{m} \beta_i x_i + \sum_{i=1}^{m} \sum_{j=1}^{m} \beta_{ij} x_i x_j + \sum_{i=1}^{m} \beta_{ii} x_i^2 + \varepsilon \quad (3)$$

where x_i and x_j are the coded parameter variables; β_0 is a constant; β_i, β_{ij}, and β_{ii} represent the coefficients of the linear, quadratic, and interactive effects, respectively. These coefficients can be obtained through the fitting of experimental data [29]. The fitting precision of the mathematical model was evaluated using, the coefficient of determination R^2, adjusted R^2, predicted R^2, and adequate precision. The F-test and p-value were applied to check the significance of the regression coefficients.

2.4.2. Experimental Design

Based on previous single-factor experiments, the laser power (A), scanning line interval (B), scanning speed (C), and laser pulse frequency (D), were chosen as parameter variables and the adhesion strength as the response variable. A standard RSM design with CCD was applied to evaluate the effects of the laser process parameter variables (A, B, C, and D) on the response variable (Y_M). Incorporating the five coded levels (−2, −1, 0, 1, 2) for each of the four variables resulted in a total of 30 experiments. The laser process parameters and their levels investigated in this study are shown in Table 1.

Table 1. Laser process parameters and their levels used in RSM design.

Parameter	Unit	Notation	Level				
			−2	−1	0	1	2
Laser power	W	*A*	1.0	2.5	4.0	5.5	7.0
Scanning line interval	μm	*B*	20	35	50	65	80
Scanning speed	mm/s	*C*	200	600	1000	1400	1800
Pulse frequency	kHz	*D*	20	40	60	80	100

The statistical software Design-Expert V11.0 was employed to establish the correlations between the variation in the laser process parameters and the adhesion strength (Y_M). The results obtained from experiments are listed in Table 2.

Table 2. Design matrix and measured responses.

Std Order	Run Order	Laser Process Parameter				Adhesion Strength Y_M (MPa)
		A (W)	*B* (μm)	*C* (mm/s)	*D* (kHz)	
1	21	2.5	35	600	40	7.48
2	8	5.5	35	600	40	7.78
3	18	2.5	65	600	40	7.32
4	4	5.5	65	600	40	8.53
5	7	2.5	35	1400	40	6.79
6	28	5.5	35	1400	40	8.14
7	15	2.5	65	1400	40	5.55
8	16	5.5	65	1400	40	7.56
9	11	2.5	35	600	80	7.71
10	1	5.5	35	600	80	9.28
11	19	2.5	65	600	80	6.45
12	2	5.5	65	600	80	10.11
13	24	2.5	35	1400	80	6.03
14	30	5.5	35	1400	80	9.48
15	22	2.5	65	1400	80	5.19
16	6	5.5	65	1400	80	9.63
17	20	1.0	50	1000	60	4.81
18	25	7.0	50	1000	60	10.43
19	26	4.0	20	1000	60	7.38
20	27	4.0	80	1000	60	6.88
21	9	4.0	50	200	60	8.51
22	10	4.0	50	1800	60	5.57
23	29	4.0	50	1000	20	6.60
24	3	4.0	50	1000	100	6.43
25	12	4.0	50	1000	60	9.91
26	23	4.0	50	1000	60	9.87
27	5	4.0	50	1000	60	10.01
28	14	4.0	50	1000	60	9.69
29	13	4.0	50	1000	60	10.31
30	17	4.0	50	1000	60	9.48

3. Results and Discussion

3.1. Analysis of Variance

The analysis of variance is a well-known and powerful statistical analysis method [30]. In this study, ANOVA was applied to test whether the laser process parameters had a significant effect on the adhesion strength to investigate the prediction ability of the developed regression models in the design space. The ANOVA table for the adhesion strength (Y_M) of electroless copper on CFCs is shown in Table 3.

The associated *p*-value for the model was less than 0.05, indicating that the model term was statistically significant [31]. In Table 3, the ANOVA results show that the laser power (*A*), scanning speed (*C*), the quadratic effect of the laser power(A^2), scanning line interval

(B^2), scanning speed (C^2), and laser pulse frequency (D^2), along with the interaction effect of laser power and laser pulse frequency (AD), were the significant model terms associated with adhesion strength. The other terms of the mathematical model were not significant, and they needed to be eliminated through a backward elimination process to improve the adequacy of the model.

Table 3. ANOVA for the adhesion strength model before stepwise elimination.

Source	Sum of Squares	*df*	Mean Square	*F*-Value	*p*-Value Prob > *F*	
Model	79.09	14	5.65	18.06	<0.0001	significant
A	35.60	1	35.60	113.84	<0.0001	
B	0.47	1	0.47	1.50	0.2403	
C	6.17	1	6.17	19.73	0.0005	
D	0.80	1	0.80	2.57	0.1299	
AB	1.35	1	1.35	4.32	0.0621	
AC	1.27	1	1.27	4.07	0.1116	
AD	4.25	1	4.25	13.60	0.0022	
BC	0.45	1	0.45	1.42	0.2512	
BD	7.563×10^{-4}	1	7.563×10^{-4}	2.418×10^{-3}	0.9614	
CD	1.406×10^{-3}	1	1.406×10^{-3}	4.497×10^{-4}	0.9474	
A^2	5.86	1	5.86	18.75	0.0006	
B^2	9.38	1	9.38	30.00	<0.0001	
C^2	10.12	1	10.12	32.36	<0.0001	
D^2	14.96	1	14.96	47.85	<0.0001	
Residual	4.69	15	0.31			
Lack of fit	4.29	10	0.43	2.45	0.2100	Not significant
Pure error	0.40	5	0.080			
Core total	83.78	29				
Standard deviation = 0.56					R^2 = 0.9440	
Mean = 7.96					Adjusted R^2 = 0.8918	
Coefficient of variation = 7.02					Predicted R^2 = 0.6981	
Predicted residual error of sum of squares (PRESS) = 25.30					Adequate precision = 14.968	

The ANOVA table for the simplified quadratic model is shown in Table 4. The fitness of the adhesion strength model R^2 was 0.9074, the adjusted R^2 was 0.8657 and the predicted R^2 was 0.7149. The values of all three terms are close to one, which indicates the established model had excellent predictive performance [32]. The adequate precision, which represents the signal-to-noise ratio, is greater than four, implying adequate model discrimination [33]. The *F*-value of the lack of fit is 2.18, which means that lack of fit was not significant by comparison with pure error.

Table 4. ANOVA for the adhesion strength model after stepwise elimination.

Source	Sum of Squares	*df*	Mean Square	*F*-Value	*p*-Value Prob > *F*	
Model	76.02	9	8.45	21.77	<0.0001	Significant
A	35.60	1	35.60	91.74	<0.0001	
B	0.47	1	0.47	1.20	0.2854	
C	6.17	1	6.17	15.90	0.0007	
D	0.80	1	0.80	2.07	0.1658	
AD	4.25	1	4.25	10.96	0.0035	
A^2	5.86	1	5.86	15.11	0.0009	
B^2	9.38	1	9.38	24.18	<0.0001	
C^2	10.12	1	10.12	26.08	<0.0001	
D^2	14.96	1	14.96	38.56	<0.0001	
Residual	7.76	20	0.39			
Lack of fit	7.36	15	0.49	2.18	0.2273	Not significant
Pure error	0.40	5	0.080			
Core total	83.78	29				
Standard deviation = 0.62					R^2 = 0.9074	
Mean = 7.96					Adjusted R^2 = 0.8657	
Coefficient of variation = 7.82					Predicted R^2 = 0.7149	
Predicted residual error of sum of squares (PRESS) = 23.89					Adequate precision = 14.394	

The final reduced mathematical predicted model in terms of coded factors for the adhesion strength (Y_M), which was obtained after eliminating the insignificant terms, is given as follows:

$$Y_M = 9.88 + 1.22A - 0.14B - 0.51C + 0.18D + 0.52AD - 0.46A^2 - 0.58B^2 - 0.61C^2 - 0.74D^2 \quad (4)$$

where the final reduced empirical model in terms of actual factors is as follows:

$$\begin{aligned} Y_M = & -8.29124 + 1.42477A + 0.25065B + 6.32474 \times 10^{-3}C + 0.16199D + 0.017188AD \\ & - 0.20551A^2 - 2.59954 \times 10^{-3}B^2 - 3.79622 \times 10^{-6}C^2 - 1.84661 \times 10^{-3}D^2 \end{aligned} \quad (5)$$

In addition, according to the sum of squares of the laser process parameter variables in Table 4, the effect of the laser process parameters on adhesion strength are as follows: laser power (A) > scanning speed (C) > pulse frequency (D) > scanning line interval (B).

3.2. Effect of Process Parameters on the Responses

Figure 6 is a perturbation plot that illustrates the effect of four key laser process parameters at the center point on the adhesion strength in the design space. It can be clearly seen from the figure that the laser power has a large positive effect on adhesion strength, while scanning speed has a large negative effect on adhesion strength. This is consistent with the previous work reported by Xu et al. [34]. This phenomenon can be explained as follows: The adhesion strength between electroless copper and CFCs largely depends on the surface roughness of the CFC sheet, and a greater roughness results in a higher adhesion strength [35]. The surface roughness is directly related to the laser energy density and the irradiation time. As shown in Figure 3a, the pristine CFC sheet had a very smooth surface. Following nanosecond laser ablation, numerous microcavity structures (Figure 3b) interspersed with irregular granular nanoprotrusions (Figure 3c) emerged on the CFC surface. Additionally, a higher laser power meant that more energy was absorbed by the CFC sheet, leading to the formation of more micro/nanostructures on the surface of the CFC sheet (Figure 3d–f) and an increased adhesion strength between the electroless copper and CFCs. However, a higher scanning speed shortened the laser irradiation time, causing less energy to be input to the CFC sheet and reduced adhesion strength. At the same time, it can be observed from Figure 6 that the scanning line interval (B) had a little effect on the adhesion strength. Simultaneously, the adhesion strength increased with the scanning line interval until it reached its central value and then started to decrease as scanning line interval increased beyond its center limit; this result is consistent with the previous results reported by Qin et al. [36]. It is well known that the laser line-by-line scanning method usually creates a groove structure on the sample [37]. For low scanning line intervals, closer areas of laser line-by-line scanning were generated, resulting in an incomplete groove structure and a reduced surface roughness, thus causing weak adhesion strength. Further increasing the scanning line interval towards the center value, the adhesion strength increased with the roughness of the CFC surface. However, increasing the scanning line interval beyond the central value resulted in a smaller laser ablation area, leading to reduced surface roughness, subsequently decreasing the adhesion strength. In Figure 6, it can also be seen that both too-low or too-high pulse frequency caused low adhesion strength, and the change trend was consistent with the effect of the scanning line interval on adhesion strength.

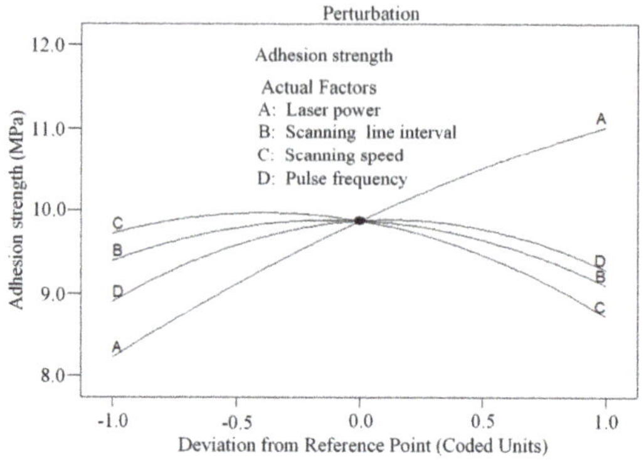

Figure 6. Perturbation plot showing the effect of all factors on the adhesion strength.

Figure 7a,b show the interaction effect of laser power and scanning line interval on adhesion strength. It is evident that the adhesion strength between the metal layer and CFCs tended to increase the higher the laser power. This is because the increase in the laser power led to an increase in laser energy density. Then, the CFC sheet absorbed enough laser energy and formed more micro/nanostructures on the CFC surface. Therefore, the adhesion strength increased gradually until the laser power increased to the limit. Moreover, it can be seen from Figure 7a,b that the adhesion strength increased first and then decreased with the increase in the scanning line interval. A lower scanning line interval reduced the roughness of the CFC surface. This may have been due to the cumulative thermal effects resulting from the high overlap percentage of the laser scanning line interval. At the same time, a higher scanning line interval meant that more areas between adjacent scanning lines were not ablated by the laser, so the roughness of CFC surface was naturally lower, and the adhesion strength was poor.

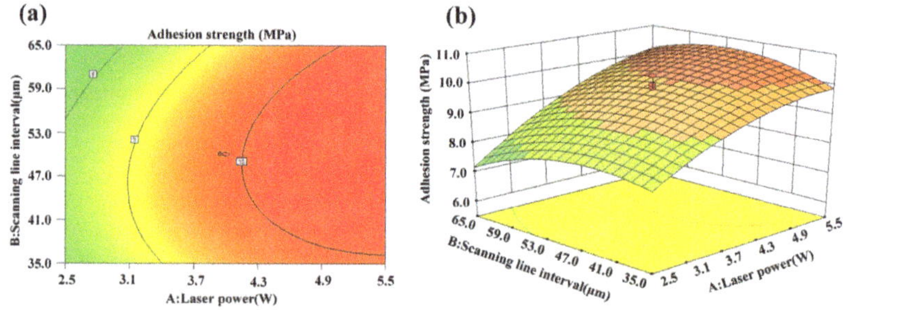

Figure 7. Interaction effect of laser power and scanning line interval on adhesion strength: (**a**) contour plot and (**b**) 3D surface plot.

The interaction effect of laser power and scanning speed on adhesion strength is presented in Figure 8a,b. It is clear that the adhesion strength tended to increase for higher laser powers and lower scanning speeds. The increase in laser power and the decrease in scanning speed resulted in the increase in the laser energy density, which effectively improved the roughness of the CFC surface and accordingly increased the adhesion strength of the electroless copper.

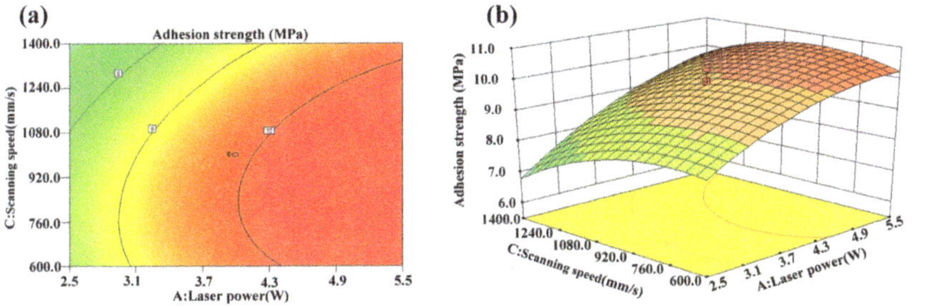

Figure 8. Interaction effect of laser power and scanning speed on adhesion strength: (**a**) contour plot and (**b**) 3D surface plot.

The interaction effect of laser power and pulse frequency on adhesion strength is illustrated in Figure 9a,b. The optimal adhesion strength was attained with an appropriate pulse frequency and high laser power. Higher laser power caused more micro/nanostructures to form on the surface of the CFC sheet, which was beneficial for the bond strength.

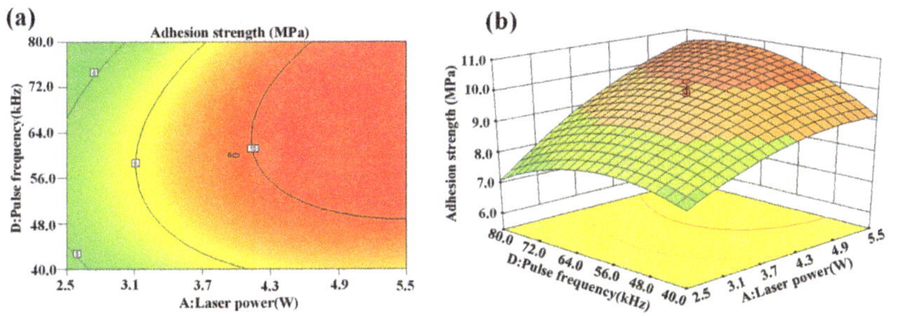

Figure 9. Interaction effect of laser power and pulse frequency on adhesion strength: (**a**) contour plot and (**b**) 3D surface plot.

Figure 10a,b show the interaction effect of scanning line interval and scanning speed on adhesion strength. The optimal scanning line interval and appropriate scanning speed are advantageous for achieving higher adhesion strength. The reasons are the same as those discussed above.

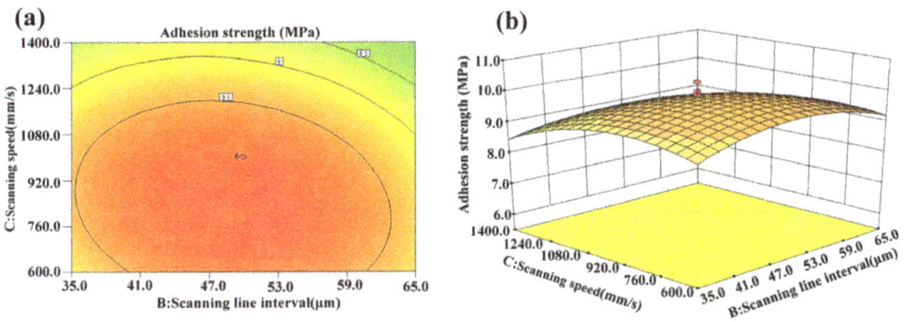

Figure 10. Interaction effect of scanning line interval and scanning speed on adhesion strength: (**a**) contour plot and (**b**) 3D surface plot.

Figure 11a,b present the interaction effect of scanning speed and pulse frequency on adhesion strength. The adhesion strength was relatively low at higher or lower scanning speed and pulse frequency. Thus, the adhesion strength was enhanced in the near-central levels of scanning speed and pulse frequency.

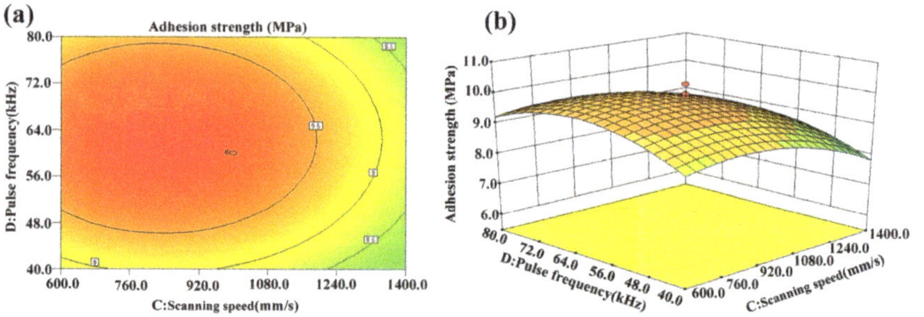

Figure 11. Interaction effect of scanning speed and pulse frequency on adhesion strength: (**a**) contour plot and (**b**) 3D surface plot.

Figure 12a,b show the interaction effect of scanning line interval and pulse frequency on adhesion strength, which is similar to that of scanning speed and pulse frequency on adhesion strength shown in Figure 11.

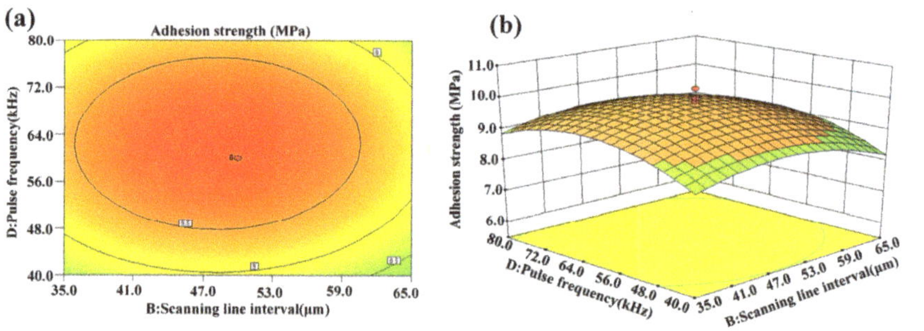

Figure 12. Interaction effect of scanning line interval and pulse frequency on adhesion strength: (**a**) contour plot and (**b**) 3D surface plot.

3.3. Validation of the Developed Model

To validate the developed model derived from multiple regression analysis, three groups of optimized laser parameters were chosen randomly within the ranges in Table 1 to conduct confirmation experiments. The actual values of the results in terms of the average of three trials were calculated. Table 5 shows the actual values, predicted results, and calculated percentage error of the confirmation experiments. It shows that the maximum relative error of the prediction for the optimal parametric combination was less than 5.0%, indicating that the developed model can yield near-accurate result. The relationship between the experimental and estimated values of adhesion strength is shown in Figure 13. It can be found that the percentage error between the actual and predicted values was small, indicating that the developed models are adequate, and the predicted data were in good agreement with the measured values. Therefore, it can be concluded that the developed model could successfully predict adhesion strength. In addition, using a numerical optimization method, Design Expert analysis revealed that the optimal laser power, scanning line interval, scanning speed, and laser pulse frequency for adhesion strength were 5.5 W, 48.2 μm, 834.0 mm/s and 69.5 kHz, respectively.

Table 5. Prediction and validation test results.

Exp. No.	A (W)	B (μm)	C (mm/s)	D (kHz)	Adhesion Strength (MPa)		∣Error∣ (%)
					Actual	Predicted	
1	2	40	500	60	6.63	6.95	4.82
2	6	70	1500	40	6.05	6.26	3.47
3	5	30	1000	70	9.26	9.71	4.86

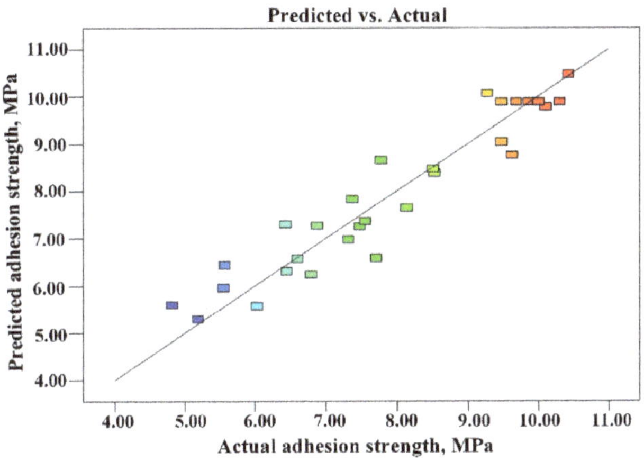

Figure 13. Plots of the actual vs. predicted on adhesion strength.

4. Conclusions

The effects of laser process parameters on the adhesion strength between electroless copper and CFCs were investigated using RSM. The following conclusions could be drawn from this study:

(1) A four-factor, five-level CCD was successfully employed to develop a mathematical model for optimizing the laser process parameters that affect adhesion strength. The ANOVA results showed that the developed model could be adequately applied to evaluate the adhesion strength at a 95% confidence level.

(2) Laser power has a large positive effect on adhesion strength, which is negatively impacted by scanning speed. The laser pulse frequency and scanning line interval have little effect on adhesion strength.

(3) The maximum relative error of the prediction at the optimal parametric combination was less than 5.0%, indicating that the developed models could adequately predict the results of the adhesion strength within the design space of the laser process parameters presented in this paper.

(4) Under optimal process conditions (laser power of 5.5 W, scanning line interval of 48.2 μm, scanning speed of 834.0 mm/s, and pulse frequency of 69.5 kHz), electroless copper on CFCs could be produced with an adhesion strength of 10.91 MPa.

Author Contributions: Investigation and methodology and writing—original draft preparation, X.W., J.L., H.L. and Z.Q.; software and writing, X.W. and Z.Q.; laser processing and chemical plating, Z.Z. and J.C.; SEM image, Z.Z. All authors have read and agreed to the published version of the manuscript.

Funding: This study was supported by the Open Project Program of Wuhan National Laboratory for Optoelectronics (2021WNLOKF017), the Fundamental Research Funds of Central Universities for South-Central Minzu University (CZQ21029), the Doctoral Research Fund of Hubei Institute of Science and Technology (BK202103), and the Innovation and Entrepreneurship Key Incubation Project of Hubei University of Science and Technology (2022, No18).

Data Availability Statement: The data that support the findings of this study are available from the corresponding author, Zhongli Qin, upon reasonable request.

Conflicts of Interest: The authors declare no conflict of interest.

References

1. Zhang, H.-Y.; Li, J.-Y.; Pan, Y.; Liu, Y.-F.; Mahmood, N.; Jian, X. Flexible carbon fiber-based composites for electromagnetic interference shielding. *Rare Met.* **2022**, *41*, 3612–3629. [CrossRef]
2. Chen, X.; Wen, K.; Wang, C.; Cheng, S.; Wang, S.; Ma, H.; Tian, H.; Zhang, J.; Li, X.; Shao, J. Enhancing mechanical strength of carbon fiber-epoxy interface through electrowetting of fiber surface. *Compos. Part B Eng.* **2022**, *234*, 109752. [CrossRef]
3. Hao, M.; Qian, X.; Zhang, Y.; Yang, J.; Li, C.; Gong, H.; Wang, X.; Wang, P.; Liu, L.; Huang, Y. Thermal conductivity enhancement of carbon fiber/epoxy composites via constructing three-dimensionally aligned hybrid thermal conductive structures on fiber surfaces. *Compos. Sci. Technol.* **2023**, *231*, 109800. [CrossRef]
4. Guo, Y.; Xu, Y.; Wang, Q.; Dong, Q.; Yi, X.; Jia, Y. Eliminating lightning strike damage to carbon fiber composite structures in Zone 2 of aircraft by Ni-coated carbon fiber nonwoven veils. *Compos. Sci. Technol.* **2019**, *169*, 95–102. [CrossRef]
5. Wang, F.S.; Ji, Y.Y.; Yu, X.S.; Chen, H.; Yue, Z.F. Ablation damage assessment of aircraft carbon fiber/epoxy composite and its protection structures suffered from lightning strike. *Compos. Struct.* **2016**, *145*, 226–241. [CrossRef]
6. Che, H.; Vo, P.; Yue, S. Metallization of carbon fibre reinforced polymers by cold spray. *Surf. Coat. Technol.* **2017**, *313*, 236–247. [CrossRef]
7. Shao, Z.; Zhang, Y.; Zhang, N.; Xia, J. Preparation and Research of Electroless Copper on Carbon Fibers. *Mater. Manuf. Process.* **2014**, *31*, 12–17. [CrossRef]
8. Guo, Z.; Sang, L.; Wang, Z.; Chen, Q.; Yang, L.; Liu, Z. Deposition of copper thin films by plasma enhanced pulsed chemical vapor deposition for metallization of carbon fiber reinforced plastics. *Surf. Coat. Technol.* **2016**, *307*, 1059–1064. [CrossRef]
9. Gillet, V.; Aubignat, E.; Costil, S.; Courant, B.; Langlade, C.; Casari, P.; Knapp, W.; Planche, M.P. Development of low pressure cold sprayed copper coatings on carbon fiber reinforced polymer (CFRP). *Surf. Coat. Technol.* **2019**, *364*, 306–316. [CrossRef]
10. Li, W.; Liu, L.; Zhong, C.; Shen, B.; Hu, W. Effect of carbon fiber surface treatment on Cu electrodeposition: The electrochemical behavior and the morphology of Cu deposits. *J. Alloys Compd.* **2011**, *509*, 3532–3536. [CrossRef]
11. Chen, H.; Liu, G.; Xu, C.; Hou, X.; Liu, Y. Copper@carbon fiber composites prepared by a simple electroless plating technique. *Mater. Lett.* **2016**, *173*, 211–213. [CrossRef]
12. Wang, F.; Arai, S.; Endo, M. Preparation of nickel–carbon nanofiber composites by a pulse-reverse electrodeposition process. *Electrochem. Commun.* **2005**, *7*, 674–678. [CrossRef]
13. Archambault, G.; Jodoin, B.; Gaydos, S.; Yandouzi, M. Metallization of carbon fiber reinforced polymer composite by cold spray and lay-up molding processes. *Surf. Coat. Technol.* **2016**, *300*, 78–86. [CrossRef]
14. Prysiazhnyi, V.; Stupavská, M.; Ráhel', J.; Kleber, C.; Černák, M.; Rafailović, L.D. A comparison of chemical and atmospheric plasma assisted copper plating on carbon fiber reinforced epoxy polymer surfaces. *Surf. Coat. Technol.* **2014**, *258*, 1082–1089. [CrossRef]
15. Liu, J.; Ma, H.; Meng, L.; Yang, H.; Yang, C.; Ruan, S.; Ruan, S.; Ouyang, D.; Mei, S.; Deng, L.; et al. Laser Powder Bed Fusion of 316L Stainless Steel: Effect of Laser Polishing on the Surface Morphology and Corrosion Behavior. *Micromachines* **2023**, *14*, 850. [CrossRef]
16. Qin, Z.; Xiang, H.; Liu, J.; Zeng, X. High-performance oil-water separation polytetrafluoroethylene membranes prepared by picosecond laser direct ablation and drilling. *Mater. Des.* **2019**, *184*, 108200. [CrossRef]
17. Tao, Y.; Lin, L.; Ren, X.; Wang, X.; Cao, X.; Gu, H., Ye, Y.; Ren, Y.; Zhang, Z. Four-Dimensional Micro/Nanorobots via Laser Photochemical Synthesis towards the Molecular Scale. *Micromachines* **2023**, *14*, 1656. [CrossRef]
18. Chen, Q.; Jing, Y.; Yin, J.; Li, Z.; Xiong, W.; Gong, P.; Zhang, L.; Li, S.; Pan, R.; Zhao, X.; et al. High Reflectivity and Thermal Conductivity Ag–Cu Multi-Material Structures Fabricated via Laser Powder Bed Fusion: Formation Mechanisms, Interfacial Characteristics, and Molten Pool Behavior. *Micromachines* **2023**, *14*, 362. [CrossRef]
19. Li, Z.; Li, H.; Yin, J.; Li, Y.; Nie, Z.; Li, X.; You, D.; Guan, K.; Duan, W.; Cao, L.; et al. A review of spatter in laser powder bed fusion additive manufacturing: In situ detection, generation, effects, and countermeasures. *Micromachines* **2022**, *13*, 1366. [CrossRef]
20. Gustke, K.; Gebauer, J.; Drehmann, R.; Lasagni, A.F.; Lampke, T. Enhancement of the Adhesion of Wire Arc Sprayed Coatings on Carbon Fiber-Reinforced Plastic by Surface Laser Structuring. *Coatings* **2021**, *11*, 467. [CrossRef]
21. Li, Y.; Meng, S.; Gong, Q.; Huang, Y.; Gan, J.; Zhao, M.; Liu, B.; Liu, L.; Zou, G.; Zhuang, D. Experimental and Theoretical Investigation of Laser Pretreatment on Strengthening the Heterojunction between Carbon Fiber-Reinforced Plastic and Aluminum Alloy. *ACS Appl. Mater. Interfaces* **2019**, *11*, 22005–22014. [CrossRef]
22. Palavra, A.; Coelho, B.N.; de Hosson, J.T.M.; Lima, M.S.F.; Carvalho, S.M.; Costa, A.R. Laser surface treatment for enhanced titanium to carbon fiber-reinforced polymer adhesion. *J. Braz. Soc. Mech. Sci. Eng.* **2017**, *39*, 2917–2924. [CrossRef]
23. Rajarajan, C.; Sonar, T.; Sivaraj, P.; Raja, S.; Mathiazhagan, N. The Effect of Resistance Spot Welding Parameters on Microstructure and Strength of DP800 Steel Joints Using Response Surface Methodology. *Adv. Mater. Sci.* **2022**, *22*, 53–78. [CrossRef]

24. Asgari, G.; Shabanloo, A.; Salari, M.; Eslami, F. *Sonophotocatalytic treatment of AB*113 dye and real textile wastewater using ZnO/persulfate: Modeling by response surface methodology and artificial neural network. *Environ. Res.* **2020**, *184*, 109367. [CrossRef]
25. Bhardwaj, T.; Shukla, M.; Paul, C.P.; Bindra, K.S. Direct energy deposition-laser additive manufacturing of titanium-molybdenum alloy: Parametric studies, microstructure and mechanical properties. *J. Alloys Compd.* **2019**, *787*, 1238–1248. [CrossRef]
26. Kuntoğlu, M.; Aslan, A.; Pimenov, D.Y.; Giasin, K.; TMikolajczyk; Sharma, S. Modeling of cutting parameters and tool geometry for multi-criteria optimization of surface roughness and vibration via response surface methodology in turning of AISI 5140 steel. *Materials* **2020**, *13*, 4242. [CrossRef]
27. Lv, M.; Liu, J.; Zeng, X.; Du, Q.; Ai, J. High-adhesion Cu patterns fabricated by nanosecond laser modification and electroless copper plating. *Appl. Surf. Sci.* **2015**, *353*, 1150–1155. [CrossRef]
28. Sashank, S.; Babu, P.D.; Marimuthu, P. Experimental studies of laser borided low alloy steel and optimization of parameters using response surface methodology. *Surf. Coat. Technol.* **2019**, *363*, 255–264. [CrossRef]
29. Huang, S.; Chen, R.; Zhang, H.; Ye, J.; Yang, X.; Sheng, J. A study of welding process in connecting borosilicate glass by picosecond laser pulses based on response surface methodology. *Opt. Laser Technol.* **2020**, *131*, 106427. [CrossRef]
30. Pongsumpun, P.; Iwamoto, S.; Siripatrawan, U. Response surface methodology for optimization of cinnamon essential oil nanoemulsion with improved stability and antifungal activity. *Ultrason. Sonochem.* **2020**, *60*, 104604. [CrossRef]
31. Bandara, P.C.; Nadres, E.T.; Rodrigues, D.F. Use of Response Surface Methodology To Develop and Optimize the Composition of a Chitosan-Polyethyleneimine-Graphene Oxide Nanocomposite Membrane Coating To More Effectively Remove Cr(VI) and Cu(II) from Water. *ACS Appl. Mater. Interfaces* **2019**, *11*, 17784–17795. [CrossRef]
32. Abdelmoneim, A.; Elshaer, R.N.; El-Shennawy, M.; Sobh, A.S. Modeling of wear resistance for TC21 Ti-alloy using response surface methodology. *Sci. Rep.* **2023**, *13*, 4624. [CrossRef] [PubMed]
33. Garg, S.; Nayyar, A.; Buradi, A.; Shadangi, K.P.; Sharma, P.; Bora, B.J.; Jain, A.; Shah, M.A. A novel investigation using thermal modeling and optimization of waste pyrolysis reactor using finite element analysis and response surface methodology. *Sci. Rep.* **2023**, *13*, 10931. [CrossRef] [PubMed]
34. Xu, H.; Zhang, J.; Feng, J.; Zhou, T. Fabrication of Copper Patterns on Polydimethylsiloxane through Laser-Induced Selective Metallization. *Ind. Eng. Chem. Res.* **2021**, *60*, 8821–8828. [CrossRef]
35. You, X.; Chen, Y.; Huang, Y.; Wang, C.; Zhou, G.; He, W.; Wang, S.; Tang, Y.; Zhang, W.; Li, Z.; et al. Surface coarsening of carbon fiber/cyanate ester composite for adhesion improvement of electroless copper plating as conductive patterns. *Mater. Chem. Phys.* **2020**, *255*, 123597. [CrossRef]
36. Qin, Z.; Ai, J.; Du, Q.; Liu, J.; Superhydrophobic, X.Z. polytetrafluoroethylene surfaces with accurately and continuously tunable water adhesion fabricated by picosecond laser direct ablation. *Mater. Des.* **2019**, *173*, 107782. [CrossRef]
37. Song, Y.; Wang, C.; Dong, X.; Yin, K.; Zhang, F.; Xie, Z.; Chu, D.; Duan, J.A. Controllable superhydrophobic aluminum surfaces with tunable adhesion fabricated by femtosecond laser. *Opt. Laser Technol.* **2018**, *102*, 25–31. [CrossRef]

micromachines

Article

Design and Optimization of Thin-Walled Main Support Structure for Space Camera Based on Additive Manufacturing

Jiahao Peng [1,2], Shijie Liu [1,*], Dong Wang [3], Anpeng Xu [1,2], Xin Huang [1,2], Tianqi Ma [1], Jing Wang [1] and Hang Li [1]

1 Changchun Institute of Optics, Fine Mechanics and Physics, Chinese Academy of Sciences, Changchun 130033, China; hellopjh@yeah.net (J.P.); xap_123@163.com (A.X.); huangxin19@mails.ucas.ac.cn (X.H.); jlspmatq@163.com (T.M.); wangjing@ciomp.ac.cn (J.W.); lixing20@mails.jlu.edu.cn (H.L.)
2 University of Chinese Academy of Sciences, Beijing 100049, China
3 China North Vehicle Research Institute, Beijing 100072, China; wangdong173@mails.ucas.ac.cn
* Correspondence: liushijie@ciomp.ac.cn

Abstract: In order to solve the design requirements of high stiffness and lightweight for the primary support structure of a wide-field auroral imager, we propose a solution for designing and optimizing a large-scale complex thin-walled structure using additive manufacturing. Firstly, we devise an integrated thin-walled structure and test material for the main support. Secondly, shape optimization is achieved via the optimization of the lateral slope angle of the primary support based on Timoshenko cantilever beam theory. Additionally, an active fitting optimization algorithm is proposed for the purpose of refining the wall thickness of the thin-walled structure. Then, we determine the structural design of the main support. This primary support is manufactured via selective laser melting (SLM). Following processing, the structure size is 538 mm × 400 mm × 384 mm, and the mass is 7.78 kg. Finally, frequency scanning experiments indicate that, in the horizontal direction, there is a natural frequency of 105.97 Hz with an error rate of approximately 3% compared to finite element analysis results. This research confirms that our large-scale complex, thin-walled main support structure design meets all design requirements.

Keywords: wide-field auroral imager; main support structure; integrated thin-walled structure; active fitting optimization algorithm; additive manufacturing

Citation: Peng, J.; Liu, S.; Wang, D.; Xu, A.; Huang, X.; Ma, T.; Wang, J.; Li, H. Design and Optimization of Thin-Walled Main Support Structure for Space Camera Based on Additive Manufacturing. *Micromachines* **2024**, *15*, 211. https://doi.org/10.3390/mi15020211

Academic Editors: Yang Liu, Jie Yin, Linda Ke and Kai Guan

Received: 14 November 2023
Revised: 27 January 2024
Accepted: 28 January 2024
Published: 30 January 2024

1. Introduction

Observational research and assurance services for the space environment can help to reduce the risks posed by extreme space weather to aerospace, navigation, communications, and national economies [1–3]. Fengyun meteorological satellites have been extensively utilized in disaster prevention and mitigation, addressing climate change, and monitoring and predicting space weather. This use has yielded significant achievements [4–6]. The wide-field auroral imager (WFAI), a principal payload on Fengyun-3 polar-orbiting meteorological satellites, is specifically designed to capture data during periods of auroral radiation intensity within the 140 nm to 180 nm wavelength range. These observational data enable the provision of forecasts for magnetic storms, magnetospheric substorms, as well as ionospheric conditions within polar regions [7].

Considering the FY-3(08) satellite (WFAI), it can be observed that the WFAI employs the swing-sweep method within the −90° to +90° range to achieve a dual-channel 130° × 130° expansive field of view, as well as to enable on-orbit calibration. The WFAI is strategically positioned in the central region of the satellite's payload bay. In order to mitigate any interference arising from other payloads on the satellite, the main support structure's dimensions, oriented perpendicularly to the satellite payload bay plate, must be adequately extensive. Furthermore, the primary optomechanical structure of the imager is situated atop the main support structure, resulting in the camera displaying an elevated center of gravity. Simultaneously, the dimensions and mass of the WFAI in other orientations are subject to stringent constraints, and they must adhere to safety margin specifications during launch conditions. Consequently, these issues present challenges in crafting a main support structure for the camera, which is simultaneously lightweight yet characterized by high stiffness and stability.

This study predominantly focuses on the application of laser additive manufacturing to create large-scale, complex, thin-walled main support structures that can serve as space cameras. Initially, operating according to the design requirements of the wide-field auroral imager, namely, lightweight and high stiffness, an integrated thin-walled structure is devised for the main support. Subsequently, the lateral slope angle of the main support is optimized in accordance with the Timoshenko cantilever beam theory. Furthermore, an active fitting optimization algorithm is proposed for the refinement of the thin-walled structure's thickness. Finally, considering the limitations of traditional methods in processing, selective laser melting (SLM) is employed to fabricate the main support. The vibration natural frequency test validates the optimal design results presented in this paper. The remaining sections of this research paper are organized as follows. In Section 2, we review the related literature. In Section 3, we summarize the design requirements and structural form of the camera. In Section 4, the original design of the main support structure is accomplished. In Section 5, the optimization-based design of the main support structure is performed, including shape optimization and size optimization. In Section 6, we introduce the additive manufacturing process of the main support structure and conduct experimental verification of the space camera. Conclusions and future research directions are given in Section 7.

2. Literature Review

Thin-walled structures play crucial roles in the aerospace field due to their performance advantages, such as being lightweight and exhibiting high specific stiffness, specific strength, and high-load-bearing capacity [8–10]. Possessing excellent temperature control characteristics and strong energy absorption performance, thin-walled structures are widely utilized in the main support structure of space cameras. The principal forms of thin-walled main support structures produced for space cameras include the thin-walled tube type, thin-walled frame type, etc. The thin-walled cylindrical structure of carbon fiber materials, characterized by lightweight and increasingly mature technology, is widely used in space cameras such as Quickbird-2 [11], ALSAT-2A [12], and KompSat-3 cameras [13]. Generally, the thin-walled frame type is integrally cast or in combination. This structural form often has high structural stiffness and sound structural stability, being generally used for small- and medium-sized off-axis space cameras with complex structural shapes. ALI camera [14] and IKONOS [15] adopt this structural form. The U-shaped frame of the wide-angle auroral imager on the Fengyun 3D satellite adopts CNC machining to achieve a thin-walled cavity structure, and its base support adopts casting methods to achieve a thin-walled frame structure [16]. With the development of space cameras, there is a growing demand for weight reductions. Therefore, it is necessary to design its supporting structural form. The traditional casting and CNC machining methods for thin-walled structures exhibit the disadvantages of long production process cycles and low material utilization, making it difficult to achieve the integrated manufacturing of large and complex thin-walled structures [17,18]. The carbon fiber material lamination process is commonly used to manufacture simple geometric components and will reduce the inherent advantages of carbon fiber composite materials when applied to more complex plate-like structures [19,20].

AM is increasingly widely used in aerospace due to its ability to create complex and customized designs, reduce component weight, improve manufacturing efficiency, shorten product development cycles, and make supply chains more flexible [21,22]. Research into AM in aerospace has also seen exponential growth over the last decade [23]. The aerospace additive manufacturing market industry is projected to grow from USD 8.29

billion in 2023 to USD 36.25 billion by 2032, exhibiting a compound annual growth rate of 20.24% during the forecast period [24]. Ishfaq K. et al. discussed the application of AM in the space field, and these applications demonstrate the feasibility and benefits of AM technology application in space environments [25]. Ransikarcum K et al. conducted a study on the business scope and supply chain scope of AM [26,27]. Wang D et al. investigated the influence of process parameters on densification, microstructure, and mechanical properties of a Ti-6Al-4V alloy printed by SLM [28]. Zheng Li et al. conducted a literature review on the in situ detection, generation, effects, and countermeasures of spatter in L-PBF [29]. Among numerous metal 3D printing technologies, the selective laser melting (SLM) method is very regarded due to its advantages in dimensional accuracy, surface quality, and component performance [30,31].

The efficiency and flexibility of additive manufacturing technology are particularly suitable for space camera structures that require precision and high performance. RUAG designed and produced an optimized version of an existing antenna bracket with AM solutions for the latest editions of the mission, namely, the Sentinel-1C and Sentinel-1D [32]. The space detector SVOM-MXT feature camera contains an enclosed aluminum shield supported by three small support feet produced via additive manufacturing [33]. NASA's Perseverance Rover carried 11 components fabricated using metal AM techniques. Of these 11 components, 6 are AM-built heat exchangers used in the MOXIE. The 3D-printed parts of the PIXL include its front and back cover, mounting frame, X-ray bench, and bench support, which are hollow, thin-wall structures. The wall thickness of these parts is between 1 mm and 1.1 mm [34]. MENG Hongtao proposes a complex outer baffle space camera based on AM [35]. Silicon carbide is used to develop a space remote-sensing camera via 3D printing technology. The outer envelope of the camera is 430 mm × 370 mm × 450 mm, with a weight of only 13.5 kg [36]. Design and manufacturing methods that integrate optimization with additive manufacturing technology are used to meet the design requirements of being lightweight and displaying high surface-shape accuracy with space mirrors [37–39].

Via our review of the above literature, we found that thin-walled structures, being among the primary support methods for space cameras, are currently produced mainly through traditional manufacturing methods such as CNC machining, casting, and composite material weaving. At present, the application of AM technology in space cameras is mainly concentrated on smaller or lower-strength structural components such as brackets, light shields, aperture plates, camera shells, reflectors, etc. Meanwhile, there is relatively little research and application of AM technology on complex thin-walled main load-bearing structural components with high stiffness and large dimensions. Gaps in the existing research are highlighted in Table 1.

Table 1. Existing studies and research gaps in thin-wall structures in space cameras.

Research Topic	Existing Research	Research Gaps
Structural type design	Thin-walled cylinder type [11–13] Thin-walled frame type [14,15]	Integrated complex thin-walled structure designs
Traditional manufacturing technology	CNC machining and casting [16] Composite material weaving [11–13]	Low-cost, high-quality, high-efficiency and pollution-free in space
Additive manufacturing applications	Small-scale functional components [32,33,35–37] Strength assessment components [30] Small-scale support structure [31,34]	Applications in large-scale, high-rigidity, and thin-wall primary load-bearing structures

This study not only provides a novel perspective for the design and manufacturing of thin-walled structures in space cameras but also expands the application scope of additive manufacturing technology in the aerospace field, providing a practical application of AM technology in the disciplinary area centering thin-walled main supports. The main innovations of this article are as follows:

- For the first time, the laser additive manufacturing of a large-size complex thin-walled structure is applied to the main bearing structure of the space camera.

- Shape optimization is accomplished by optimizing the lateral slope angle of the main support according to the Timoshenko cantilever beam theory.
- An active fitting optimization algorithm is proposed for use to refine the wall thickness of the thin-walled structure. The optimization algorithm not only reduces the number of iterations but also obtains more precise solutions.
- The utilization of additive manufacturing for large-scale, complex, thin-walled structures introduces a novel perspective in both the field of space camera support and AM.

3. Design Requirements and Structural Form of the Camera

3.1. Key Technical Indexes

The main technical specifications of the wide-field auroral imager (WFAI) are delineated in Table 2.

Table 2. Main technical specifications of the camera.

No.	Parameters	Specifications
1	Mass (kg)	$\leq$25
2	Frequency (Hz)	$\geq$100
3	Volume (mm)	$\leq$726 $\times$ 635 $\times$ 500

3.2. Camera Structure Form

The main constituents of the WFAI are depicted in Figure 1. These components primarily encompass the main support structure, four independent lens components, the scanning axis system, the stepper motor, and the on-orbit calibration component.

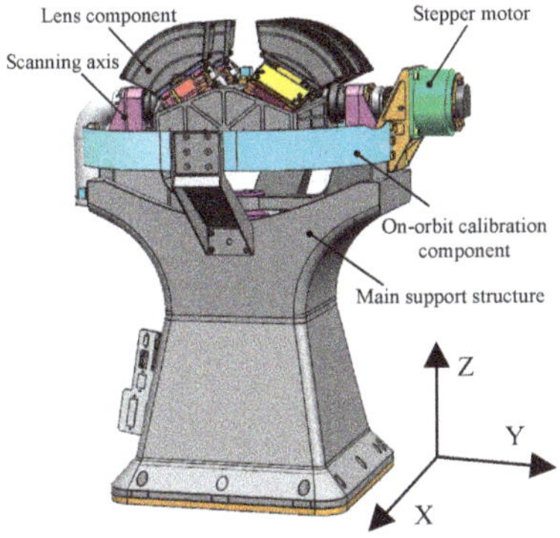

Figure 1. Sketch of the camera structure.

The support base plays a pivotal role in the wide-field auroral imager, serving as the primary fixture for various camera components, either directly or indirectly. Notably, the scanning axis system is affixed atop the main support, spanning two relatively independent planes. Additionally, the assembly of four independent lenses, forming the imaging system, is conducted, and the product is positioned within the midsection of the bracket. A motor is responsible for propelling the scanning axis system along the designated track, thereby facilitating the scanning imaging process. Hence, the stability of the main support determines the precision of the shaft, governing the imaging quality of the imaging system and influencing the operational longevity of the camera. Furthermore, in adherence to the specifications outlined in Table 2, the mass of the main support constitutes a substantial 30% portion of the entire camera's weight. This component also functions as the principal structural element that dictates the camera's rigidity, considering that other constituents have already undergone extensive lightweight design processes. The primary support structure further benefits from a versatile array of material options and structural configurations within the optical system, rendering it better suited for achieving a high degree of lightweight design when compared to the remaining components.

4. Main Support Structure Design

4.1. Material Selection

The selection of materials for space cameras primarily revolves around key properties, including excellent structural stiffness and strength, as well as thermal stability [38]. Table 3 illustrates the principal performance indicators of materials frequently employed in space cameras.

Table 3. Material parameters of space support structure [40].

Name	Densities ρ(g·cm^{-3})	Modulus E (Gpa)	E/ρ (kN·m/g)	Expansivity (10^{-6} K^{-1})	Tensile Strength	Specific Strength
TC4	4.44	110	24.7	8.8	802	180.6
Al alloy	2.71	69	25.5	23	410	146.4
Alloy steel	7.83	210	26.9	12	780	100
Invar	8.1	145	17.9	2.4	302	37.3

It can be seen from the parameters in the table that the specific stiffness of titanium alloy, aluminum alloy, and alloy steel is similar, that the coefficient of linear expansion of aluminum alloy is excessively large, and that the specific strength of alloy steel is small. Although the Invar exhibits superior thermal stability, its specific stiffness and strength are lower. Analyzing the above factors together, it is apparent that the overall performance of titanium alloy is the best. As such, titanium alloy is chosen as the laser additive manufacturing material.

4.2. Integrated Design

The traditional one-dimensional rotational main support structure comprises two primary components, namely, the U-frame and the base, as depicted in Figure 2. The U-frame serves as the mounting interface for the rotational axis system, while the base acts as the connecting element between the space camera and the satellite platform. The bolted connections used in assembly can compromise the stiffness near the connections, thereby affecting the overall structural dynamics. In terms of manufacturing, in order to achieve lightweight, compact, and intricate structures, CNC machining is commonly employed [41], resulting in material waste and extended processing times. Casting is traditionally used for larger, more complex structures [42], but it often leads to defects, reduced strength, and increased waste rates.

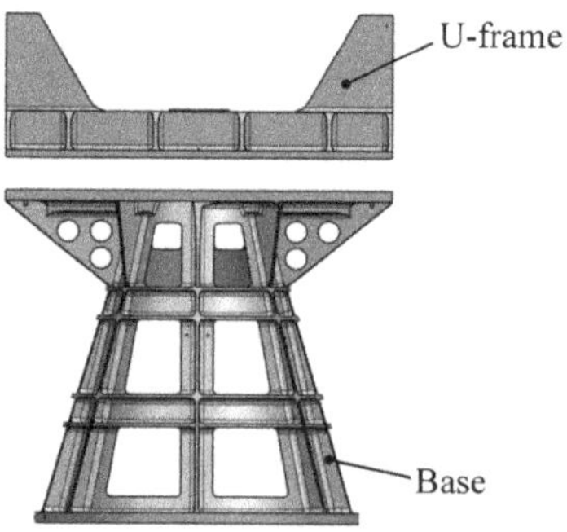

Figure 2. Traditional structural form of main support.

4.3. Structural Form Design

The WFAI achieves a substantial field of view using a swing-sweep mechanism. Consequently, it possesses significant dimensions in the Z-direction, resulting in a relatively high-payload center of mass. Additionally, due to the eccentric load on the scanning mechanism, the main support is subjected to bending and torsional loads, primarily during launch. On occasions when the cross-sectional area is the same, the bending section modulus and torsional section modulus of a thin-walled hollow structure are greater than those of a solid structure. In order to meet the requirements for mass, stiffness, and stability in space cameras, the main support employs a thin-walled box-type structure. Thin-walled box structures are lightweight, offering excellent bending and torsional performances. Internally, these boxes house electronic components, shielding them from radiation, while the exterior surface of the thin-walled box structure provides effective thermal control. The preliminary structural diagram is depicted in Figure 3.

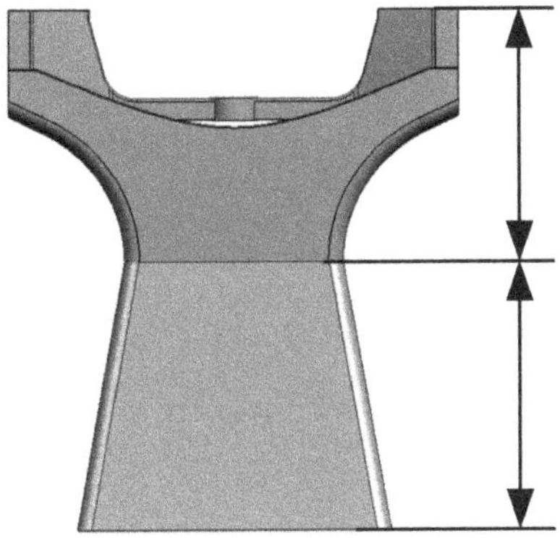

Figure 3. Preliminary structural sketch.

As shown in Figure 3, considering the satellite space size requirements, rotation profile, and optical requirements for on-orbit calibration, the changeable area of Part I of the thin-walled beam is small. As such, the optimizable section is Part II. For this reason, this section is designed as a trapezoidal structure. While the overall mass and total height remain unchanged, Part II sees evolution from an equal-section rectangle into a variable-section rectangle.

5. Optimization

Once the structural form of the main support has been determined, its shape and size should be optimized to achieve the goals of the highest stiffness and lightest mass.

5.1. Shape Optimization

5.1.1. Optimization Objective and Variable

Based on the structural design results in shown Section 4.3, the thin-walled main support structure is simplified into a variable cross-section Timoshenko cantilever beam structure. The differential equation for the free vibration of the variable cross-section Timoshenko beam is as follows [43]:

$$\begin{cases} \frac{\partial}{\partial x}\left[\chi GA(x)\left[\varphi(x,t)-\frac{\partial y(x,t)}{\partial x}\right]\right]+\rho A(x)\frac{\partial^2 y(x,t)}{\partial t^2}=0 \\ \frac{\partial}{\partial x}\left[EI(x)\frac{\partial \varphi(x,t)}{\partial x}\right]+\chi GA(x)\left[\frac{\partial \varphi(x,t)}{\partial x}-\varphi(x,t)\right]-\rho I(x)\frac{\partial^2 \varphi(x,t)}{\partial t^2}=0 \end{cases} \quad (1)$$

This is the coupled equation of motion for a Timoshenko beam, where A(x) is the cross-sectional area, E is the modulus of elasticity, G is the shear modulus, χ is a coefficient related to the shape of the cross-section, ρ is the density of the beam, I(x) is the moment of inertia of the cross-section, y(x,t) is the displacement of the beam, and φ(x,t) is the angle of rotation of the beam cross-section.

Adopt the separated variable approach and bring the following equation into Equation (1).

$$\begin{cases} y(x,t) = \overline{Y}(x)\sin(\overline{\omega}t + \gamma) \\ \varphi(x,t) = \overline{\Phi}(x)\sin(\overline{\omega}t + \gamma) \end{cases} \quad (2)$$

Then, derive the second part of Equation (1) for x. The principal mode function of the vibration of the beam can be expressed as:

$$\begin{cases} -\frac{\partial}{\partial x}\left[\chi GA(x)\left(\overline{\Phi} - \overline{Y}'\right)\right] + \overline{\lambda}\rho A(x)\overline{Y} = 0 \\ \frac{\partial^2}{\partial x^2}\left(EI(x)\overline{\Phi}'\right) - \frac{\partial}{\partial x}\left[\chi GA(x)\left(\overline{\Phi} - \overline{Y}'\right)\right] + \overline{\lambda}\frac{\partial}{\partial x}\left(\rho I(x)\overline{\Phi}\right) = 0 \end{cases} \quad (3)$$

where the eigenvalue $\overline{\lambda} = \omega^2$ for a variable cross-section Timoshenko beam. Thus, the principal factors affecting the natural frequency are I(x) and A(x). For an arbitrary-variable cross-section beam, solving the analytical solution of Equation (3) is usually difficult because Equation (3) is a system of differential equations with variable coefficients.

The objective underlying shape optimization is to obtain a first-order natural frequency as high as possible. From the above equation and Figure 4, it can be seen that Part I of the thin-walled beam possesses a small changeable area, meaning that the optimizable section is Part II, which are in the yellow square. In Part II, any cross-section shape is a thin-walled rectangular cross-section, and as the angle θ between the side of the main support and the vertical plane (lateral slope angle) changes, the thin-walled rectangular cross-section dimensions evolve also. Therefore, the cross-sectional moment of inertia also exhibits alterations. For this reason, the lateral slope angle θ is the optimization variable of shape optimization.

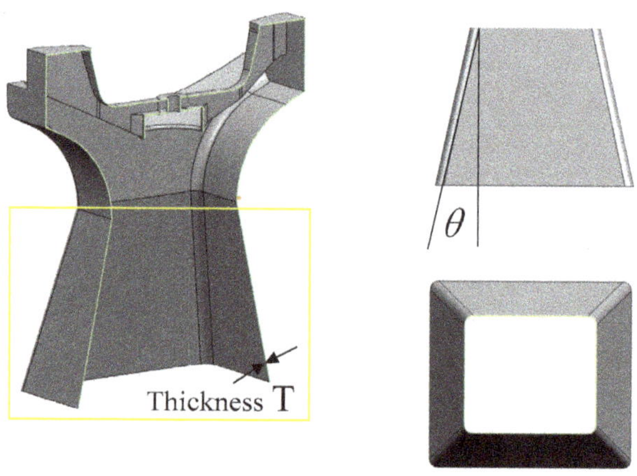

Figure 4. Main support optimization variables.

5.1.2. Optimization Strategy

The optimization of lateral slope angle θ is conducted with the objective of obtaining the highest first-order natural frequency of the support base for a given base mass and height. In this paper, the optimization strategy is an ergodic method. This involves determining the range of values of θ and then choosing the ergodic step size Δθ. Then, it is necessary to calculate the different wall thickness Ti corresponding to different θi with the given mass Md of the support base. Next, researchers should derive the first-order natural frequency ωi corresponding to θi according to the free vibration equation of the support base. Finally, the relationship between the lateral slope angle θ and the first-order natural frequency ωi is to be obtained by traversing all the values of θ, thus determining the optimum lateral slope angle θ and the first-order natural frequency ωi.

5.1.3. Optimization Process and Result

The optimization process of the lateral slope angle θ and the results thereof are as follows. Initially, the permissible range for θ, based on the satellite installation space, lies between 0° and 12°. Then, a step size of 2° is selected for the exploration. The mass of Part II of the main support structure is set at 4 kg. The corresponding wall thickness T can be calculated from the total mass and θ. Subsequently, for each θ value within the specified range, the corresponding free vibration equation for Part II of the main support structure is formulated. By solving this equation, the first-order natural frequency for each θ value is obtained. The optimization results are illustrated in Figure 5. As observed in the graph, the optimal lateral slope angle (θ) is determined to be 12°, with a corresponding wall thickness of 1.779 mm.

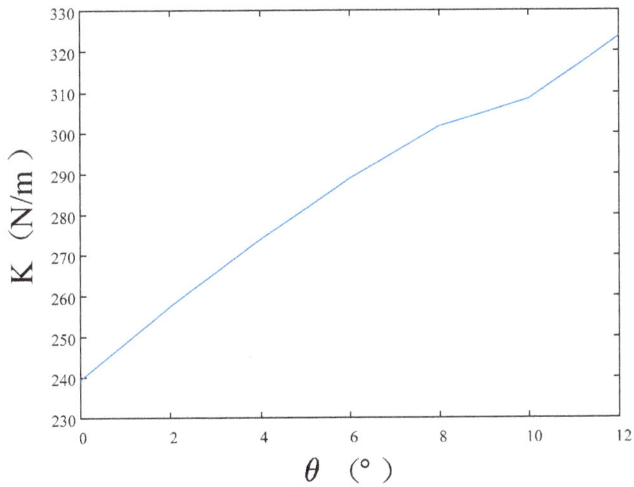

Figure 5. Optimization results of the lateral slope angle.

5.2. Structural Reinforcement Design

The primary photomechanical structure of the space camera is arranged on Part I of the main support, and its strength, stiffness, and stability must overcome the vibration of the mechanical environment of the rocket launching stage. Otherwise, the accuracy of the shaft system cannot be guaranteed, which in turn affects the imaging quality of the optical system and the working life of the camera. Therefore, the cavity of Part I is structurally strengthened, as shown in Figure 6. The inside cavity of Part I mainly consists of the center force-bearing cylinder, eight ribs, etc., which together form the main force transmission path. The scanning axis system, stepping motor, and calibration components transfer the load to the main force transmission path. Meanwhile, for the purpose of preventing stress concentration, the shaft mounting surface and satellite connection surface are thickened to 3 mm, and other wall thicknesses are designed according to the optimization result of 1.779 mm. At this time, the mass of the main support is 10,615 Kg. In order to conduct additive manufacturing, an auxiliary support plate is designed, and a rounded design is performed in the thin-wall connection to prevent cracking during laser printing. Simultaneously, the center bearing cylinder also plays the role of supporting the calibration of the deuterium lamp assembly and the role of the shading cylinder.

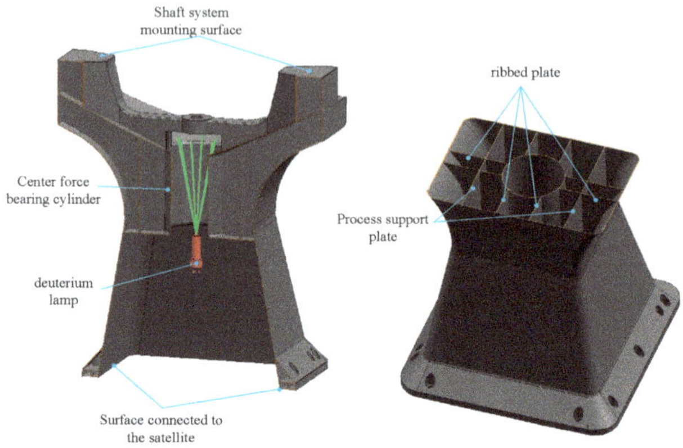

Figure 6. Structural reinforcement of the main support.

5.3. *Size Optimization*

5.3.1. Optimization Objectives and Variables

Shape optimization ensures the high stiffness of the main support, while size optimization accomplishes the goal of lowering the weight of the support base.

The objective of size optimization is to minimize the mass of the main support base. Simultaneously, for the purpose of avoiding resonance phenomena during the rocket's launch that could potentially damage the structure, the first-order natural frequency of the camera is required to be no less than 120 Hz. Therefore, the constraint for the optimization problem is that the first-order natural frequency of the main support base must exceed 120 Hz. The wall thickness of thin-walled structures is the most important factor affecting the mass of the main support. As a consequence, the wall thickness T is used as an optimization variable for dimensional optimization.

5.3.2. Optimization Strategy

The wall thickness T of the support base is used as the optimization variable, the minimum mass is used as the optimization objective, and the first-order natural frequency of the main support is kept higher than 120 Hz as a constraint for optimization.

The conventional dimensional optimization method involves giving an initial wall thickness T_0 and then increasing the thickness Δt each time until the resulting thickness meets the set requirement. The next step involves outputting the wall thickness at a minimum volume, denoted as follows:

$$\text{s.t}\begin{cases} \text{Mass}_{\text{min}}(T_0 + \Delta t \times N) \\ f_1(T_0 + \Delta t \times N) \geq 120\ \text{Hz} \end{cases} \tag{4}$$

where T_0 is the initial wall thickness, Δt is the wall thickness increment, and N is the number of iterations.

However, there remain some problems with this approach. If Δt takes a large value, although the number of iterations can be reduced, the solution may be rough; conversely, if Δt takes a small value, although a detailed solution can be obtained, the number of iterations will be increased, consuming a lot of computational resources.

For this reason, we propose an active fitting optimization algorithm, i.e., the active fitting of the polynomial function is performed after each calculation, the wall thickness at the target frequency is solved numerically, and the minimum wall thickness is selected for calculation. These steps not only reduce the number of iterations but also achieve a detailed solution. The theory of the active fitting optimization algorithm is illustrated in Algorithm 1.

5.3.3. Optimization Process and Result

The optimization process and results of the wall thickness T are as follows. The principal support structure was divided into shell elements. For the sake of simplification, all structural components at the cantilever end were condensed into a single mass node known as M. This approach took into consideration the eccentricity of the mass node and incorporated moments of inertia in three directions. Care was taken to maintain precise connection arrangements, resulting in the simplified finite element model depicted in Figure 7.

Algorithm 1: Active Fitting Optimization Algorithm

Input: T is the wall thickness of the model, FRE is the 1st-order mode of the model, N is the maximum number of iterations, M is a polynomial number, FRE_{target} is the target value of the 1st-order mode of the model, R^2_{target} is the fit threshold, and ΔFRE is the residual value.
Output: T.

$T_0 = \varnothing, T_1 = \varnothing, T_1 \neq T_0$; calculations using the finite element method $FRE_0(T_0), FRE_1(T_1)$ //Step 1. Calculation of 1st-order modes FRE_0 and FRE_1 for T_0 and T_1 wall thicknesses using finite element methods
$N = \varnothing; M = \varnothing; FRE_{target} = \varnothing$;//Step 2. The variable assignment cannot be 0.
for (i = 1 to N) do
From $[T_0FRE_0] \sim [T_iFRE_i]$, calculate the
$F_i(t) = p_M t^M + \cdots + p_2 t^2 + p_1 t + p_0 = \sum_{j=0}^{M} p_j t^j$;
//Step 3. Constructing the fitting function
  Calculate the degree of fit R^2;//Step 4. Calculating goodness of fit
  if $(R^2 \geq R^2_{target})$ then
    when $F_i(t) = FRE_{target}$ solve $(t_0 t_1 \cdots t_n)$;
$T^{min}_{i+1} = (\min(t_0 t_1 \cdots t_n) \cup t > 0)$;
Set the wall thickness of the model as T_{i+1}, and use the finite element method to calculate the 1st-order mode as FRE_{i+1};
    if $|FRE_{i+1} - FRE_{target}| \leq \Delta FRE$) then
  end for; else
      $[T_{i+1}FRE_{i+1}] = [T^{min}_{i+1}FRE_{i+1}]$;
    end if
else
    $i = i + 1$;
    $M = M + 1$;
end if
end for
return T_{i+1}, i, N

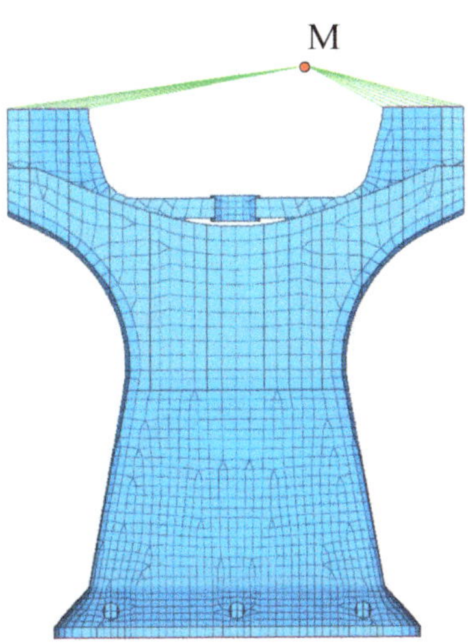

Figure 7. Finite element model.

The active fitting optimization algorithm is used to calculate the optimization of the model wall thickness. Firstly, the 1st-order mode for wall thickness $T_0 = 0.5$, $T_1 = 0.55$ is calculated as FRE_0 $(T_0) = 77$, FRE_1 $(T_1) = 80.8$, the number of cycles is set to be N = 10, the number of polynomials is set to be M = 1, the target value for the 1st-order mode of the model is set to be $FRE_{target} = 120$, the goodness of fit is set to be $R^2_{target} = 0.9999$, and $\Delta FRE = 0.01$ is the residual value. After six cycles and four finite element calculations, the 1st-order mode FRE = 119.9927 Hz of the model is obtained for a wall thickness of T = 1.19 mm. The variation of the fitting curve is shown in Figure 8, and the frequency residual curve during convergence is shown in Figure 9.

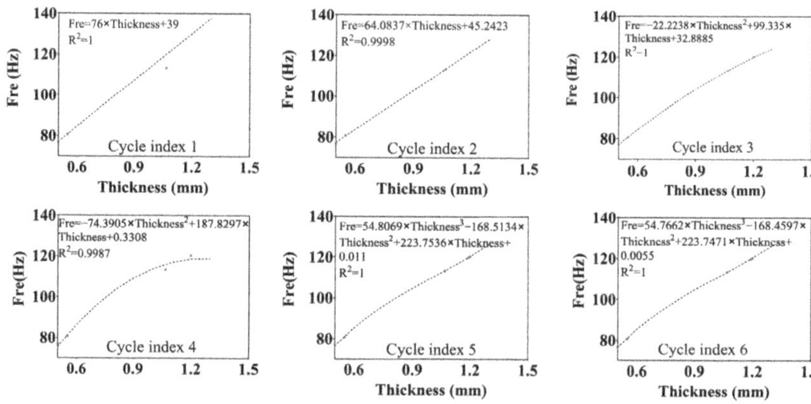

Figure 8. Changes in the fitted curves during the optimization process.

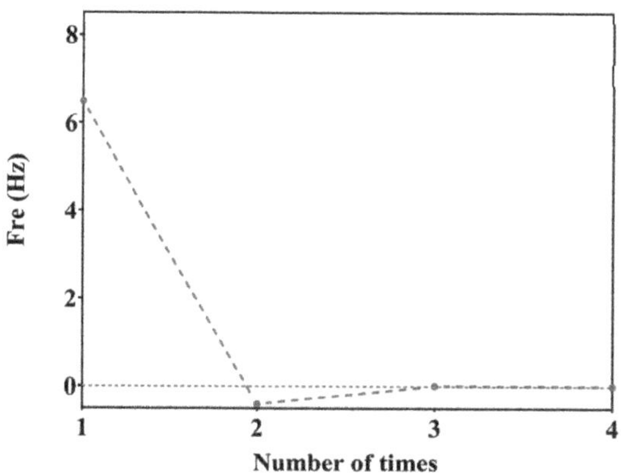

Figure 9. Frequency residuals in the convergence process.

5.4. Mechanical Analysis

The structural design of the support base was completed based on the optimized wall thickness obtained in the previous section. The total mass of the main support structure was 7.56 kg, resulting in a 28% reduction in camera mass. Such a measure ensures that the structure's natural frequency remains within the optimization constraints while also helping to significantly reduce overall camera weight. Following the optimization, local reinforcement design was applied to the camera's mounting locations, and stress concentration design modifications were conducted at threaded connection points.

In the optimization design, the load was considered as a point mass, and the rotational inertia of the load was not taken into account. However, in the actual system, there is some flexibility in the load, shaft system, and their connections. As a result, the natural frequencies of the actual system are lower than the optimized results. To obtain a more accurate representation of the system's natural frequencies, it was necessary to conduct a modal analysis of the entire system. The 11 connection holes at the bottom of the main support were constrained, and the system's natural frequencies in two horizontal directions were calculated to be 109.2 Hz and 110.5 Hz, as illustrated in Figure 10. The results indicate that the first-order natural frequency, calculated using a finite element model with realistic loads, is 10.8 Hz lower than the results obtained via consideration of point masses.

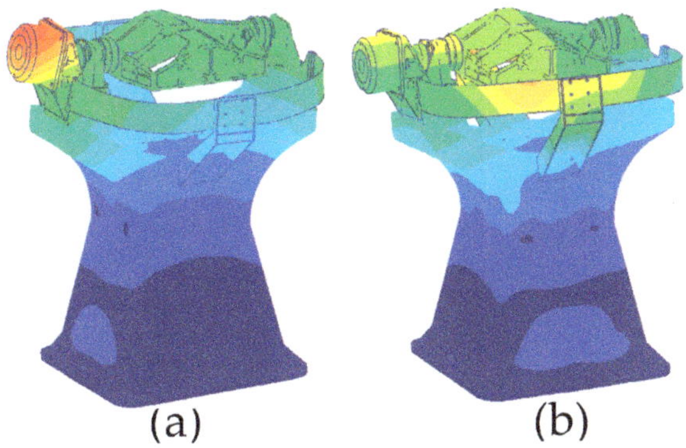

Figure 10. Horizontal modal vibration pattern of space camera. (**a**) Y direction natural frequency, (**b**) X direction natural frequency.

6. Laser Additive Manufacturing and Experimental Verification

6.1. Laser Additive Manufacturing

Compared with the traditional one-dimensional scanning main support structure, which consists of a U-frame and a base, the optimized main support structure adopts integrated thin-wall structures, such as the thin-wall structures shown in Figures 4 and 6, whose overall size reaches 538 mm × 400 mm × 384 mm. The wall thickness is 1.19 mm, and there are some closed cavity structures. It is almost impossible to complete the work using conventional processing means. Additive manufacturing can overcome the problems of chattering in traditional thin-wall CNC machining and quality defects in thin-wall casting. As such, this paper chooses selective laser melting technology to manufacture this structure.

Given that the main support structure was constructed from titanium alloy Ti6Al4V material and possesses substantial dimensions, with an outer envelope measuring 538 mm × 400 mm × 384 mm, a comprehensive assessment of the available large-scale 3D printing equipment was conducted. After thorough consideration, the decision was made to employ the BLT-S615 multi-metal 3D printing system manufactured by Bright Laser Technologies (Xi'an, China). The 3D model of this 3D printing device is shown in Figure 11, the specific parameters of which are outlined in Table 4.

Table 4. Parameters of the BLT-S615 multi-metal 3D printing system.

Material Support	Titanium Alloy, Aluminum Alloy, High-Temperature Alloy, Stainless Steel, High-Strength Steel, Tool Steel
Forming size	600 mm × 600 mm × 1500 mm (W × D × H), 650 mm × 650 mm × 1300 mm (W × D × H)
Power of the laser	500 W × 4, 500 W × 6
Layering thickness	20 μm~100 μm
Maximum scan speed	7 m/s
Efficiency of forming	100 cm^3/h
Preheating temperature	RT + 20~200 °C
Beam quality	M2 < 1.1
Optical structure	F-θ footage
Powder laying agencies	Single/two-way spreading of powder
Power wastage	≤18 KW
Size	4700 mm × 5100 mm × 3800 mm (W × D × H)
Weight	≈14,900 kg

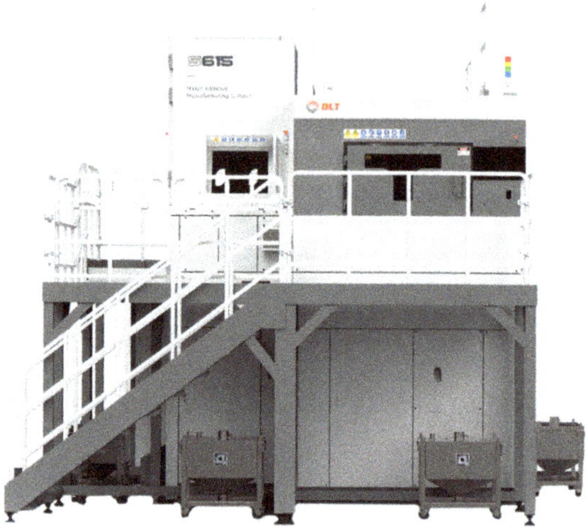

Figure 11. The BLT-S615 multi-metal 3D printing system.

BLT-S615 equipment possesses a maximum size of 600 mm × 600 mm × 1500 mm, meeting the required dimensions for the main support structure. It features a laser power of 500 W × 4, a laser wavelength in the range of 1060 nm to 1080 nm, and a maximum scanning speed of 7 m/s. Additionally, achievable part accuracy can be controlled within ± 0.05 mm. Therefore, the BLT-S615 was selected for the manufacturing of the main support structure using metal additive manufacturing technology. The quality met design specifications. The specific experimental print parameters of the thin-walled main support structure by SLM in this paper are shown in Table 5. After the completion of printing, the ultrasonic thickness gauge is utilized to measure part of the wall thickness. The thickness range is from 1.21 mm to 1.25 mm, and the error is within the allowable range. The mass of the main support structure is 7.78 Kg, which is slightly greater than the design weight but still meets the design requirements.

Table 5. Process parameters of the main support structure by SLM.

Device Model	Laser Power	Spot Diameter	Powder Layer Thickness	Scanning Velocity	Hatch Spacing	Powder Materials
BLT-S615	4 × 340 W	80–85 μm	60 μm	1250 mm/s	120 um	Ti6Al4V

6.2. Experimental Verification

To test the system's inherent frequency in the horizontal direction, a horizontal 0.2 g sweep frequency vibration test was conducted using an electromagnetic vibration table, as depicted in Figure 12. The results of the camera's horizontal sweep frequency test are shown in Figure 13. The camera's inherent frequencies in the horizontal direction were determined to be 105.97 Hz, with an error of 3% compared to the simulated result of 109.2 Hz, validating the optimized design results presented in this paper.

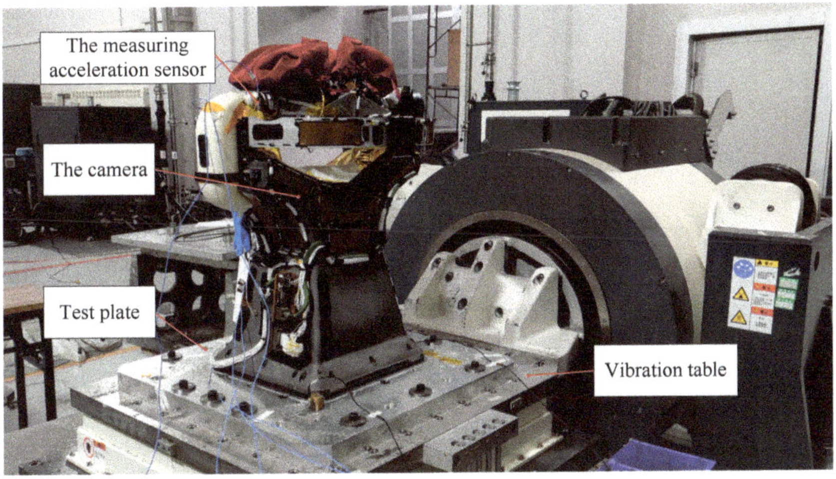

Figure 12. Sweep frequency vibration test site.

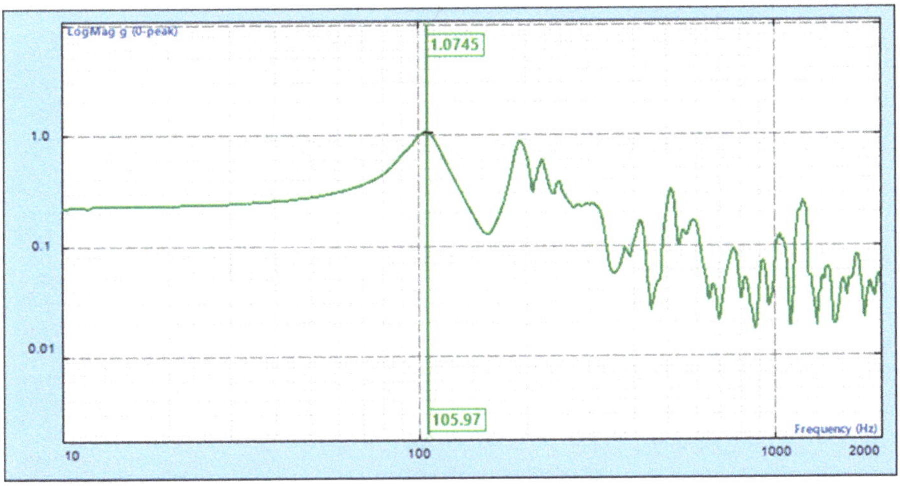

Figure 13. X-direction sweep test results.

7. Conclusions and Future Research

In this paper, we designed the main support structure for a wide-angle auroral imager with an integrated thin-wall structure via structural optimization. We then manufactured this via selective laser melting (SLM). First, this marked the first application of laser additive manufacturing of large-size complex thin-walled structures to the production of the main bearing structure of a space camera. The processed size of the main support structure reached 538 mm × 400 mm × 384 mm, and the wall thickness of the thin-walled structure was 1.19 mm. The vibration test proved that the structure exhibited good stiffness with lighter mass. Second, the initial design of the thin-walled main support was simplified to have a Timoshenko cantilever beam structure. The study analyzed the principal parameters affecting the natural frequency according to Timoshenko beam theory, subsequently optimizing the angle between the side of the main support and the vertical plane. Finally, an active fitting optimization algorithm was proposed for the refinement of the thin-walled structure's wall thickness. The optimization algorithm not only reduced the number of iterations but also obtained more precise solutions. The proposed methodology provides valuable insights and references for the future design and manufacturing of similar space optical camera sensors. Furthermore, the approach of utilizing additive manufacturing to produce large-scale complex thin-walled structures introduces a novel perspective in both the field of space camera support and AM.

This paper presented a case study of a large-scale complex thin-wall main support structure based on additive manufacturing in a space camera. The primary focus of the research lay in the design and optimization of the main support structure. During the research process, several other aspects deserving analysis were identified. Firstly, it was crucial to determine precisely how process parameters used in laser additive manufacturing, such as powder materials, laser power, scanning velocity, hatch spacing, and powder layer thickness, impact both the quality and frequency of the space camera as well as its structural stress. Secondly, because of the thin wall thickness and large size of the main support structure, the thermal stress in the laser AM process can easily cause deformation of the overall structure. Therefore, this article also discusses how to reduce thin-walled deformation by designing and controlling process parameters in laser AM. Lastly, relevant experiments must be conducted to verify their thermal stability and mechanical stability due to thin-walled structures being prone to deformation, buckling, and other problems in environmental tests.

Author Contributions: Conceptualization, J.P. and S.L.; methodology, J.P.; software, J.P.; validation, J.P., S.L. and D.W.; formal analysis, J.P.; investigation, J.P.; resources, J.P.; data curation, J.P., A.X. and H.L.; writing—original draft preparation, J.P.; writing—review and editing, J.P., X.H., T.M. and J.W.; visualization, J.P.; supervision, S.L.; project administration, S.L. All authors have read and agreed to the published version of the manuscript.

Funding: This research received no external funding.

Data Availability Statement: The data presented in this study are publicly available in the article.

Conflicts of Interest: The authors declare no conflicts of interest.

References

1. De Lourdes González, G. Storm-time variability of ionospheric irregularities over South America. *J. Atmos. Sol.-Terr. Phys.* **2022**, *241*, 105980. [CrossRef]
2. Wang, L.; Yang, X.; Dai, L.; Wang, C.; Zhang, H.; Chang, Z.; Jing, T. An on-orbit cross-calibration between the relativistic electron observations from BeiDou M04 and GPS ns63. *Adv. Space Res.* **2022**, *70*, 2805–2817. [CrossRef]
3. Smith, A.W.; Forsyth, C.; Rae, I.J.; Garton, T.M.; Jackman, C.M.; Bakrania, M.; Johnson, J.M. On the Considerations of Using Near Real Time Data for Space Weather Hazard Forecasting. *Space Weather* **2022**, *20*, e2022SW003098. [CrossRef]
4. Hong, J.; Mao, F.; Gong, W.; Gan, Y.; Zang, L.; Quan, J.; Chen, J. Assimilating Fengyun-4A observations to improve WRF-Chem PM2. 5 predictions in China. *Atmos. Res.* **2022**, *265*, 105878. [CrossRef]
5. Yang, Z.; Zhang, P.; Gu, S.; Hu, X.; Tang, S.; Yang, L.; Bai, W. Capability of Fengyun-3D satellite in earth system observation. *J. Meteorol. Res.* **2019**, *33*, 1113–1130. [CrossRef]
6. Zhu, Z.; Shi, C.; Gu, J. Characterization of bias in Fengyun-4B/AGRI infrared observations using RTTOV. *Remote Sens.* **2023**, *15*, 1224. [CrossRef]
7. Zhang, X.X.; Chen, B.; He, F.; Song, K.F.; He, L.P.; Liu, S.J.; Guo, Q.-F.; Li, J.-W.; Wang, X.-D.; Zhang, H.-J.; et al. Wide-field auroral imager onboard the Fengyun satellite. *Light Sci. Appl.* **2019**, *8*, 47. [CrossRef] [PubMed]
8. Dehaghani, M.R.; Tang, Y.; Panicker, S.; Wu, D.; Coatanea, E.; Wang, G.G. Modeling and optimization of height-related geometrical parameters for thin wall structures manufactured by metal additive manufacturing. *Int. J. Adv. Manuf. Technol.* **2023**, *129*, 4663–4675. [CrossRef]
9. Ding, D.; Zhao, R.; Lu, Q.; Pan, Z.; Li, H.; Wang, K.; He, K. A shape control strategy for wire arc additive manufacturing of thin-walled aluminium structures with sharp corners. *J. Manuf. Process.* **2021**, *64*, 253–264. [CrossRef]
10. Kumar, M.B.; Sathiya, P. Methods and materials for additive manufacturing: A critical review on advancements and challenges. *Thin-Walled Struct.* **2021**, *159*, 107228. [CrossRef]
11. Novack, T.; Esch, T.; Kux, H.; Stilla, U. Machine learning comparison between WorldView-2 and QuickBird-2-simulated imagery regarding object-based urban land cover classification. *Remote Sens.* **2011**, *3*, 2263–2282. [CrossRef]
12. Kameche, M.; Benmostefa, S. In-flight MTF stability assessment of ALSAT-2A satellite. *Adv. Space Res.* **2016**, *58*, 117–130. [CrossRef]

13. Kim, H.O.; Kim, H.S.; Lim, H.S.; Choi, H.J. Space-based earth observation activities in South Korea [Space Agencies]. *IEEE Geosci. Remote Sens. Mag.* **2015**, *3*, 34–39. [CrossRef]
14. Middleton, E.M.; Ungar, S.G.; Mandl, D.J.; Ong, L.; Frye, S.W.; Campbell, P.E.; Pollack, N.H. The earth observing one (EO-1) satellite mission: Over a decade in space. *IEEE J. Sel. Top. Appl. Earth Obs. Remote Sens.* **2013**, *6*, 243–256. [CrossRef]
15. Haverkamp, D. Extracting straight road structure in urban environments using IKONOS satellite imagery. *Opt. Eng.* **2002**, *41*, 2107–2110. [CrossRef]
16. Guo, Q.; Chen, B.; Liu, S.; Song, K.; He, L.; He, F.; Shi, G. Optomechanical design of a wide-field auroral imager on Fengyun-3D. *Appl. Opt.* **2022**, *61*, 3349–3356. [CrossRef] [PubMed]
17. Wang, K.; Wang, L.; Zheng, K.; He, Z.; Politis, D.J.; Liu, G.; Yuan, S. High-efficiency forming processes for complex thin-walled titanium alloys components: State-of-the-art and perspectives. *Int. J. Extrem. Manuf.* **2022**, *2*, 032001. [CrossRef]
18. Zhou, X.; Ma, J.; Zhou, W.; Welo, T. Forming-based geometric correction methods for thin-walled metallic components: A selective review. *Int. J. Adv. Manuf. Technol.* **2023**, *128*, 17–39. [CrossRef]
19. Sayam, A.; Rahman, A.M.; Rahman, M.S.; Smriti, S.A.; Ahmed, F.; Rabbi, M.F.; Faruque, M.O. A review on carbon fiber-reinforced hierarchical composites: Mechanical performance, manufacturing process, structural applications and allied challenges. *Carbon Lett.* **2022**, *32*, 1173–1205. [CrossRef]
20. Liu, Y.; Zhu, Y. 3D-Printed Soft Wearable Electronics: Techniques, Materials, and Applications. In *Additive Manufacturing: Materials, Functionalities and Applications*; Springer International Publishing: Cham, Switzerland, 2022; pp. 1–49.
21. Khorasani, M.; Ghasemi, A.; Rolfe, B.; Gibson, I. Additive manufacturing a powerful tool for the aerospace industry. *Rapid Prototyp. J.* **2022**, *28*, 87–100. [CrossRef]
22. Hoffmann, M.; Elwany, A. In-space additive manufacturing: A review. *J. Manuf. Sci. Eng.* **2023**, *145*, 020801. [CrossRef]
23. Blakey-Milner, B.; Gradl, P.; Snedden, G.; Brooks, M.; Pitot, J.; Lopez, E.; Du Plessis, A. Metal additive manufacturing in aerospace: A review. *Mater. Des.* **2021**, *209*, 110008. [CrossRef]
24. Market Research Future, Global Aerospace Additive Manufacturing Market Research Report—Forecast 2023–2032. 2021. Available online: https://www.marketresearchfuture.com/reports/aerospace-additive-manufacturing-market-1551 (accessed on 10 November 2023).
25. Ishfaq, K.; Asad, M.; Mahmood, M.A.; Abdullah, M.; Pruncu, C. Opportunities and challenges in additive manufacturing used in space sector: A comprehensive review. *Rapid Prototyp. J.* **2022**, *28*, 2027–2042. [CrossRef]
26. Ransikarbum, K.; Pitakaso, R.; Kim, N. A Decision-Support Model for Additive Manufacturing Scheduling Using an Integrative Analytic Hierarchy Process and Multi-Objective Optimization. *Appl. Sci.* **2020**, *10*, 5159. [CrossRef]
27. Ransikarbum, K.; Khamhong, P. Integrated fuzzy analytic hierarchy process and technique for order of preference by similarity to ideal solution for additive manufacturing printer selection. *J. Mater. Eng. Perform.* **2021**, *30*, 6481–6492. [CrossRef]
28. Wang, D.; Wang, H.; Chen, X.; Liu, Y.; Lu, D.; Liu, X.; Han, C. Densification, Tailored Microstructure, and Mechanical Properties of Selective Laser Melted Ti–6Al–4V Alloy via Annealing Heat Treatment. *Micromachines* **2022**, *13*, 331. [CrossRef] [PubMed]
29. Li, Z.; Li, H.; Yin, J.; Li, Y.; Nie, Z.; Li, X.; You, D.; Guan, K.; Duan, W.; Cao, L.; et al. A review of spatter in laser powder bed fusion additive manufacturing: In situ detection, generation, effects, and countermeasures. *Micromachines* **2022**, *13*, 1366. [CrossRef]
30. Razavykia, A.; Brusa, E.; Delprete, C.; Yavari, R. An overview of additive manufacturing technologies—A review to technical synthesis in numerical study of selective laser melting. *Materials* **2020**, *13*, 3895. [CrossRef] [PubMed]
31. Wang, Z.; Ummethala, R.; Singh, N.; Tang, S.; Suryanarayana, C.; Eckert, J.; Prashanth, K.G. Selective laser melting of aluminum and its alloys. *Materials* **2020**, *13*, 4564. [CrossRef]
32. RUAG. Sentinel Antenna Bracket. Available online: https://www.eos.info/01_parts-and-applications/case_studies_applications_parts/_case_studies_pdf/en_cases/cs_m_aerospace_ruag_en.pdf (accessed on 2 February 2021).
33. Manil, P.; Jan, Y.; Nunio, F.; Lomello, F.; Arhancet, A.; Lacroix, M.; Lapresle, J. Structural optimization, additive manufacturing and vibration testing of titanium alloy supports based on the space detector SVOM-MXT. In *Advances in Optical and Mechanical Technologies for Telescopes and Instrumentation IV*; SPIE: Bellingham, WA, USA, 2021; Volume 11451, pp. 526–544.
34. NASA. NASA's Perseverance Rover Bringing 3D-Printed Metal Parts to Mars. 2020. Available online: http://www.nasa.gov/feature/jpl/nasas-perseverance-rover-bringing-3d-printed-metal-parts-to-mars (accessed on 19 February 2021).
35. Meng, H.A. Baffle Manufacturing Technology based on 3D printing for the secondary mirror space cameras. *Spacecr. Recovery Remote Sens.* **2023**, *5*, 46–53. (In Chinese)
36. Zhang, N.; Qinglin, L.; Junlei, C. Application and verification of 3D printing technology in the development of space remote sensing camera. *AOPC 2021 Opt. Sens. Imaging Technol.* **2021**, *12065*, 348–353.
37. Fan, Y.; Dong, D.; Li, C.; Sun, Y.; Zhang, Z.; Wu, F.; Guan, Y. Research and experimental verification on topology-optimization design method of space mirror based on additive-manufacturing technology. *Machines* **2021**, *9*, 354. [CrossRef]
38. Jia, X.; Hu, B.; Wang, F.; Li, S.; Sun, L.; Wu, J. Research progress and core technologies of optical-mechanical system based on additive manufacturing. *AOPC 2021 Adv. Laser Technol. Appl.* **2021**, *12060*, 276–285.
39. Chen, W.; Sun, L.; Li, S.; Wu, J.; Zhang, Z. Precision manufacturing of metal mirrors based on additive manufacturing. *Eighth Symp. Nov. Photoelectron. Detect. Technol. Appl.* **2022**, *12169*, 1982–1988.
40. Yoder, P.R., Jr. *Opto-Mechanical Systems Design*; CRC Press: Boca Raton, FL, USA, 2005.
41. Xiong, Q.; Zhou, Q. Development Trend of NC Machining Accuracy Control Technology for Aeronautical Structural Parts. *World J. Eng. Technol.* **2020**, *8*, 266–279. [CrossRef]

42. Oguntuyi, S.D.; Nyembwe, K.; Shongwe, M.B.; Mojisola, T. Challenges and recent progress on the application of rapid sand casting for part production: A review. *Int. J. Adv. Manuf. Technol.* **2023**, *126*, 891–906. [CrossRef]
43. Timoshenko, S.; Young, D.H.; Weaver, W. *Vibration Problems in Engineering*, 4th ed.; John Wiley & Sons, Inc.: New York, NY, USA, 1974.

micromachines

Article

High Reflectivity and Thermal Conductivity Ag–Cu Multi-Material Structures Fabricated via Laser Powder Bed Fusion: Formation Mechanisms, Interfacial Characteristics, and Molten Pool Behavior

Qiaoyu Chen [1,2], Yongbin Jing [1,2], Jie Yin [1,2,*], Zheng Li [1,2], Wei Xiong [1,3], Ping Gong [1,4], Lu Zhang [1,2], Simeng Li [1,2], Ruiqi Pan [1,2], Xiya Zhao [1,2] and Liang Hao [1,2,*]

1 Gemmological Institute, China University of Geosciences, Wuhan 430074, China
2 Hubei Gem & Jewelry Engineering Technology Research Center, Wuhan 430074, China
3 School of Mechanical Engineering and Electronic Information, China University of Geosciences, Wuhan 430074, China
4 Academy of Fine Arts, Guizhou Normal University, Guiyang 550001, China
* Correspondence: yinjie@cug.edu.cn (J.Y.); haoliang@cug.edu.cn (L.H.)

Abstract: Ag and Cu have different advantages and are widely used in key fields due to their typical highly electrical and thermal conductive (HETC) properties. Laser powder bed fusion (LPBF), an innovative technology for manufacturing metallic multi-material components with high accuracy, has expanded the application of Ag–Cu in emerging high-tech fields. In this study, the multi-material sandwich structures of Ag7.5Cu/Cu10Sn/Ag7.5Cu were printed using LPBF, and the formation mechanism, interface characteristics, and molten pool behavior of the Ag7.5Cu/Cu10Sn (A/C) and Cu10Sn/Ag7.5Cu (C/A) interfaces were studied to reveal the influence of different building strategies. At the A/C interface, pre-printed Ag7.5Cu promoted Marangoni turbulence at a relatively low energy density ($E_{A/C}$ = 125 J/mm^3). Due to the recoil pressure, the molten pool at the A/C interface transformed from a stable keyhole mode to an unstable keyhole mode. These phenomena promoted the extensive migration of elements, forming a wider diffusion zone and reduced thermal cracking. At the C/A interface, the molten pool was rationed from the conduction mode with more pores to the transition mode with fewer defects due to the high energy density ($E_{C/A}$ = 187.5 J/mm^3). This work offers a theoretical reference for the fabrication of HETC multi-material structures via LPBF under similar conditions.

Keywords: highly reflective and thermally conductive metals; multi-material structures; laser powder bed fusion; interfacial characteristics; Ag alloy; Cu alloy

Citation: Chen, Q.; Jing, Y.; Yin, J.; Li, Z.; Xiong, W.; Gong, P.; Zhang, L.; Li, S.; Pan, R.; Zhao, X.; et al. High Reflectivity and Thermal Conductivity Ag–Cu Multi-Material Structures Fabricated via Laser Powder Bed Fusion: Formation Mechanisms, Interfacial Characteristics, and Molten Pool Behavior. *Micromachines* **2023**, *14*, 362. https://doi.org/10.3390/mi14020362

Academic Editor: Francesco Ruffino

Received: 11 November 2022
Revised: 10 January 2023
Accepted: 13 January 2023
Published: 31 January 2023

1. Introduction

The typical highly electrically and thermally conductive metals (HETCMs) (e.g., Ag, Au, Cu) are widely used in various fields because of the special characteristics [1]. Ag has been used in smart electronics [2], green energy [3], wearables [4], jewelry [5], medicine and health care [6], and other emerging fields due to its excellent electrical [4], thermal [7], reflective [8], and antibacterial [9] properties. Cu has high conductivity [10], heat conduction [11], corrosion resistance [12] and toughness [13], and is widely used in electric power [14], heat dissipation [15], aerospace [16], marine applications [17], and other fields. However, the relatively high price, low hardness, low yield strength, and poor wear resistance of Ag limit its further application [18]. While Cu is widely used in a pure and an alloy form due to its machinability and relatively low cost.

Metallic additive manufacturing (AM) is an emerging technology that can efficiently produce functionally graded materials (FGMs), resulting in improved performance through gradual changes in the composition or microstructure [19]. Multi-material structures are typical FGMs [20]. By integrating the structures and functions of multiple materials, the combined properties of the multiple materials (e.g., local wear resistance, high thermal conductivity, thermal insulation, chemical corrosion resistance, etc.) [21] can be achieved to meet the growing demands of the aerospace, biomedical, automotive, and jewelry industries, among others [22]. Laser powder bed fusion (LPBF) is a common metal additive manufacturing process [23,24] that has the advantages of precision manufacturing and multi-scale precision control [25]. In the LPBF process, the combination of laser rapid solidification and high thermal conductivity metal contributes to promote grain refinement [17], which provides the possibility to improve the comprehensive performance of HETCMs [1]. The smaller spots and thinner layers enable LPBF to fabricate metallic multi-material components with more complex structures and higher geometric resolutions [21], which are otherwise difficult or impossible to fabricate with the traditional manufacturing techniques [26] such as powder metallurgy [27], welding [28], casting [29], cold pressing [30], etc. Through LPBF technology, precious metal materials can be used efficiently, and its green and flexible manufacturing characteristics are also helpful in expanding the application of Ag materials in emerging high-tech fields.

Although the mentioned studies highlight the potential advantages of Ag and Cu, there are few articles related to the LPBF fabrication of Ag–Cu because of the expensive price of Ag and the difficulty of printing HETCMs [31,32]. For example, Robinson et al. [32] printed 99% pure Ag using single-tracks fabrication on Cu and steel substrates and found that the Cu substrates with high thermal conductivity and high reflectivity were not suitable for LPBF fabrication of pure Ag single-tracks. Xiong et al. [18] found that the grain refinement caused by the LPBF process contributed to the Vickers hardness of the LPBFed Ag alloy three times higher than that of the casting Ag alloy. John et al. [33] found that when using LPBF in situ alloying to fabricate Cu–Ag structures with different Ag contents, the pore number and average pore size of both the prefabricated and the annealed samples decreased with the increasing Ag content. Temel et al. [34] fabricated Ag-coated Cu cores through LPBF to obtain materials with higher conductivity and oxidation resistance at a relatively low cost. The above studies demonstrated the feasibility of fabricating Ag–Cu multi-material via LPBF. However, existing studies mainly use in situ alloying and pre-alloying. The fabrication of Ag–Cu multi-material LPBF using a sandwich structure has not been investigated.

Some LPBF-based metallic multi-material structures (MMS) have been investigated, such as copper–steel [20], copper–nickel [35], titanium–copper [36], aluminum–copper [37], etc. The research contents involve the interface characteristics of multiple materials, including the interfacial microstructure, the elemental diffusion mechanism, the interfacial melting mode, etc. Chen et al. [38] printed 316L–Cu10Sn MMS using LPBF and found that many dendritic cracks in the diffusion zone came from the different physical properties of 316L and Cu10Sn. Wei et al. [36] printed Cu10Sn and Ti6Al4V with different construction strategies using LPBF and found that the thin elements reaction zone at the interface contributed to the delamination of dissimilar materials, while more extensive element migration enhanced the bonding strength of multi-material interfaces. Tan et al. [39] found that due to the high cooling rate of the copper substrate, gradient/fine dendrite grains were generated at the interface of the LPBFed copper–steel MMS, resulting in an ultra-high copper–steel bonding strength. Liu et al. [20] investigated the correlation of energy density with the Marangoni convection-driven molten pool flow behavior in 316L–Cu10Sn multi-material LPBF fabrication.

Previous studies have shown that different building strategies in metallic multi-material LPBF fabrication affect interface behavior [40,41], but there is no related study on Ag–Cu multi-material. The multi-material used in the current research are mainly dissimilar materials with certain thermo-physical properties (e.g., thermal conductivity, thermal expansion coefficient, laser absorptivity, melting temperature, etc.), but the HETCMs MMS fabricated by LPBF has not been reported yet. Therefore, a study of the bonding characteristics of the interface between Ag alloy and Cu alloy is helpful in breaking down the barriers of HETCMs MMS additive manufacturing and expanding the application of HETCMs in various industries.

In this study, the Ag7.5Cu/Cu10Sn/Ag7.5Cu multi-material sandwich structures were printed by LPBF. The sandwich structure had two interfaces, the Ag7.5Cu/Cu10Sn interface (Cu10Sn was printed on Ag7.5Cu), and the Cu10Sn/Ag7.5Cu interface (Ag7.5Cu was printed on Cu10Sn). The interfacial characteristics (such as morphology, defect, microstructure, and element diffusion), interfacial melting mode, and their formation mechanisms were investigated.

2. Methods and Experiments

2.1. Experimental Materials

The spherical gas-atomized powders used in the experiment included Ag7.5Cu alloy powders (Figure 1a) and Cu10Sn alloy powders (Figure 1b), supplied by the Legor Group. Previous studies have shown that different powder particle sizes not only increased packing density but also improved the powder absorptivity of highly reflective metal powders through multiple reflections [42,43]. The particle size distribution of the powders was measured by a Mastersizer 2000 laser particle size analyzer. The Ag7.5Cu powders had a D10 of 14.7μm, a D50 of 22.8 μm, and a D90 of 33.0 μm, and the Cu10Sn powders had a D10 of 24.5 μm, a D50 of 36.1 μm, and a D90 of 57.5 μm, as shown in Figure 1c and 1d, respectively. The oxygen level of the powders was measured by a Leco ONH836 Oxygen/Nitrogen/Hydrogen elemental analyzer. The chemical compositions of two powders are shown in Table 1.

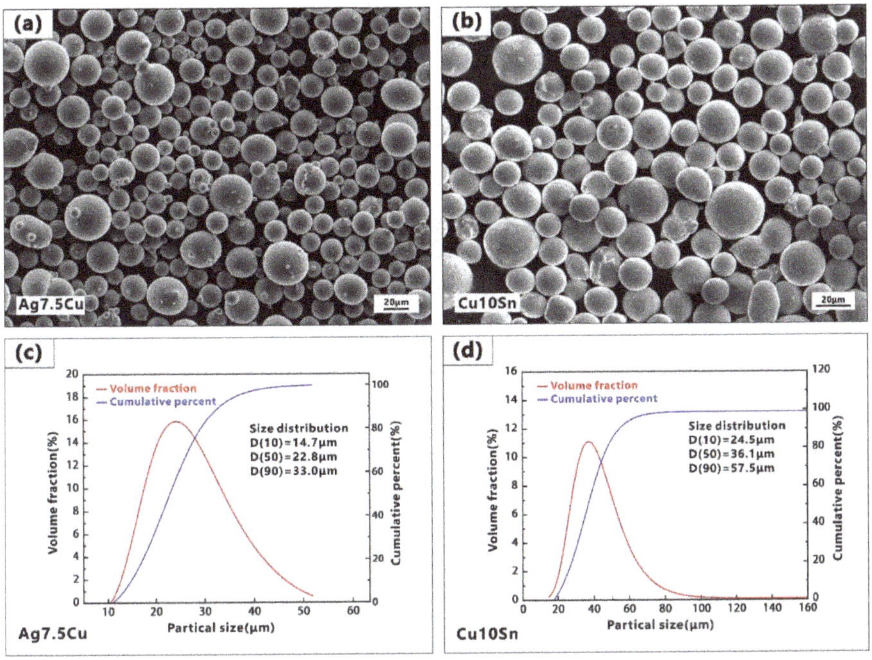

Figure 1. (**a**,**b**) SEM images of Ag7.5Cu and Cu10Sn powders, respectively; particle size distribution of (**c**) Ag7.5Cu powders and (**d**) Cu10Sn powders.

Table 1. Chemical compositions of Ag7.5Cu and Cu10Sn alloy powders (wt.%).

Materials	Ag	Cu	Sn	O	Others
Ag7.5Cu	Bal.	7.56	—	0.016	Less than 0.05
Cu10Sn	—	Bal.	10.20	0.038	Less than 0.05

2.2. LPBF Process of Ag7.5Cu–Cu10Sn Multi-Material Structure

In this work, the sample was produced by LPBF, using a SISMA MYSINT100 system and an Nd: YAG fiber laser with wavelength of 1.06 μm, high power (up to 200 W), and small focus (30 μm). The sample was built in a nitrogen environment (residual oxygen content of 0.5 vol%). A schematic illustration of the multi-material LPBF process as shown in Figure 2. The sandwich structure sample was a 6mm x 6mm x 6mm cube consisting of Ag7.5Cu alloy on the bottom and top, and Cu10Sn alloy in the middle, with a 2 mm height of each metal (see Figure 2a). Due to the sandwich structure, two interfaces with different printing strategies could be observed in the same sample. The Ag7.5Cu/Cu10Sn interface and the Cu10Sn/Ag7.5Cu interface were referred to as the "A/C interface" and "C/A interface", respectively.

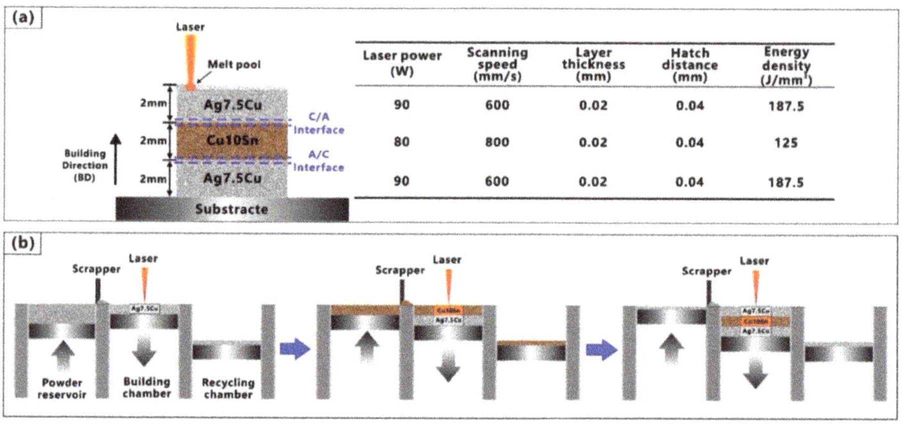

Figure 2. Schematic illustration of LPBF process for fabricating Ag7.5Cu/Cu10Sn/Ag7.5Cu multi-material samples: (**a**) building strategy and parameters; (**b**) printing process.

Differences in the melting point and thermal conductivity of different materials would lead to the insufficient melting of metals with higher melting points in the powder mixing region [19]. As metals with high reflectivity (Ag, Cu > 95%, 1064 nm) [44] and high thermal conductivity(Ag: 429 W/(m·K), Cu: 401 W/(m·K)) [45] require more laser energy to melt [23]. Based on our previous experiments [18], compared with Cu10Sn alloy, a higher energy density was used when printing the Ag7.5Cu alloy to avoid the insufficient melting of the Ag alloy. The optimized parameters for manufacturing the Ag7.5Cu alloy and Cu10Sn alloy were listed on the right side of Figure 2a, and a chessboard scanning strategy was applied to reduce the residual stress. The volume energy densities (E_d) were calculated based on Equation (1) [46].

Figure 2b shows the LPBF printing strategies for fabricating the samples. After the first part of the Ag7.5Cu printing was completed, the Ag7.5Cu powders in the powder reservoir and powder recycling chamber were removed and cleaned, and then, the Cu10Sn powders were re-filled into the powder reservoir to continue the printing. This step was repeated until the entire multi-material sample was printed.

$$E_d = \frac{P}{v \times h \times t} \quad (1)$$

where v is the scanning speed; h is the hatch distance; and t is the layer thickness.

2.3. Microscopic Feature

The side surface of the Ag7.5Cu/Cu10Sn/Ag7.5Cu sample was polished by the common metallographic method, up to 0.3 μm silica colloidal. The sample was polished mechanically to obtain mirror-like surface and etched with two different etchants. One etchant (50 mL ammonia water + 50 mL H_2O_2 (3 vol%) + 50 mL distilled water) was used for the Ag7.5Cu part. The Cu10Sn part was etched with 5 g $FeCl_3$, 10 mL HCI, and 100 mL distilled H_2O mixed solution. An optical microscope (OM) with a Leica M205 was used to characterize the surface morphology and defects. The interfacial microstructure and element distribution were observed and analyzed with a ZEISS Gemini 300 scanning electron microscope (SEM) equipped with an Smartedx energy dispersive spectroscope (EDS).

3. Results and Discussion

3.1. Interfacial Morphology and Defects

Figure 3 shows the optical mesoscopic images of the two interfacial regions of the Ag7.5Cu/Cu10Sn/Ag7.5Cu sandwich structure sample fabricated using LPBF. Three regions could be observed in the interface: the Ag7.5Cu region, the diffusion zone, and the Cu10Sn region. The diffusion zone of the A/C interface (~90 μm) was wider than that of the C/A interface (~40 μm), as indicated by the yellow dash lines in Figure 3(a1,a2) and Figure 3(b1,b2), respectively. No obvious mesoscopic defects (delamination or cracking) were observed in the binding regions (see Figure 3(a1,b1)), and the Ag7.5Cu was firmly metallurgically bonded to the Cu10Sn part. On the one hand, Ag and Cu were metallurgically compatible. As the Ag–Cu alloy system was an eutectic [47], the maximum solid solubilities of Cu in Ag at the eutectic temperature (779.1 °C) and at room temperature (20 °C) were 8.27 wt% and <1 wt%, respectively [47]. On the other hand, the large difference in the coefficient of thermal expansion (CTE) between dissimilar metals was one of the main causes of interfacial bonding defects [48,49], while the CTEs of Ag (19 ppm/K) and Cu (17 ppm/K) were relatively close [50], resulting in good metallurgical bonding in the LPBF manufacturing process.

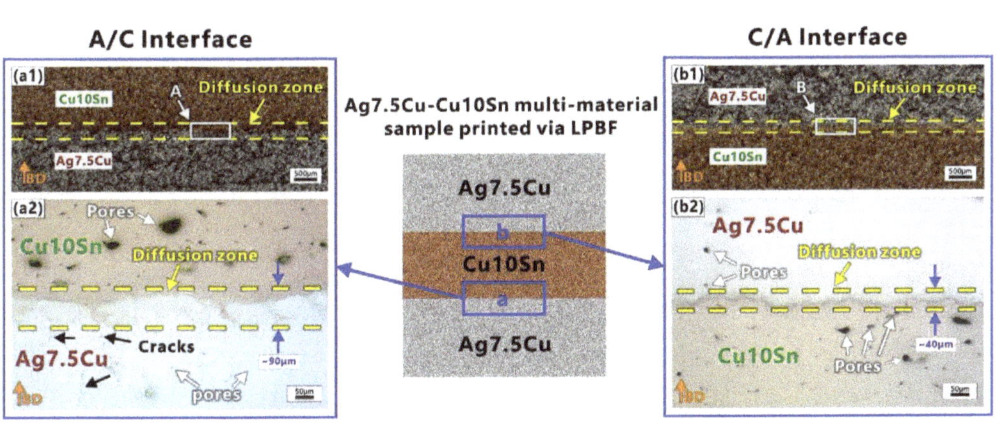

Figure 3. Optical images of two interfaces: (**a1**) macroscopic images of the A/C interface; (**a2**) enlarged view of area A in (**a1**); (**b1**) macroscopic images of C/A interface; (**b2**) enlarged view of area (B) in (**b1**). Orange arrows indicate the building direction (BD).

Figure 4 shows the microstructure of the two interfaces after etching. The purpose of the etching was to better reveal the morphology of the molten pool. The overlapped molten pools with 'fish scale' [51] solidification lines could be clearly identified in the two interfaces, as shown in Figure 4. During LPBF fabrication, as new layers were deposited on top, a certain amount of the previously deposited layers were remelted [52,53], and the changes in the depth and shape of the molten pool resulted in irregular, wavy-shaped tracks [54]. Micro cracks and micro pores were observed in the Ag7.5Cu region and the Cu10Sn region respectively, while fewer defects were observed in the two diffusion zones of two interfaces, as shown in Figures 3 and 4. Lack of fusion (LoF) [55] pores could be seen in the Cu10Sn region, mainly at the edge of the molten pool (see white arrows in Figure 4a,c). Microcracks were observed mainly in the Ag7.5Cu region (see black arrows in Figure 4a,c). Some cracks extended from the edge of the diffusion zone to the Ag7.5Cu side at the A/C interface (see black arrows in Figure 4b). These microcracks occurred at the molten pool boundary and extend in the same direction as the columnar crystal growth (parallel to the direction of the maximum thermal gradient). Microcracks were considered to be thermal cracks caused by high thermal stress due to high energy density, and they usually occurred and propagated along the grain boundaries [56]. The layer-by-layer melting and solidification characteristics of the LPBF process facilitated epitaxial growth, and it tended to induce columnar grains [57]. The columnar grains provided a long and straight path for crack propagation along the grain boundary [58], while the high solidification rate of Ag was not conducive to liquid backfill, making LPBFed Ag more prone to cracking. In addition, the unique solidification morphology caused by Marangoni convection was observed at the A/C interface, as shown in Figure 4b. In the LPBF process, the Marangoni convection, recoil pressure, and surface tension of the molten pool promoted elemental mixing and drove the molten pool into a keyhole mode [59,60]. Previous studies have shown that the large thermal stress induced by the high thermal gradients increased the possibility of crack generation [61]. Therefore, cracks were more likely to occur in the heat-affected zone (HAZ) below the keyhole mode molten pool [62]. This also explains why there were more microcracks below the diffusion zone of the A/C interface.

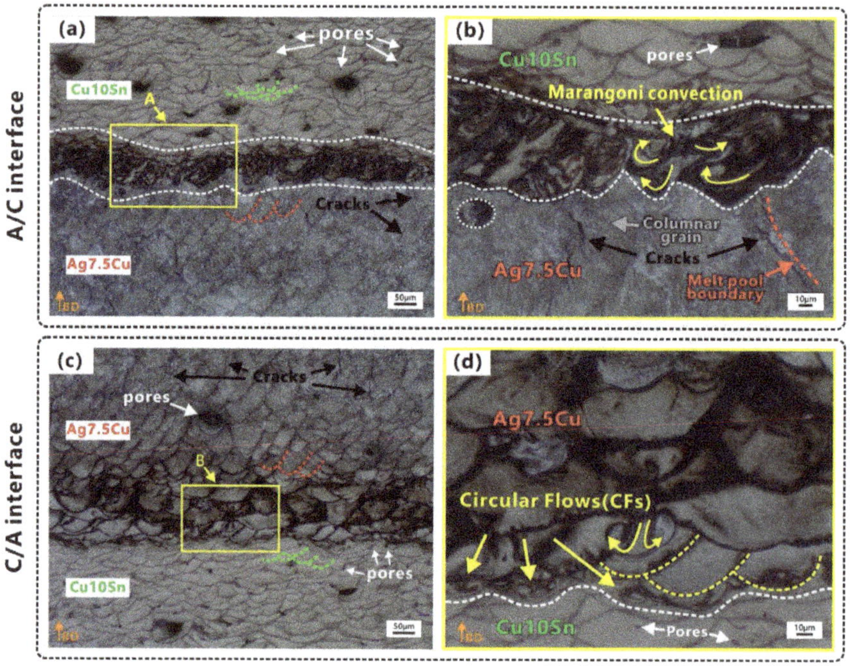

Figure 4. Characteristics of two interfaces after etching: (**a**) mesoscopic morphology at A/C interface; (**b**) enlarged view of area A in (**a**); (**c**) mesoscopic morphology at C/A interface; (**d**) enlarged view of area B in (**c**). Orange arrows indicate the building direction (BD).

3.2. Element Diffusion

The main element distributions of the A/C and C/A interfaces are shown in Figures 5 and 6, respectively. Since both materials contained Cu elements, EDS mapping and line scan result (Figure 7) were combined for a comprehensive evaluation in order to better determine the width of the diffusion zone.

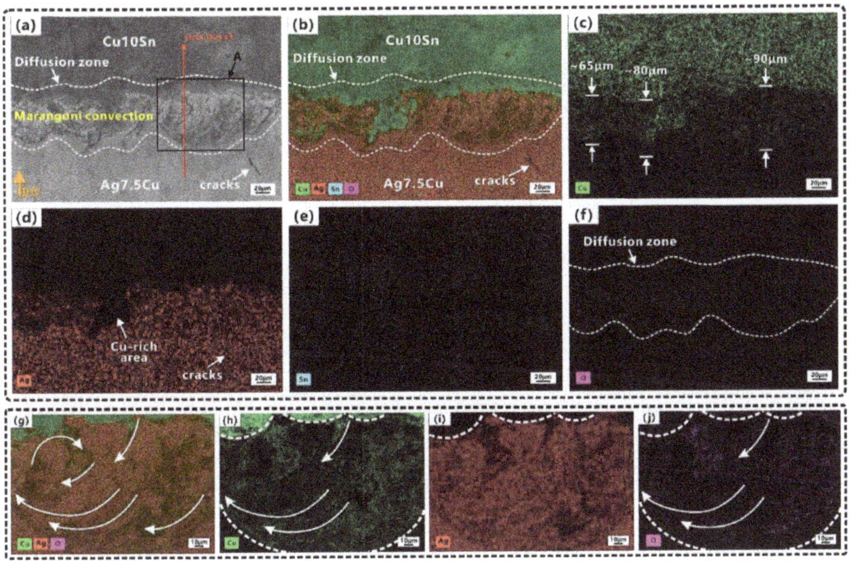

Figure 5. Elemental distributions at A/C interface: (**a**) SEM image; (**b**) EDS mapping of (**a**); (**c–f**) showing elemental distributions of Cu, Ag, Sn, O, respectively; (**g**) EDS mapping of area A in (**a**); (**h–j**) showing elemental distributions of Cu, Ag, O, respectively. Orange arrow indicates the building direction (BD).

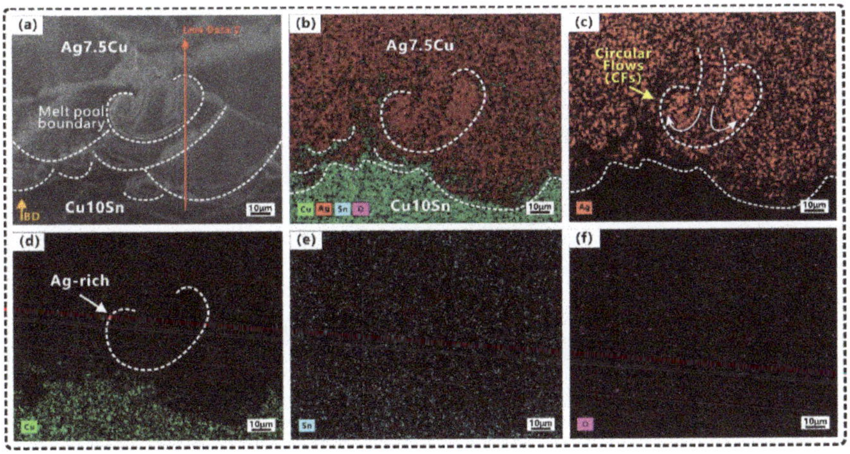

Figure 6. Elemental distributions at C/A interface: (**a**) SEM image; (**b**) EDS mapping of (**a**); (**c–f**) showing elemental distributions of Ag, Cu, Sn, O, respectively. Orange arrow indicates the building direction (BD).

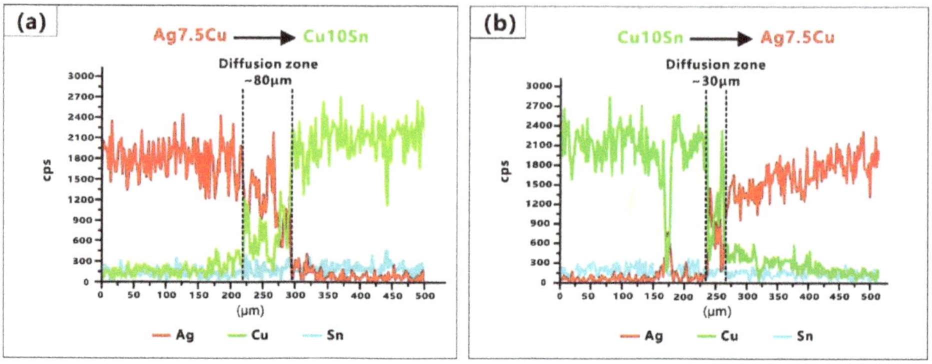

Figure 7. EDS line scan results along the scan path 1 (**a**) and 2 (**b**) presented in Figures 5a and 6a, respectively.

The mixing and dilution of Ag, Cu, and Sn were evident in the molten pool at two interfaces (Figures 5 and 6). The element diffusion distance of A/C interface (65–90 μm) was wider than that of C/A interface (30–40 μm). Some strong Marangoni convection could be observed at the A/C interface (Figures 4b and 5a), and circular flows (CFs) were observed at the C/A interface (Figures 4d and 6c). Tan et al. [48] also found that when printing copper–steel multi-material, the high thermal conductivity of copper at the bottom resulted in a better interfacial bond strength. However, they only used one material with high thermal conductivity, and the other material was steel, which had poor thermal conductivity. During LPBF, the CFs in the molten pool were induced by Marangoni convection (ΔM), which was proportional to the surface tension (γ) and temperature gradient (ΔT) [63]. This phenomenon was enhanced by the underlying Ag/Cu because the high thermal conductivity of Ag/Cu increased ΔT, which enhanced the dynamics of the Marangoni convection of the molten pool and further facilitated element migration in the diffusion zone. This was especially the case at the A/C interface (underlying Ag) since the thermal conductivity of Ag was higher than that of Cu. This explains in one aspect why the stronger Marangoni convection and a wider diffusion zone were observed at the A/C interface.

Due to the complex non-equilibrium solidification process in the LPBF process [64], the long-term diffusion was limited during the rapid cooling process [65], which was not conducive to the generation of element portioning. Therefore, it did not have the conditions to produce large eutectic areas. Wang et al. [66] manufactured 925 Ag through LPBF and found that some submicron-sized eutectic structures were formed. The Ag–Cu phase diagram showed eutectic reaction at 28.1wt.% Cu [47]. The Cu-rich area (Figure 5d) was observed at the A/C interface. Figure 5g–j shows the mapping images of the molten pool of the area A in Figure 5a. The melting tracks with curved strips and swirls were clearly observed, which proved that Ag and Cu formed a good interpenetrating interface. The circular movement of the molten metal (Marangoni convection) in the molten pool contributed to the diffusion and homogenization of elements [67], which also promoted more eutectic formation in the diffusion zone. Due to the good liquidity and castability of eutectic alloys, thermal cracks were effectively reduced or even eliminated by increasing the amount of the eutectic phase [68]. The low melting eutectic structure backfilled the cracks and increased the grain boundary area to prevent crack propagation [69], and the alloy composition near the eutectic point of the phase diagram was also conducive to achieve high castability [70]. This also explains why there were fewer cracks in the diffusion zone.

Comparing Figure 5f and j with Figure 6f, the oxygen content in the strong Marangoni convection zone at the A/C interface was higher, and the mapping area of the oxygen content was consistent with that of Cu. Since the sample was constructed in a nitrogen atmosphere (0.5 vol% residual oxygen) and stored powders were used in the experiments.

On the one hand, this phenomenon was attributed to the higher oxidation degree of Cu10Sn powders than that of Ag7.5Cu powders (see Table 1). On the other hand, powder oxidation may have had an effect on the Marangoni convection. As the shape of the molten pool was determined by the melting mode of the LPBF process, the different degrees of Marangoni convection at the A/C and C/A interfaces could lead to differences in the melting modes at the two interfaces. This will be discussed in Section 3.3.

3.3. Melting Mode in Multi-Material LPBF

The melting mode in the multi-material interfacial region led to different molten pool morphologies and significantly affected the distance of the element migration [51]. Therefore, different melting modes were very important for the interfacial bonding of multi-material.

In this study, the diffusion zone of the A/C interface (~80 μm) was wider than that of the C/A interface (~30 μm) (Figure 7). The migration distance of the elements at the A/C interface was about twice that at the C/A interface. The different width of the diffusion zone was due to the different melting modes in the manufacture of LPBF-fabricated multi-material. Qi et al. [59] believed that there were three modes in the LPBF fabrication of Al7050: the keyhole, the transition, and the conduction modes. The keyhole mode produced a deep and narrow V-shaped or goblet molten pool; the conduction mode produced a semicircular molten pool; and the transition mode was between the first two. These three modes could be identified by the ratio of molten pool depth (D) to width (W) (D/W, the aspect ratio of the molten pool cross-section), the keyhole mode ($D/W \geq 0.55$), the conduction mode ($D/W < 0.25$), and the transition mode ($0.25 \leq D/W < 0.55$). In comparison, Ag and Cu had higher reflectivity and thermal conductivity than Al, and the solidification rates of Ag and Cu were faster. Therefore, there would be differences in the formation mechanism of the Ag–Cu molten pool. It was necessary to make a comprehensive judgment based on the morphology of the molten pool, the ratio of molten pool depth to width (D/W), and the ratio of molten pool depth to layer thickness (D/L). The molten pool morphology gave a direct indication of the melting mode, while D/W could be used to assist in judging the transformation of the molting mode, and D/L indicated the penetration depth.

3.3.1. Melting Mode at the A/C Interface

At the A/C interface, it could be observed that the molten pools of the Ag7.5Cu region had goblet or V shapes (see the red dash line in Figure 4a); the depth of the molten pool was about 62 μm; and the D/L was ~3.1 (deep penetration). Therefore, the melting mode of the Ag7.5Cu region belonged to the keyhole mode. When using a laser input p = 80 W, a scan speed v = 800 mm/s, and a volume energy density $E_{A/C}$ = 125 J/mm^3 (Figure 2a), strong Marangoni convection (Figure 4b) and strong element mixing (Figure 5b) were produced in the diffusion zone of the A/C interface. The depth of the keyhole would be further deepened when Cu10Sn was printed on Ag7.5Cu, since the laser beam reflected several times along the keyhole could significantly increase the laser absorption rate [71]. This contributed to the more sufficient melting of highly reflective metals and further promoted the intense mixing of materials, resulting in good metallurgical bonding at the A/C interface (see the schematic diagram in Figure 8). The molten pool subsequently transitioned from the stable keyhole mode (D/W~0.59) to the deeper and unstable keyhole mode (D/W~0.75, D/L~3.5). Due to the high cooling rate of the LPBF process (up to 10^6–10^8 K/s) and the high thermal conductivity of Ag, the solidification time was short. On the one hand, the instability of the critical keyhole caused by recoil pressure [72] released acoustic waves that pushed the keyhole tip away from the keyhole [73], which were trapped by the solidification front and formed an isolated area (Figure 8a,b). On the other hand, the strong recoil pressure promoted the deeper penetration of Cu10Sn by the Marangoni convection, which in turn resulted in more extensive element migration (~80 μm) (Figure 7a). In addition, during LPBF, increasing oxygen content changed the temperature coefficient of the surface tension from negative to positive, thus changing the Marangoni convection from an inward centrifugal flow to an outward cardiac flow [74]. The comparison of the oxygen content in the stronger (Figure 5f) and weaker (Figure 6f) Marangoni convection regions indicated that the higher oxygen content in the powder could intensify the Marangoni convection and promote more agitation in the molten pool.

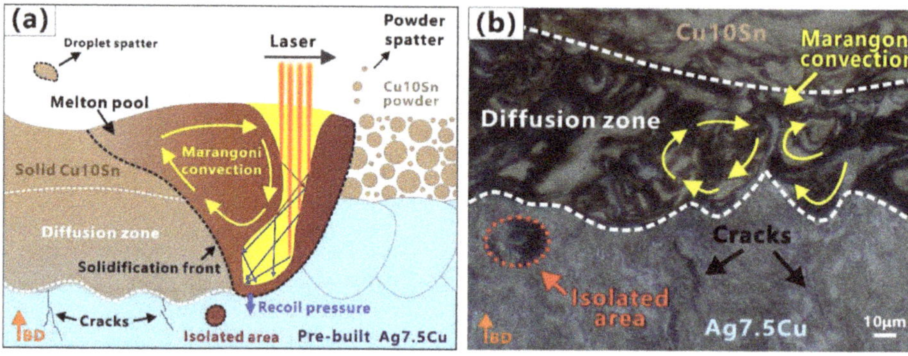

Figure 8. Melting mode of A/C interface in multi-material LPBF: (**a**) schematic illustration of molten pool behavior; (**b**) molten pool morphology at A/C interface. Orange arrows indicate the building direction (BD).

3.3.2. Melting Mode at the C/A Interface

In the Cu10Sn region of the C/A interface, the D/W ratio of the molten pool was calculated to be ~0.41. According to Qi's article, the molten mode of this proportion belonged to the transition mode. In this study, the molten pool in the Cu10Sn region had semicircular shapes (see green dash line in Figure 4c), the D/L was ~1.3 (thin penetration), and there was a presence of LoF. Therefore, we believed that the melting mode of the Cu10Sn region was more inclined to the conduction mode. In the diffusion zone of the C/A interface, higher laser power (p = 90 W), lower scanning speed (v = 600 mm/s), and higher volume energy density ($E_{C/A}$ = 187.5 J/mm^3) were used to print the Ag7.5Cu on the pre-printed Cu10Sn (Figure 2a). The molten pool transformed from the conduction mode to the transition mode (D/W ~0.44, D/L ~2), forming a deeper molten pool in the diffusion zone (see the yellow dash line in Figure 4d). In the transition mode, part of the recoil pressure was used to maintain the keyhole shape, while the other part pushed the molten pool in the opposite direction to the scanning direction [73], thus forming the CFs (see the schematic diagram in Figure 9a). Therefore, the shapes of molten pool in the diffusion zone were between the V shape and the semicircular shape, accompanied by a small range of CFs (see Figures 4d and 9b).

During the LPBF process, when the powder bed absorbed sufficient laser energy, the previously solidified layer would remelt due to heat conduction. The longer the liquid time of the molten pool, the longer the diffusion time of the elements. Simulated Scheil–Gulliver solidification curves (Figure 10) showed that Ag7.5Cu had a narrower solidification temperature range than Cu10Sn. Due to the high solidification rate of Ag, the diffusion time was short, and the molten pool had already solidified rapidly before Ag was evenly distributed in the diffusion zone (see the Ag-rich zone in Figure 6d), resulting in the nonuniform and narrower diffusion zone (~30 μm) (see Figure 7b). Furthermore, when printing the upper layer of Ag7.5Cu, the pre-solidified diffusion zone could be covered, and some existing features such as CFs and defects (holes and cracks) could even be covered since remelting could repaired some defects [62]. This is one of the reasons why few CFs and defects were observed in the diffusion zone at the C/A interface.

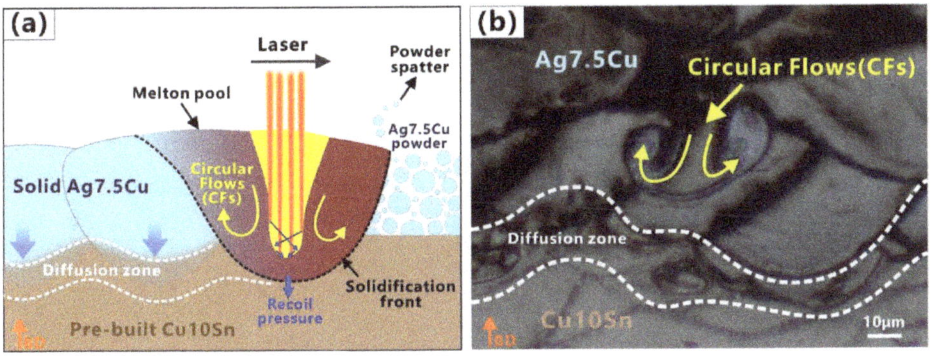

Figure 9. Melting mode of C/A interface in multi-material LPBF: (**a**) schematic illustration of molten pool behavior; (**b**) molten pool morphology at C/A interface. Orange arrows indicate the building direction (BD).

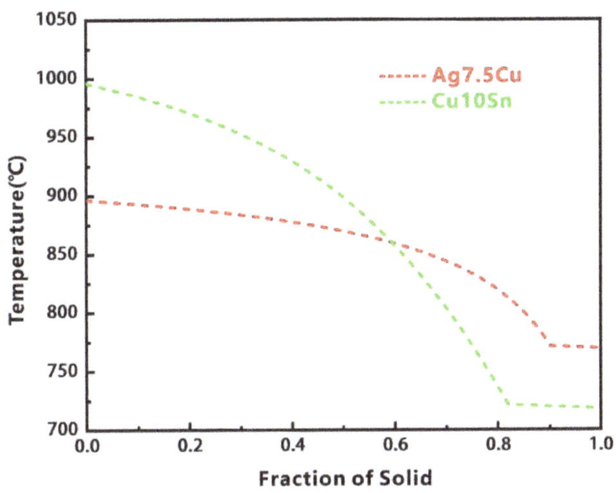

Figure 10. Scheil–Gulliver solidification curves of Ag7.5Cu and Cu10Sn.

In summary, on the one hand, Marangoni turbulence was promoted when metals with high reflectivity were used as substrates. Marangoni convection facilitated eutectic formation in the diffusion zone, which could mitigate the crack defects. The multiple laser reflections inside the keyhole increased the laser absorption of the highly reflective metals. Marangoni convection and the recoil pressure facilitated the transition of the molten pool from the stable keyhole mode to the deeper and unstable keyhole mode. These promoted more extensive migration of the elements, resulting in a wider diffusion zone, and enhanced the metallurgical bonding at the A/C interface. Furthermore, the higher oxygen content in the Cu alloy powders affected Marangoni convection, resulting in more intense agitation in the molten pool and enhanced Ag–Cu interpenetration. On the other hand, using a higher energy density to print the C/A interface, CFs and recoil pressure changed the molten pool from the conduction mode to the transition mode and reduced defects in the diffusion zone. Due to the high cooling rate of LPBF and the rapid solidification rate of Ag, the diffusion time of elements was short. When the upper layer of Ag7.5Cu was printed, CFs and pore defects in the underlying diffusion zone were covered and reduced, respectively. These resulted in a narrower diffusion zone at the C/A interface than that at the A/C interface.

4. Conclusions

In this study, the Ag–Cu multi-material structures were fabricated via LPBF. The characteristics and micromorphology (molten pool morphology, defects, element diffusion, etc.) of the A/C (printing Cu10Sn on Ag7.5Cu) and C/A interfaces (printing Ag7.5Cu on Cu10Sn) were investigated. The different melting modes and formation mechanisms of the two interfaces were discussed.

The following is a summary of the main conclusions:

(1) **The Ag7.5Cu/Cu10Sn/Ag7.5Cu sandwich structure was proposed and successfully printed using LPBF for the first time.** It was successfully printed with no obvious mesoscopic defects in the binding regions. Ag7.5Cu and Cu10Sn were firmly metallurgically bonded to each other. Some thermal cracks and LoF pores were observed in the Ag7.5Cu region and the Cu10Sn region, respectively, but fewer microscopic defects were found in the diffusion zone.

(2) **The high thermal conductivity of the substrate enhanced the Marangoni convection, which strengthened interfacial bonding strength and reduced the defects.** The diffusion zone of the A/C interface (~80 μm) was wider than that of the C/A interface (~30 μm). The strong Marangoni convection was observed at the A/C interface; it promoted an intense mixing of elements, resulting in more extensive elemental migration and a wider diffusion zone. The energy density used in the diffusion zone at the A/C interface ($E_{A/C}$ = 125 J/mm^3) was lower than that at the C/A interface ($E_{C/A}$ = 187.5 J/mm^3), indicating that the material with high thermal conductivity as a substrate could significantly promote Marangoni convection. The strong convection of the molten pool increased the amount of eutectic formation in the diffusion zone, thus mitigating the crack defects. Furthermore, the higher oxygen content in the Cu10Sn powders resulted in a stronger Marangoni convection in the powder mixing zone, enhancing the Ag–Cu interpenetration.

(3) **Defining the melting mode in the Ag–Cu interface region.** At the A/C interface, the keyhole mode was further enhanced by the Marangoni convection and the recoil pressure. The stable keyhole mode (D/W ~0.59, D/L ~3.1) enhanced the laser energy absorption and provided a large penetration depth. The unstable keyhole mode (D/W ~ 0.75, D/L ~ 3.5) was accompanied by more turbulent molten pools, which promoted elemental migration and resulted in better interfacial bonding strength. At the C/A interface, the molten pool transformed from the conduction mode (D/W~0.41, D/L~1.3) to the transition mode (D/W ~0.44, D/L ~2) at a higher energy density, reducing the defects. The underlying diffusion zone was covered by the upper Ag7.5Cu layer, resulting in a narrower diffusion zone and reduced the interfacial bonding strength.

Author Contributions: Conceptualization, Z.L. and Q.C.; methodology, L.H., J.Y. and W.X.; investigation, Y.J., W.X., P.G. and X.Z.; resources, L.H. and J.Y.; writing—original draft preparation, Q.C.; writing—review and editing, J.Y., Q.C.; Y.J., L.Z., S.L. and R.P.; supervision, J.Y. and L.H.; project administration, L.H. and J.Y.; funding acquisition, L.H., J.Y. and Z.L. All authors have read and agreed to the published version of the manuscript.

Funding: This research is funded by the National Natural Science Foundation of China (No. 61805095, No. 51675496, No. 51902295); the Science and Technology Project of the Hubei Province (2021BEC010); the Wuhan Applied Foundational Frontier Project from the Wuhan Science and Technology Bureau Project, China (No. 2020010601012172); Knowledge Innovation Program of Wuhan-Basic Research (No.2022020801010196); and the Fundamental Research Funds for the Central Universities, China University of Geosciences (Wuhan).

Data Availability Statement: The data that support the findings of this study are available from the corresponding author upon reasonable request.

Acknowledgments: The authors gratefully appreciate Haisheng Xu from Hubei Sanjiang Space Jiangbei Machinery Engineering Co., Ltd. for his help.

Conflicts of Interest: The authors declare no conflict of interest.

References

1. Xiong, W.; Hao, L.; Peijs, T.; Yan, C.; Cheng, K.; Gong, P.; Cui, Q.; Tang, D.; Al Islam, S.; Li, Y. Simultaneous strength and ductility enhancements of high thermal conductive Ag7.5Cu alloy by selective laser melting. *Sci. Rep.* **2022**, *12*, 4250. [CrossRef] [PubMed]
2. Fantino, E.; Chiappone, A.; Roppolo, I.; Manfredi, D.; Bongiovanni, R.; Pirri, C.F.; Calignano, F. 3D Printing: 3D Printing of Conductive Complex Structures with In Situ Generation of Silver Nanoparticles (Adv. Mater. 19/2016). *Adv. Mater.* **2016**, *28*, 3711. [CrossRef] [PubMed]
3. Bradley, D. Every silver-lined solar cell: Energy. *Mater. Today* **2009**, *12*, 10. [CrossRef]
4. Ali, A.; Baheti, V.; Militky, J. Energy harvesting performance of silver electroplated fabrics. *Mater. Chem. Phys.* **2019**, *231*, 33–40. [CrossRef]
5. Yu, Q.; Meng, K.; Guo, J. Research on innovative application of silver material in modern jewelry design. *MATEC Web Conf.* **2018**, *176*, 02013. [CrossRef]
6. Ciacotich, N.; Din, R.U.; Sloth, J.J.; Møller, P.; Gram, L. An electroplated copper–silver alloy as antibacterial coating on stainless steel. *Surf. Coatings Technol.* **2018**, *345*, 96–104. [CrossRef]
7. Bernasconi, R.; Hart, J.L.; Lang, A.C.; Magagnin, L.; Nobili, L.; Taheri, M.L. Structural properties of electrodeposited Cu-Ag alloys. *Electrochim. Acta* **2017**, *251*, 475–481. [CrossRef]
8. Kato, K.; Omoto, H.; Tomioka, T.; Takamatsu, A. Visible and near infrared light absorbance of Ag thin films deposited on ZnO under layers by magnetron sputtering. *Sol. Energy Mater. Sol. Cells* **2011**, *95*, 2352–2356. [CrossRef]
9. Silver, S.; Phung, L.T.; Silver, G. Silver as biocides in burn and wound dressings and bacterial resistance to silver compounds. *J. Ind. Microbiol. Biotechnol.* **2006**, *33*, 627–634. [CrossRef]
10. Lu, L.; Shen, Y.; Chen, X.; Qian, L.; Lu, K. Ultrahigh Strength and High Electrical Conductivity in Copper. *Science* **2004**, *304*, 422–426. [CrossRef] [PubMed]
11. Bachmaier, A.; Pfaff, M.; Stolpe, M.; Aboulfadl, H.; Motz, C. Phase separation of a supersaturated nanocrystalline Cu–Co alloy and its influence on thermal stability. *Acta Mater.* **2015**, *96*, 269–283. [CrossRef]
12. Tuthill, A. *Guidelines for the Use of Copper Alloys in Seawater*; NiDI Publication: The Hague, The Netherlands, 1987; pp. 1–2.
13. Tylecote, R.F. *A History of Metallurgy*; The Metals Society: London, UK, 1992.
14. Walsh, F.; Low, C. A review of developments in the electrodeposition of tin-copper alloys. *Surf. Coat. Technol.* **2016**, *304*, 246–262. [CrossRef]
15. Singer, F.; Deisenroth, D.; Hymas, D.; Ohadi, M. Additively manufactured copper components and composite structures for thermal management applications. In Proceedings of the 2017 16th IEEE Intersociety Conference on Thermal and Thermomechanical Phenomena in Electronic Systems (ITherm), Orlando, FL, USA, 30 May–2 June 2017; pp. 174–183.
16. Blakey-Milner, B.; Gradl, P.; Snedden, G.; Brooks, M.; Pitot, J.; Lopez, E.; Leary, M.; Berto, F.; du Plessis, A. Metal additive manufacturing in aerospace: A review. *Mater. Des.* **2021**, *209*, 110008. [CrossRef]
17. Scudino, S.; Unterdörfer, C.; Prashanth, K.; Attar, H.; Ellendt, N.; Uhlenwinkel, V.; Eckert, J. Additive manufacturing of Cu–10Sn bronze. *Mater. Lett.* **2015**, *156*, 202–204. [CrossRef]
18. Xiong, W.; Hao, L.; Li, Y.; Tang, D.; Cui, Q.; Feng, Z.; Yan, C. Effect of selective laser melting parameters on morphology, microstructure, densification and mechanical properties of supersaturated silver alloy. *Mater. Des.* **2019**, *170*, 107697. [CrossRef]
19. Wei, C.; Gu, H.; Li, Q.; Sun, Z.; Chueh, Y.-H.; Liu, Z.; Li, L. Understanding of process and material behaviours in additive manufacturing of Invar36/Cu10Sn multiple material components via laser-based powder bed fusion. *Addit. Manuf.* **2021**, *37*, 101683. [CrossRef]
20. Liu, L.; Wang, D.; Deng, G.; Yang, Y.; Chen, J.; Tang, J.; Wang, Y.; Liu, Y.; Yang, X.; Zhang, Y. Interfacial Characteristics and Formation Mechanisms of Copper–steel Multimaterial Structures Fabricated via Laser Powder Bed Fusion Using Different Building Strategies. *Chin. J. Mech. Eng. Addit. Manuf. Front.* **2022**, *1*, 100045. [CrossRef]
21. Wang, D.; Liu, L.; Deng, G.; Deng, C.; Bai, Y.; Yang, Y.; Wu, W.; Chen, J.; Liu, Y.; Wang, Y.; et al. Recent progress on additive manufacturing of multi-material structures with laser powder bed fusion. *Virtual Phys. Prototyp.* **2022**, *17*, 329–365. [CrossRef]
22. Wei, C.; Zhang, Z.; Cheng, D.; Sun, Z.; Zhu, M.; Li, L. An overview of laser-based multiple metallic material additive manufacturing: From macro- to micro-scales. *Int. J. Extrem. Manuf.* **2020**, *3*, 012003. [CrossRef]
23. Gu, D.D.; Meiners, W.; Wissenbach, K.; Poprawe, R. Laser additive manufacturing of metallic components: Materials, processes and mechanisms. *Int. Mater. Rev.* **2012**, *57*, 133–164. [CrossRef]
24. Li, Z.; Li, H.; Yin, J.; Li, Y.; Nie, Z.; Li, X.; You, D.; Guan, K.; Duan, W.; Cao, L.; et al. A Review of Spatter in Laser Powder Bed Fusion Additive Manufacturing: In Situ Detection, Generation, Effects, and Countermeasures. *Micromachines* **2022**, *13*, 1366. [CrossRef]
25. Gu, D.; Shi, X.; Poprawe, R.; Bourell, D.L.; Setchi, R.; Zhu, J. Material-structure-performance integrated laser-metal additive manufacturing. *Science* **2021**, *372*, eabg1487. [CrossRef] [PubMed]
26. Zhang, C.; Chen, F.; Huang, Z.; Jia, M.; Chen, G.; Ye, Y.; Lin, Y.; Liu, W.; Chen, B.; Shen, Q.; et al. Additive manufacturing of functionally graded materials: A review. *Mater. Sci. Eng. A* **2019**, *764*, 138209. [CrossRef]
27. Chauhan, P.K.; Khan, S. Microstructural examination of aluminium-copper functionally graded material developed by powder metallurgy route. *Mater. Today Proc.* **2020**, *25*, 833–837. [CrossRef]
28. Safari, M.; Yaghooti, H.; Joudaki, J. Laser welding of titanium and stainless steel sheets using Ag-Cu interlayers: Microstructure and mechanical characterization. *Int. J. Adv. Manuf. Technol.* **2021**, *117*, 3063–3073. [CrossRef]

29. Soderhjelm, C.J. Multi-Material Metal Casting: Metallurgically Bonding Aluminum to Ferrous Inserts. Ph.D. Thesis, Worcester Polytechnic Institute, Worcester, MA, USA, 2017.
30. Mousavi, Z.; Pourabdoli, M. Physical and chemical properties of Ag–Cu composite electrical contacts prepared by cold-press and sintering of silver-coated copper powder. *Mater. Chem. Phys.* **2022**, *290*, 126608. [CrossRef]
31. Klotz, U.E.; Tiberto, D.; Held, F. Optimization of 18-karat yellow gold alloys for the additive manufacturing of jewelry and watch parts. *Gold Bull.* **2017**, *50*, 111–121. [CrossRef]
32. Robinson, J.; Stanford, M.; Arjunan, A. Stable formation of powder bed laser fused 99.9% silver. *Mater. Today Commun.* **2020**, *24*, 101195. [CrossRef]
33. Robinson, J.; Arjunan, A.; Stanford, M.; Lyall, I.; Williams, C. Effect of silver addition in copper-silver alloys fabricated by laser powder bed fusion in situ alloying. *J. Alloy. Compd.* **2021**, *857*, 157561. [CrossRef]
34. Varol, T.; Hacısalihoğlu, İ.; Kaya, G.; Güler, O.; Yıldız, F.; Aksa, H.C.; Akçay, S.B. The Effect of Selective Laser Melting Process on the Microstructure, Density, and Electrical Conductivity of Silver-Coated Copper Cores. *J. Mater. Eng. Perform.* **2021**, *30*, 5216–5226. [CrossRef]
35. Zhang, M.; Yang, Y.; Wang, D.; Song, C.; Chen, J. Microstructure and mechanical properties of CuSn/18Ni300 bimetallic porous structures manufactured by selective laser melting. *Mater. Des.* **2019**, *165*, 107583. [CrossRef]
36. Wei, C.; Liu, L.; Cao, H.; Zhong, X.; Xu, X.; Gu, Y.; Cheng, D.; Huang, Y.; Li, Z.; Guo, W.; et al. Cu10Sn to Ti6Al4V bonding mechanisms in laser-based powder bed fusion multiple material additive manufacturing with different build strategies. *Addit. Manuf.* **2022**, *51*, 102588. [CrossRef]
37. Sing, S.; Lam, L.; Zhang, D.; Liu, Z.; Chua, C. Interfacial characterization of SLM parts in multi-material processing: Intermetallic phase formation between AlSi10Mg and C18400 copper alloy. *Mater. Charact.* **2015**, *107*, 220–227. [CrossRef]
38. Chen, J.; Yang, Y.; Song, C.; Zhang, M.; Wu, S.; Wang, D. Interfacial microstructure and mechanical properties of 316L /CuSn10 multi-material bimetallic structure fabricated by selective laser melting. *Mater. Sci. Eng. A* **2019**, *752*, 75–85. [CrossRef]
39. Tan, C.; Chew, Y.; Bi, G. Additive manufacturing of steel–copper functionally graded material with ultrahigh bonding strength. *Mater. Sci. Technol.* **2021**, *72*, 217–222. [CrossRef]
40. Bai, Y.; Zhang, J.; Zhao, C.; Li, C.; Wang, H. Dual interfacial characterization and property in multi-material selective laser melting of 316L stainless steel and C52400 copper alloy. *Mater. Charact.* **2020**, *167*, 110489. [CrossRef]
41. Chen, J.; Yang, Y.; Wang, D.; Liu, Z.; Song, C. Effect of manufacturing steps on the interfacial defects of laser powder bed fusion 316L/CuSn10. *Mater. Lett.* **2021**, *292*, 129377. [CrossRef]
42. Boley, C.D.; Khairallah, S.A.; Rubenchik, A.M. Calculation of laser absorption by metal powders in additive manufacturing. In *Additive Manufacturing Handbook*; CRC Press: London, UK, 2017; pp. 507–509.
43. Attar, H.; Bönisch, M.; Calin, M.; Zhang, L.-C.; Scudino, S.; Eckert, J. Selective laser melting of in situ titanium–titanium boride composites: Processing, microstructure and mechanical properties. *Acta Mater.* **2014**, *76*, 13–22. [CrossRef]
44. Paquin, R.A. Properties of metals. In *Handbook of Optics*; McGraw-Hill Companies, Inc.: New York, NY, USA, 1995; Volume 2, pp. 35–49.
45. Ashby, M.F.; Jones, D.R. *Engineering Materials 1: An Introduction to Properties, Applications and Design*; Elsevier: Waltham, MA, USA, 2011; Volume 1.
46. Caiazzo, F.; Alfieri, V.; Casalino, G. On the relevance of volumetric energy density in the investigation of Inconel 718 laser powder bed fusion. *Materials* **2020**, *13*, 538. [CrossRef]
47. Subramanian, P.; Perepezko, J.H. The ag-cu (silver-copper) system. *J. Phase Equilibria Diffus.* **1993**, *14*, 62–75. [CrossRef]
48. Tan, C.; Zhou, K.; Ma, W.; Min, L. Interfacial characteristic and mechanical performance of maraging steel-copper functional bimetal produced by selective laser melting based hybrid manufacture. *Mater. Des.* **2018**, *155*, 77–85. [CrossRef]
49. Song, C.; Hu, Z.; Xiao, Y.; Li, Y.; Yang, Y. Study on Interfacial Bonding Properties of NiTi/CuSn10 Dissimilar Materials by Selective Laser Melting. *Micromachines* **2022**, *13*, 494. [CrossRef] [PubMed]
50. Zweben, C. 4.16 Metal Matrix Composite Thermal Management Materials. In *Comprehensive Composite Materials II*; Elsevier: Oxford, UK, 2018. [CrossRef]
51. Yao, L.; Huang, S.; Ramamurty, U.; Xiao, Z. On the formation of "Fish-scale" morphology with curved grain interfacial microstructures during selective laser melting of dissimilar alloys. *Acta Mater.* **2021**, *220*, 117331. [CrossRef]
52. Tan, C.; Zhou, K.; Ma, W.; Zhang, P.; Liu, M.; Kuang, T. Microstructural evolution, nanoprecipitation behavior and mechanical properties of selective laser melted high-performance grade 300 maraging steel. *Mater. Des.* **2017**, *134*, 23–34. [CrossRef]
53. Mutua, J.; Nakata, S.; Onda, T.; Chen, Z.-C. Optimization of selective laser melting parameters and influence of post heat treatment on microstructure and mechanical properties of maraging steel. *Mater. Des.* **2018**, *139*, 486–497. [CrossRef]
54. Yan, C.Z.; Hao, L.; Hussein, A.; Young, P.; Huang, J.T.; Zhu, W. Microstructure and mechanical properties of aluminium alloy cellular lattice structures manufactured by direct metal laser sintering. *Mater. Sci. Eng. A* **2015**, *628*, 238–246. [CrossRef]
55. Tang, M.; Pistorius, P.C.; Beuth, J.L. Prediction of lack-of-fusion porosity for powder bed fusion. *Addit. Manuf.* **2017**, *14*, 39–48. [CrossRef]
56. Tang, Y.T.; Panwisawas, C.; Ghoussoub, J.N.; Gong, Y.; Clark, J.W.; Németh, A.A.; McCartney, D.G.; Reed, R.C. Alloys-by-design: Application to new superalloys for additive manufacturing. *Acta Mater.* **2021**, *202*, 417–436. [CrossRef]
57. Bermingham, M.; StJohn, D.; Krynen, J.; Tedman-Jones, S.; Dargusch, M. Promoting the columnar to equiaxed transition and grain refinement of titanium alloys during additive manufacturing. *Acta Mater.* **2019**, *168*, 261–274. [CrossRef]

58. Han, Q.; Mertens, R.; Montero-Sistiaga, M.L.; Yang, S.; Setchi, R.; Vanmeensel, K.; Van Hooreweder, B.; Evans, S.L.; Fan, H. Laser powder bed fusion of Hastelloy X: Effects of hot isostatic pressing and the hot cracking mechanism. *Mater. Sci. Eng. A* **2018**, *732*, 228–239. [CrossRef]
59. Qi, T.; Zhu, H.; Zhang, H.; Yin, J.; Ke, L.; Zeng, X. Selective laser melting of Al7050 powder: Melting mode transition and comparison of the characteristics between the keyhole and conduction mode. *Mater. Des.* **2017**, *135*, 257–266. [CrossRef]
60. Yin, J.; Zhang, W.; Ke, L.; Wei, H.; Wang, D.; Yang, L.; Zhu, H.; Dong, P.; Wang, G.; Zeng, X. Vaporization of alloying elements and explosion behavior during laser powder bed fusion of Cu–10Zn alloy. *Int. J. Mach. Tools Manuf.* **2021**, *161*, 103686. [CrossRef]
61. Stopyra, W.; Gruber, K.; Smolina, I.; Kurzynowski, T.; Kuźnicka, B. Laser powder bed fusion of AA7075 alloy: Influence of process parameters on porosity and hot cracking. *Addit. Manuf.* **2020**, *35*, 101270. [CrossRef]
62. Fu, J.; Li, H.; Song, X.; Fu, M. Multi-scale defects in powder-based additively manufactured metals and alloys. *J. Mater. Sci. Technol.* **2022**, *122*, 165–199. [CrossRef]
63. Arafune, K.; Hirata, A. Thermal and solutal Marangoni convection in In–Ga–Sb system. *J. Cryst. Growth* **1999**, *197*, 811–817. [CrossRef]
64. Qin, H.; Dong, Q.; Fallah, V.; Daymond, M.R. Rapid Solidification and Non-equilibrium Phase Constitution in Laser Powder Bed Fusion (LPBF) of AlSi10Mg Alloy: Analysis of Nano-precipitates, Eutectic Phases, and Hardness Evolution. *Met. Mater. Trans. A* **2020**, *51*, 448–466. [CrossRef]
65. Fan, H.; Yang, S. Effects of direct aging on near-alpha Ti–6Al–2Sn–4Zr–2Mo (Ti-6242) titanium alloy fabricated by selective laser melting (SLM). *Mater. Sci. Eng. A* **2020**, *788*, 139533. [CrossRef]
66. Wang, Z.; Xie, M.; Li, Y.; Zhang, W.; Yang, C.; Kollo, L.; Eckert, J.; Prashanth, K.G. Premature failure of an additively manufactured material. *NPG Asia Mater.* **2020**, *12*, 30. [CrossRef]
67. Yao, C.; Wang, Z.; Tay, S.L.; Zhu, T.; Gao, W. Effects of Mg on microstructure and corrosion properties of Zn–Mg alloy. *J. Alloy. Compd.* **2014**, *602*, 101–107. [CrossRef]
68. Hatch, J.E. (Ed.) *Aluminum: Properties and Physical Metallurgy 1984*; American Society for Metals: Metals Park, OH, USA, 2000; pp. 110–111.
69. Xiao, R.; Zhang, X. Problems and issues in laser beam welding of aluminum–lithium alloys. *J. Manuf. Process.* **2014**, *16*, 166–175. [CrossRef]
70. He, L.; Wu, S.; Dong, A.; Tang, H.; Du, D.; Zhu, G.; Sun, B.; Yan, W. Selective laser melting of dense and crack-free AlCoCrFeNi2.1 eutectic high entropy alloy: Synergizing strength and ductility. *J. Mater. Sci. Technol.* **2022**, *117*, 133–145. [CrossRef]
71. King, W.E.; Barth, H.D.; Castillo, V.M.; Gallegos, G.F.; Gibbs, J.W.; Hahn, D.E.; Kamath, C.; Rubenchik, A.M. Observation of keyhole-mode laser melting in laser powder-bed fusion additive manufacturing. *J. Mater. Process. Technol.* **2014**, *214*, 2915–2925. [CrossRef]
72. Yin, J.; Wang, D.; Yang, L.; Wei, H.; Dong, P.; Ke, L.; Wang, G.; Zhu, H.; Zeng, X. Correlation between forming quality and spatter dynamics in laser powder bed fusion. *Addit. Manuf.* **2020**, *31*, 100958. [CrossRef]
73. Zhao, C.; Shi, B.; Chen, S.; Du, D.; Sun, T.; Simonds, B.J.; Fezzaa, K.; Rollett, A.D. Laser melting modes in metal powder bed fusion additive manufacturing. *Rev. Mod. Phys.* **2022**, *94*, 045002. [CrossRef]
74. Leung, C.L.A.; Marussi, S.; Towrie, M.; Atwood, R.C.; Withers, P.J.; Lee, P.D. The effect of powder oxidation on defect formation in laser additive manufacturing. *Acta Mater.* **2019**, *166*, 294–305. [CrossRef]

MDPI AG
Grosspeteranlage 5
4052 Basel
Switzerland
Tel.: +41 61 683 77 34

Micromachines Editorial Office
E-mail: micromachines@mdpi.com
www.mdpi.com/journal/micromachines

www.ingramcontent.com/pod-product-compliance
Lightning Source LLC
Chambersburg PA
CBHW041947260926
51850CB00018B/1954
* 9 7 8 3 7 2 5 8 1 7 2 0 7 *